ARCHITECTS OF GROWTH?

The **Institute of Southeast Asian Studies (ISEAS)** was established as an autonomous organization in 1968. It is a regional centre dedicated to the study of socio-political, security and economic trends and developments in Southeast Asia and its wider geostrategic and economic environment. The Institute's research programmes are the Regional Economic Studies (RES, including ASEAN and APEC), Regional Strategic and Political Studies (RSPS), and Regional Social and Cultural Studies (RSCS).

ISEAS Publishing, an established academic press, has issued more than 2,000 books and journals. It is the largest scholarly publisher of research about Southeast Asia from within the region. ISEAS Publishing works with many other academic and trade publishers and distributors to disseminate important research and analyses from and about Southeast Asia to the rest of the world.

ARCHITECTS OF GROWTH?

Sub-national Governments and Industrialization in Asia

Edited by
Francis E. Hutchinson

ISEAS INSTITUTE OF SOUTHEAST ASIAN STUDIES
Singapore

First published in Singapore in 2014 by
ISEAS Publications
Institute of Southeast Asian Studies
30 Heng Mui Keng Terrace, Pasir Panjang
Singapore 119614

E-mail: publish@iseas.edu.sg *Website*: bookshop.iseas.edu.sg

The responsibility for facts and opinions in this publication rests exclusively with the authors and their interpretations do not necessarily reflect the views or the policy of the publisher or its supporters.

ISEAS Library Cataloguing-in-Publication Data

Architects of Growth? : sub-national governments and industrialization in Asia / edited by Francis E. Hutchinson.

1. Economic development—Political aspects—Asia.
2. Subnational governments—Asia.
3. Industrialization—Asia.
4. Electronic industries—Asia.

I. Hutchinson, Francis E.

HC412 A66 2014

ISBN 978-981-4414-53-1 (soft cover)
ISBN 978-981-4414-54-8 (E-book PDF)

Typeset by International Typesetters Pte Ltd
Printed in Singapore by Mainland Press Pte Ltd

CONTENTS

LIST OF TABLES

LIST OF FIGURES

FOREWORD

The chapters in this book are based on papers presented at a remarkable conference organized by the Institute of Southeast Asian Studies on 7 October 2011. They are groundbreaking in several ways.

For one thing, they discuss territories instead of countries. This approach is highly significant, and promises to lead ISEAS research in new directions in our effort to understand Southeast Asian dynamics beyond national divides.

The chapters also study major actors in the political economy of many nations, who are often overlooked. The focus is tactically moved towards politics and key processes found at levels below the national conceptual umbrella.

Furthermore, the book as a whole examines the nexus between politics and economics and in the process not only gives a deeper meaning to regionalism, but also inspects the geography and geopolitics of developmental processes today.

These are important strands that when interwoven, amount to a unique innovation. The case studies do not come only from within Southeast Asia, but also cover industrial areas external to the region as well.

I congratulate Dr Francis Hutchinson for conceiving of and organizing this conference and bringing to bear, in the process, his wide range of international contacts, and his penchant for comparative studies in an area that is necessarily complex.

Heartfelt thanks to Konrad Adenauer Stiftung for helping to sponsor the conference.

Ooi Kee Beng
Deputy Director
ISEAS

PREFACE AND ACKNOWLEDGEMENTS

Conventional academic and practical approaches to economic policy-making focus on the national level. However, over the last three decades, this nation-centred approach has been called into question by the re-scaling of economic, political and social processes. Decreasing barriers to trade, ever further-reaching production networks, greater flows of information, and demands by citizens for more participation in public life have challenged the primacy of the nation-state.

In addition, many national-level governments now have to contend with energized and proactive sub-national governments. In part, this is due to policies enacted by national governments themselves. Over the past decades, the "Silent Revolution" of decentralization has swept the globe, as central governments (with some notable exceptions) have devolved responsibilities to state, provincial, and municipal governments.

Furthermore, economic globalization seems to be heightening the relationship between location and economic activity. Despite investment being mobile, economic activity and, in particular, innovation have remained very sensitive to geography and the local institutional environment. This is seen in the emergence of high-performing regional economies in industrialized as well as industrializing countries.

These trends have implications for sub-national governments. While state and provincial governments may have gained in importance, the challenges facing them have also increased. More mobile investment along with heightened competition in many sectors means that policy-makers must now, not only attract investment, but also work hard to retain it. Firms that have problems getting reliable suppliers, securing appropriate labour, or obtaining licences will relocate to more amenable locations — often in a neighbouring state or province. Thus, competition between sub-national governments for investment is also on the rise, as they strive to outdo each other by providing incentives and specialized infrastructure.

However, while state and provincial governments have acquired new visibility and additional responsibilities, they are different from their national counterparts in that they do not have a wide range of tools or a large quantity of resources at their disposal. In addition, they are embedded in a specific power relationship with their national governments, and must reconcile central priorities with those of their constituents.

Conversely, despite their relative paucity of revenue and governmental responsibilities, state and provincial governments are usually responsible for a range of public services that are important to firms. In addition, they may be in a better position to leverage their proximity to the local economy to support new, innovative activities, particularly those that require social capital and ground-level knowledge. Indeed, scarce resources may preclude the rent-seeking often associated with state-supported enterprises, forcing firms and sub-national governments to engage in more disciplined value-enhancing dialogue.

Despite their unique characteristics and challenges, there is little research available on what state and provincial governments can or should do to effectively steward their economies — particularly at this juncture. To this end, the Institute of Southeast Asian Studies organized a conference on "State Policy and Industrialization at the Sub-national Level in Asia" on 7 October 2011 in Singapore. Researchers looking at economic developments at the sub-national level in Southeast Asia, Northeast Asia, South Asia and Europe were invited to present papers on emerging state and provincial economies.

This book is the end-result of that conference, both of which have sought to deepen our understanding of whether, how, and under what circumstances sub-national governments can play a significant role in promoting industrialization. Following the conference, the papers were revised by the authors based on comments arising during the discussion, as well as during the subsequent editing process.

The conference was generously funded by the Konrad Adenauer Stiftung, and institutional support was provided by Ambassador K. Kesavapany, who was ISEAS Director at the time. Heartfelt thanks go to both for making the event and this book possible.

I would also like to thank the conference participants for their insights. Special thanks go to Ooi Kee Beng, Faizal bin Yahya, Vikram Khanna, Omkar Shrestha and Carlos Kuriyama for their work as Chairs and Moderators.

Logistical and organizational support for the conference was provided and gratefully received by Karthi Nair, Loh Joo Yong, Ramlee Othman, and Razali.

The prompt and efficient work of Reema Bhagwan Jagtiani in the last stages of the manuscript preparation is most appreciated. Marcel Jäggi and Hans Hortig of the Future Cities Laboratory very kindly drew the maps for the case studies.

Thanks are also due to the ISEAS Publications Unit staff under the leadership of Triena Ong for their efficient and professional help.

I am extremely grateful to the contributors of this volume, who have generously given their time, insights, and support through the various rounds of revision and editing.

Francis E. Hutchinson

CONTRIBUTORS

Do Chai Chung is Researcher at the Institute for Korea Regional Studies, Seoul National University. He received his Ph.D. in economic geography at the Seoul National University in 2011. His ongoing research interests focus on the restructuring of old industrial regions, and the economic and social impacts of regional industrial policy.

Leo van Grunsven is Associate Professor in the Department of Human Geography and Planning, Faculty of Geosciences, Utrecht University, The Netherlands. His research interests focus on globalization, urban-industrial dynamics and regional development in Southeast and East Asia, and local enterprise development. Currently, his research in the Asian region focuses on the evolution and resilience of export manufacturing complexes in Southeast Asia, and the rise of inland second tier cities in China. Dr van Grunsven has published on these issues in edited books as well as journals such as: *Zeitschrift für Wirtschaftsgeographie; Journal of Development Alternatives and Area Studies*; *Environment and Planning A*; *European Planning Studies*; *Third World Planning Review*; and *Singapore Journal of Tropical Geography.*

Bram van Helvoirt is Programme Manager for Market Intelligence at the Centre for the Promotion of Imports from Developing Countries (CBI) in the Netherlands. He has a Ph.D. in International Development Studies and an MSc in Human Geography and Urban and Regional Planning, with a specialization in International Economics and Economic Geography from the University of Utrecht. Dr van Helvoirt has published articles in the field of development geography and economic development in journals such as *Asia-Pacific Viewpoint* and *International Development Planning Review.*

Francis E. Hutchinson is the Coordinator of the Regional Economic Studies Programme and Fellow at the Institute of Southeast Asian Studies, Singapore. He has a Ph.D. in Public Policy and Administration from the Australian National University and degrees from the Universities of Cambridge and Sussex. Dr Hutchinson's research focuses on governance, federalism, decentralization, and economic policy-making at the sub-national level in the Southeast Asian region. He is currently writing a sole-authored monograph on the influence of institutions on economic outcomes at the state level in Malaysia, with specific reference to Penang and Johor. Dr Hutchinson has published on sub-national issues in Malaysia and India in *Southeast Asian Affairs, Economic and Political Weekly,* and *Journal of Contemporary Asia.*

P. Vigneswara Ilavarasan is Associate Professor at the Department of Management Studies, Indian Institute of Technology Delhi, India. His research interests include: Work, Industry and Society; and Technology and Society, with a special focus on Information and Communications Technologies (ICTs) and India. Dr Ilavarasan has a Ph.D. in Sociology from IIT-Kanpur, and is the recipient of the Outstanding Young Faculty Fellowship Award from IIT-Delhi and the Professor M.N. Srinivas Memorial Prize (Indian Sociological Society). He has served as consultant to the Ministry of Labour and Employment (India), World Bank, Oxford Analytica, and Indicus Analytics. He has received research grants from IDRC, Canada and the Ministry of Science and Technology (India).

Ting-Lin Lee is Associate Professor and Chair in the Department of Asia-Pacific Industrial and Business Management, National University of Kaohsiung, Taiwan. She obtained her Ph.D. degree in Science and Technology Policy at PREST, University of Manchester, UK and an MBA at Sun Yat-Sen University, Taiwan. She worked for more than 12 years in Taiwan's central and local governments, rising to a Section Chief. Her current research centres on science and technology policy, innovation and National/Regional Innovation Systems. She has published articles in *Research Policy*, *Technology in Society*, and *International Journal of Technology, Policy, and Management.*

K.J. Joseph is the Ministry of Commerce Chair Professor at the Centre for Development Studies, Trivandrum. He has authored *Industry under Economic Liberalization: the Case of Indian Electronics* and *Information Technology and Innovation System and Trade Regime in Developing Countries: India and ASEAN.*

In addition, he has published more than 60 papers on technology, innovation, and export competitiveness. Professor Joseph is Editor-in-Chief of the journal *Innovation and Development*. He has held positions at: Jawaharlal Nehru University; Research and Information System for Developing Countries; UNESCAP; and Economic Growth Centre, Yale University.

Chandra-nuj Mahakanjana is Assistant Professor and Associate Dean for Planning and Development at the Graduate School of Public Administration, National Institute of Development Administration, Thailand. Her research focuses on decentralization and local government, in particular: balancing democratic accountability in Thai local government; the impact of decentralization on local executives in Thailand; and the role of women in local politics. She is currently working as a consultant on Central-Regional-Local relations for the UNDP.

Olaf Merk is Administrator/Economist in the Regional Development Policies Division at the Organisation for Economic Cooperation and Development (OECD), France. He is currently employed as Programme Manager of the Port-Cities Programme. At the OECD, where he began work in 2005, he directed several Territorial Reviews, including on Toronto, Copenhagen, Randstad-Holland, Switzerland and the Netherlands. He has collaborated on a dozen other OECD publications, with contributions on various subjects such as regional economic development, metropolitan governance, fiscal federalism, urban finance, public-private partnerships and urban infrastructure. Prior to the OECD, he worked at the Netherlands Ministry of Finance, where his last position was Acting Head of the Division for Financing Sub-national Governments. Olaf Merk has a Master's Degree in Political Science from the University of Amsterdam and diplomas from London School of Economics and University of Uppsala.

Sam Ock Park is Professor Emeritus of Economic Geography, Department of Geography, Seoul National University (SNU); and Distinguished Professor, Department of Public Administration, Gachon University. He has published 19 books on the locational dynamics of economic activities, regional development, high tech industries, industrial clusters and regional innovation systems. Professor Park has written 160 articles in English and Korean including papers in international journals such as: *Papers in Regional Science; Progress in Human Geography; Economic Geography; Environment and Planning A; Geoforum;* and *Journal of Economic and Social Geography (TESG).* He has also served as the Pacific Editor of *Papers in Regional*

Science (1995–97) and is currently an editorial board member of *Regional Studies, Journal of Economic Geography*, and *Papers in Regional Science* and Co-editor-in-Chief of *The Korea Spatial Planning Review*.

Toh Mun Heng is Associate Professor at the School of Business, National University of Singapore. He obtained his doctoral degree in Economics and Econometrics from the London School of Economics, University of London. His research interests and publications are in the areas of general equilibrium and econometric modelling, input-output analysis, international trade and investment, human resource development, productivity measurement, and household economics and development strategies of emerging economies in the Asia-Pacific. He has co-authored and edited several titles such as: *The Economics of Education and Manpower Development*; *Challenge and Response: Thirty Years of the Economic Development Board*; *Public Policies in Singapore: A Decade of Changes*; *ASEAN Growth Triangles*; and *Production Networks and Industrial Clusters: Integrating Economies of Southeast Asia*.

Tran Ngoc Ca is Director of the Secretariat for the National Council for Science and Technology Policy; Personal Assistant to the Science and Technology Minister; and Deputy Director, National Institute for Science and Technology Policy and Strategy Studies, Vietnam. He is the author of *Technological Capability and Learning in Firms: Vietnamese Industries in Transition* (Ashgate Publishing, UK). He has worked on projects pertaining to science, technology, and innovation policy for agencies such as UNDP, UNIDO, the World Bank, the European Commission, and International Development Research Centre, among others.

Thanut Tritasavit is a Research Associate in the Regional Economic Studies Programme at the Institute of Southeast Asian Studies, Singapore. He obtained his MSc in Environmental Economics and Environmental Management from the University of York, UK. Mr Tritasavit has contributed to the *Regional Outlook: Southeast Asia 2012–13* published by the Institute of Southeast Asian Studies.

Daniel Unger is Associate Professor of Political Science in the Department of Political Science at Northern Illinois University. He has published in the areas of international relations theory, Japanese foreign policy, comparative political economy, and comparative politics. Dr Unger has a Ph.D. from the University of California, Berkeley and is the author of *Building Social*

Capital in Thailand: Fibers, Finance, and Infrastructure published by Cambridge University Press. He is currently finishing a book on politics in Thailand.

Cassandra C. Wang is Assistant Professor in the Department of Earth Sciences of Zhejiang University, Hangzhou, China. She received her Ph.D. from the Department of Geography, University of Hong Kong. Her research interests include China's ICT industry and its dynamics of innovation, industrial clusters and the uneven growth of regional economies in China, and the upgrade of Chinese indigenous firms in the global value chain. Dr Wang has published on these issues in journals such as: *Journal of Economic Geography; Environment and Planning A; Urban Studies; Eurasion Geography and Economics;* and *Issues and Studies.*

ABBREVIATIONS

AFTA	ASEAN Free Trade Agreement
AIADMK	All India Anna Dravida Munnetra Kazhagam
ASEAN	Association of Southeast Asian Nations
AV	Audio-Visual
A*STAR	Agency for Science, Technology, and Research
BOI	Board of Investment
CEM	Contract Electronics Manufacturing
DMK	Dravida Munnetra Kazhagam
EDB	Economic Development Board
E&E	Electrical and Electronics Sector
EEI	Electrical and Electronics Industry Institute
ELCOT	Electronics Corporation of Tamil Nadu
EMS	Electronic Manufacturing Services
EPZ	Export Processing Zone
ER	Economic Region
ERSO	Electronics Research and Service Organization
EU	European Union
FAI	Fixed Asset Investment
FDI	Foreign Direct Investment
FIS	Foreign Invested Sector
FTA	Free Trade Agreement
GDP	Gross Domestic Product
GERI	Gumi Electronics & Information Technology Research Institute
GIN	Global Innovation Network
GNP	Gross National Product
GPN	Global Production Network
GRP	Gross Regional Product
HDD	Hard Disk Drive
HDDI	Hard Disk Drive Institute
HSIP	Hsinchu Science-based Industrial Park

HTCE	High Tech Campus Eindhoven
IC	Integrated Circuits
IDEMA	International Drive Equipment and Manufacturers' Association
IIT	Indian Institute of Technology
INR	Indian Rupee
IRDA	Iskandar Regional Development Authority
IT	Information Technology
ITA	Information Technology Agreement
ITRI	Industrial Technology Research Institute
JCS	Johor Civil Service
JSEPU	Johor State Economic Planning Unit
JSIC	Johor State Investment Centre
KEIC	Korea Electronics Industrial Corporation
KMIT	King Mongkut Institute of Technology
KRW	Korean Won
LCD	Liquid Crystal Display
LED	Light Emitting Diode
LGU	Local Government Unit
LIUP	Local Industry Upgrading Programme
MEMS	Microelectromechanical Systems
MEPZ	Mactan Export Processing Zone
MNCs	Multinational Corporations
MoE	Ministry of Education
MoEA	Ministry of Economic Affairs
NBDA	North Brabant Development Agency
NECTEC	National Electronic and Computer Technology Centre
NICS	Newly-Industrialized Countries
NIS	National Innovation System
NSTDA	National Science and Technology Development Agency
NTD	New Taiwan Dollar
OBM	Original Brand Manufacturers
ODM	Original Design Manufacturers
OECD	Organization for Economic Cooperation and Development
OEM	Original Equipment Manufacturers
OFDI	Outward Foreign Direct Investment
PAO	Provincial Administrative Organs
PC	Personal Computers
PCB	Printed Circuit Board
PPP	Purchasing Power Parity

R&D	Research and Development
RIS	Regional Innovation System
RM	Malaysian Ringgit
RMB	Renminbi
SER	Supra-Economic Region
SEZ	Special Economic Zone
SIPCOT	State Promotional Corporation of Tamil Nadu
SIRIM	Standards and Industrial Research Institute of Malaysia
SME	Small and Medium Enterprise
SOE	State-owned Enterprise
TEEMA	Taiwan Electrical and Electronic Manufacturing Association
TIDCO	Tamil Nadu Industrial Corporation
TIIC	Tamil Nadu Industrial Investment Corporation
UNCTAD	United Nations Convention on Trade and Development
UNIDO	United Nations Industrial Development Organization
VCCI	Vietnam Chamber of Commerce and Industry
VND	Vietnamese Dong
WTO	World Trade Organization

SECTION I

Introduction and Industry Overview

1

INTRODUCTION

Francis E. Hutchinson

For many senior managers in multinational corporations, visits to Malaysia, the Philippines, or China do not involve lengthy stays in Kuala Lumpur, Manila, or Beijing. Rather, they involve site visits to facilities in rather more distant regions, such as Johor, Cebu, or Chengdu — states or provinces that are far-removed from the national seat of power.

These sub-national regions have created environments that seem different — and slightly separate — from the rest of the country. This might be due to a particularly vibrant business community, a strongly-rooted outward orientation, a tradition of effective public administration, or an unusually deep pool of skilled workers. In these areas, investors do not liaise with the central government, but rather provincial or state government agencies for permits, licences or tax incentives. And, they source components, equipment, or design work from a variety of dynamic firms nearby. Finished goods and services are sold to neighbouring firms, or dispatched from the local air- or sea-port to international markets.

In the most successful up-and-coming regions, firm clusters attain technological capabilities far ahead of those in more established centres of industry. And, some of these states and provinces have managed to spawn activities in new sectors, even within national-level policy environments that are not conducive to innovation and higher value-added activities.

Over the past decades, Asia's geography of production has undergone a number of important changes. Technological change, growing numbers of multinational corporations, and the liberalization of trade and investment regimes are allowing production processes to be broken down and relocated to diverse locations in the pursuit of specific capabilities and cost differentials (Dicken 2003).

These changes signal heightened competition, but also increased opportunities for economies formerly on the periphery. More finely-dissected production processes, and a more hospitable policy environment have led to the development of regional production networks, which have enabled the proliferation of manufacturing activities across Asia. This "flying geese" phenomenon has seen lead countries climb the value chain, delegating less technologically-sophisticated activities to other countries in the region (Hamaguchi 2009).

On one level, this has made incursions into manufacturing easier, as countries or regions can begin producing specific components as opposed to entire finished goods (Joseph, this volume). On the other, while manufacturing activity has proliferated across Asia, over the past two decades it has also become more concentrated, tending to cluster in specific locations (Huang and Bocchi 2009).

In some cases, traditional centres of industry have benefited. However, in others, centres in relatively remote states or provinces have come to the fore. Thus, as subsequent chapters in this book will show, Cebu in the Philippines, Tamil Nadu in India, and Chengdu in China have emerged as important new centres for sophisticated manufacturing activities, negotiating their own entry into global production networks.

Uneven patterns of growth within countries are likely to be more prevalent in the future. Those regions that are more linked to the global economy through better "hard" and "soft" infrastructure, and are able to create and maintain an enabling environment for business will benefit. In national contexts where these assets constitute an exception, these regions may become "enclaves", with better international than local connections (Hill 2008).

This dynamic is further accentuated by measures undertaken by national governments. The pursuit of free trade and national competitiveness has been accompanied by moves to reduce regulatory obstacles to the movement of goods and services, capital, technology and workers. In addition, there is an emerging consensus that larger urban centres are important catalysts for growth. Thus, national governments are moving away from striving to minimize intra-regional inequalities towards seeking to capitalize on agglomeration economies and emerging centres of excellence (Williams et al. 2005; Merk, this volume).

Furthermore, the "Silent Revolution" of decentralization has seen the wide-spread devolution of responsibilities and revenue sources to lower levels of government. This has been a particularly important development for Asia, where many countries opted for strong central governments and top-down planning in the early post-independence period. In some cases, as in the cases of the Philippines and Indonesia, specific decentralization measures have been implemented. In others, national-level policies to deregulate have allowed sub-national governments to move in. India's liberalization measures and the subsequent revitalization of its state governments is a case in point.

These developments, then, entail a new set of challenges for sub-national governments. On the economic front, economic globalization offers greater opportunities for aspiring regions to break into production networks. However, this same flexibility entails less secure "tenure" for those states or provinces that have managed to break into these same networks, and in many cases, the most intense competition now comes from neighbouring regions (Tendler 2002). In addition, greater decentralization as well as fewer commitments from national level governments to "balanced" growth entails greater autonomy and stakes in this new policy environment.

However, while these developments do imply a heightened role for sub-national governments, relatively little has been written about what aspiring state and provincial governments can do to create and maintain a sustainable competitive advantage. Disciplines such as economics and political science have traditionally focused on national-level issues and imperatives (Montero 2002; Paul 2002).

That said, sub-national governments are not mere replicas of their national equivalents. They are embedded in a specific hierarchical position vis-à-vis their national counterparts, and need to reconcile national and local priorities. In addition, they have a reduced range of policy instruments at their disposal. Macroeconomic and trade policy is always the preserve of national governments, which may extend to encompass many aspects of infrastructure, education, and technology policy. Budgets are smaller, and sub-national governments may also be subject to hard budget constraints imposed by central governments.

However, sub-national governments also have important attributes that lend themselves to effective policy-making. State and provincial governments are often entrusted with unglamorous, but critical responsibilities for businesses such as land management, local infrastructure, and basic services. In addition, although budgets are smaller in an absolute sense, sub-national governments do have some financial wherewithal to make strategic investments in new sectors or to support promising emerging ones. Sub-national governments can also capitalize on their greater proximity to firms, businesses, and supporting infrastructure such as universities to steer collective efforts as well as mould the institutional environment in a way that supports economic activity in general and innovation in particular.

Research on selected cases in China, Russia, India, and Brazil has documented how, by selectively implementing national decrees, establishing links with the local private sector, and — where necessary — innovating, sub-national governments can foster the emergence and subsequent growth of promising economic activities (Segal and Thun 2001; Remick 2002; Sinha 2005; Tendler 2002). For example, based on their work on the municipal governments of Beijing and Shanghai, Segal and Thun make the following statement (to which the name of this book is due):

> Local governments do not simply try to reproduce and catch-up with development efforts initiated by the central government, but are often the actual architects of growth, designing and implementing development policies that are conducive to local institutional frameworks and specific development needs (2001, p. 558).

However, the path towards this goal is not straight-forward. National-level priorities have to be reconciled with local aspirations, social capital has to be built, and policies have to be formulated effectively. In particular, policy-makers at the sub-national level need to avoid a "race to the bottom", where promising sectors are solicited on the basis of incentives and infrastructure, rather than a genuine competitive advantage (Gray and Dunning 2002). Indeed, the challenge is to construct an enduring competitive advantage that is not easily replicated.

Drawing on cases largely — but not exclusively — from Asia, *Architects of Growth?* seeks to compare and contrast the experiences of ten up-and-coming sub-national regions that have sought or are seeking entry into the electronics sector. In exploring the issues of agency and effective policy approaches at the sub-national level, this book aims to shed light on a vital, but overlooked topic.

The sections ahead set out in greater detail the conceptual framework and research goals that guide this book and its constituent chapters.

ECONOMIC ACTIVITY AND LOCATION

Mapping out a nation's economy reveals that its industries are not evenly spread throughout its territory, but rather concentrated in a small number of locations. For example, Chicago and Detroit were synonymous with manufacturing in the past, and Silicon Valley in California and Route

128 in Massachusetts are currently known for their IT industries. Regions in Northern Italy are reputed centres for high quality fashion-wear; and Tokyo, London, and Hong Kong are centres for international finance. Specific areas or regions seem to be good at producing large numbers of new firms in certain sectors as well as attracting others from elsewhere.

What, then, is known about economic activity and location?

The theory of comparative advantage argues that "patterns of location, specialization, and trade will be driven by the geographical distribution of factors of production" (Storper 2002, p. 242). However, this framework is static and does not explain how regions with scarce endowments of capital and labour come to host new, more complex types of production such as manufacturing and services.

Indeed, until recently, many regions in Asia were characterized by scarce capital, and relatively abundant land and labour. It is thus pertinent to ask what enabled these regions to alter their comparative advantage, moving away from hosting simple, labour-intensive tasks towards more complex, capital-intensive ones.

There are well-known reasons why firms tend to agglomerate or cluster together.[1] Marshall contended that they seek to benefit from externalities or spillover effects, which benefit all firms in a given group. These externalities arise due to "traded" and "untraded" interdependencies. Traded interdependencies refer to direct transactions between firms, and mean that firms in an established cluster are more likely to benefit from a wider range of specialized suppliers. This, through better quality inputs, quicker delivery times, and more competitive prices, is held to increase the performance of all firms in the cluster. Clusters can also offer "thicker" labour markets which have more workers with required technical competencies. Untraded interdependencies are less tangible, and include benefits such as more opportunities for the interchange of ideas, techniques, technology, and business opportunities that arise from proximity between firms (Marshall 1890).

Further work has built on this foundation in a number of ways. For example, externalities can be "generalized", meaning that they are the inevitable spillovers that result when firms cluster together, regardless of their activities. Greater aggregate demand enables the emergence of a variety of infrastructural, economic, and social services that would not exist if demand were more dispersed. This occurs in large urban centres. However, externalities can also be "specialized", which entail benefits for firms that operate in the same or similar industries. These can occur in large urban centres or, indeed, smaller centres that reach a critical mass of firms engaged in compatible activities (Dicken 2003). It must be noted, however, agglomerations can also generate negative externalities such as pollution, increasing rent, labour poaching, or intellectual property theft.

Research by the Endogenous Growth school supports much of this argumentation. Romer (1986; 1990) and Lucas (1988) argue that, in addition to labour and capital, technology and knowledge are also factors of production. Technology and the knowledge required for its successful deployment for production are taken to be "endogenous" to a firm, as they are sought based on the conscious decisions made by firm owners and subsequently internalized, rather than something that occurs exogenously permeating all aspects of a given economy.[2] Investment in acquiring technology and human capital is also cumulative, as it builds on previous episodes of learning. Indeed, seeing as most of the benefits of learning are retained by the investing firm, it can acquire a competitive edge over its rivals, which can result in an oligopolistic market.

The successful acquisition of technology and its accompanying skills is therefore something that is highly location-specific, occurring in individual firms. Clustering can boost this phenomenon because the positive effects accruing from knowledge-creation and R&D can spill-over, benefiting other firms. Following this logic, the bigger the cluster of firms, the more spill-overs can be generated, and the better for the group as a whole. In addition, this approach raises the possibility of path-dependence, as small initial differences in the characteristics of different locations may prove determinant in the long-run. Once a given location begins to pull ahead, the cumulative effect of firm-specific learning and the ensuing spill-overs will enable it to pull ahead of its competitors.

However, both neo-classical and endogenous growth approaches perceive firms solely as competitors, uniquely engaging with each other through business transactions, with externalities occurring automatically. However, work from a variety of development economists and economic geographers places more emphasis on: the agency of firms and their ability to cooperate as well as compete; and the role that local traditions or customs can play in facilitating or hindering collaborative enterprises.

For example, the "Collective Efficiency" school argues that firms can actively seek to maximize the positive aspects of clustering. Joint efforts can include: inter-firm collaboration regarding products and processes that can improve knowledge and capabilities; or the creation of intermediary organizations to lower transaction costs between members and facilitate collaborative learning (Humphrey and Schmitz 1996).

Other work has approached firms as entities embedded in particular social contexts that shape their outlooks, actions and decisions and also affect the potential for collaboration. For example, the Regional Advantage school has looked at the role of the local institutional "environment" such as local customs, traditions, and attitudes, and how they can support economic growth and innovation through fostering collective attitudes regarding issues such as product quality, acceptable business practices, and openness to inter-firm collaboration (Saxenian 1994; Storper and Scott 2003).

This more nuanced, long-term, and social nature of collaboration and the pursuit of technological capabilities is complemented by evolutionary and institutional economists, who argue that the development of new technology is a sequential and path-dependent process that is shaped by the local institutional context. This refers particularly to norms and conventions that shape how knowledge is developed and then subsequently communicated between actors (Amin and Thrift 1992; Amin 1999).

Similarly, the Regional Innovation System school argues that many of the more productive aspects of agglomeration such as collaboration, collective efficiency, and generalized diffusion of knowledge do not occur automatically. A variety of market failures may prevent firms from accessing knowledge and opportunities. Non-excludable public goods such as basic research and development, a skilled workforce, or collective facilities will be under-developed due to the "free-riding" issue. Negative aspects of agglomeration may also stunt the growth of a promising cluster, or the demand for knowledge-enhancing services may be too incipient or dispersed to make private sector involvement rational. And, producers and consumers do not always have the necessary information regarding prices, products, markets or technologies (Cooke 2001; Giuliani and Bell 2008).

Informed by these approaches, Scott and Storper propose a useful framework for understanding the geographic distribution of new industries. They contend that when a given industry first emerges, production is geographically dispersed with few or no predominant centres. This is because sector-specific markets for labour or supporting services are undeveloped. Consequently, there is little incentive for firms to agglomerate. This state of affairs is termed "an open window of locational opportunity" (Storper and Scott 2003, p. 590).

However, over time, the industry in question becomes more concentrated, as those regions with more supplier markets and spillover effects, as well as favourable institutional environments begin to pull ahead. After a while, this "process of cumulatively self-reinforcing development" will mean that only a restricted number of locations will host a specific industry. Following this, the "window of opportunity" will close, as it will be hard for firms in new regions to outcompete those in more established ones.

The above-mentioned work depicts economic activity as something that is fundamentally influenced by the local environment within which it takes place. Positive externalities or, to be more specific, the balance of positive and negative externalities attracts firms to cluster in specific locations. Firms, consortia, and intermediary associations may act in a proactive manner to further bolster the benefits offered by proximity. And, a given region's institutional context also plays a role through collective attitudes to collaboration, business practices, and so forth.

However, some have argued that, regardless of the nature or cause of such externalities, technological advances are threatening to do away with the territorially-rooted nature of economic activity. For example, Friedman (2006) argues that the Internet, workflow software, as well as new forms of organizing production such as off-shoring, out-sourcing, and supply chaining are restructuring the way firms operate and allowing entrepreneurs and professionals from across the globe to produce and sell goods and services on the international market (2006). By enabling people in many sites to collaborate, conceptualize, design, and produce goods and services, location becomes, by and large, irrelevant.

That said, the emergence of high-performing regional economies in developing countries as well as increasing inter-regional disparities within developed countries suggests that there are, as Markusen states "sticky places in slippery space" (1996, p. 293). Indeed, while the barriers of space and time have been drastically reduced, this does not mean that distance is now meaningless. Rather, economic globalization has altered the importance of distance for different types of economic activity.

On the one hand, the perpetual transformation of intricate coordination functions into simple, routine activities that can be carried out at cheaper, but more distant, locations promotes dispersion. On the other, it is highly likely that tasks that rely on "tacit" knowledge and that require large amounts of face-to-face contact will continue to cluster in specific locations. Leamer and Storper argue that while modern telecommunications allow simple routine functions to be carried out at cheaper and more distant locations, a great deal of economic activity involves complex concepts and interactions that cannot be simplified and managed from afar. Certain production processes that rely on familiarity, trust, and large amounts of knowledge that cannot be codified will cluster (Leamer and Storper 2001). Archetypical examples of these activities include the financial services and fashion industries.

Indeed, given the benefits of agglomeration economies, it is also likely that routine, de-territorialized processes will also cluster in new locations. While firms in these industries cater to far-off clients, they are similarly dependent on a local labour market, specialized infrastructure, and a range of supporting services for competitiveness. This is seen in the concentration of business process outsourcing firms in urban areas such as Manila, Cebu, and Davao in the Philippines; and Bangalore, Chennai, as well as smaller cities such as Pune and Gurgaon in India.

What does this mean for aspiring regions away from capital cities and established centres of industry?

Of course, these regions may still find it hard to compete for government resources and private investment. In addition to the capital's proximity to decision-makers, primary urban centres also have privileged access to high-end infrastructure such as airports, stock exchanges, and financial institutions. Thus, in order to attract investment, aspiring regions and their urban centres must outperform the capital as well as other contending cities (Davis and Henderson 2003).

However, large cities, while offering advantages to firms, can also be affected by diseconomies of scale. As a national economy grows, its largest cities begin to suffer from issues such as rising costs for land and labour, over-charged infrastructure, and skill shortages. After a certain point, firms begin to relocate to suburban areas in the same cities in a bid to benefit from the externalities, while minimizing the effect of the diseconomies of scale. As these cities grow even more, firms may then decide to relocate to smaller cities (Henderson 2010).

The critical point at which firms decide to leave depends on whether they benefit from generalized or specialized externalities. Available research suggests that simpler activities benefit from specialized externalities and more complex activities benefit from those that are generalized (UNIDO 2009). Thus, aspiring regions may well be able to provide hospitable contexts for smaller, newer, and less sophisticated activities that benefit from specialized economies.

While not advocating the eclipse of the nation-state, the dynamism of a growing number of sub-national regions has led some to stress the importance of these geo-economic units and the need to understand the evolving relationship between local-level institutional contexts and the drivers of economic globalization (Scott 1998; Storper 1997). The OECD and the European Commission have incorporated regions as a central component in their economic development strategies, and the World Bank and the United Nations Industrial Development Organization are seeking to understand how agglomeration, economies of scale, and urban development can be better harnessed (OECD 2004; European Commission 2006; World Bank 2009; UNIDO 2009).

This work provides an initial point from which to begin analysing the links between economic activity, location, and policies. The next section will look at these issues in the Asian context and the subsequent sections will make more explicit links to policy.

EMERGING REGIONS IN ASIA

Asia has its share of large agglomerations or firm clusters — often in capital cities or the largest urban centres. For example, Taiwan and Korea have large electronics clusters in Taipei and Seoul, respectively. "The Big Four" of Beijing, Shanghai, Shenzhen, and Guangzhou are the established centres of manufacturing in China. In Southeast Asia, Jakarta, Bangkok, Manila, and to a lesser extent, Kuala Lumpur are home to a disproportionate share of their country's economic activity. This reaches extremes in the Philippines and Thailand, where Manila and Bangkok each account for more than half of total gross domestic product (Rimmer and Dick 2009).

There are reasons for this. Unlike Europe and North America, which have a more dispersed economic structure, economic activity in many Asian countries has been concentrated in a limited number of urban areas. In North America and Europe, industrialization was accompanied by the emergence of new and larger urban centres. In contrast, many Asian countries had few urban centres when they began industrializing. The largest urban centres constituted the biggest local markets and tended to have the best available infrastructure and skilled labour (Henderson 2010). Subsequently, economies of scale accompanied by economic nationalism, political centralization, and the establishment of customs procedures and trade barriers accentuated this. For example, due to many of the policies detailed above, by the 1970s influential secondary cities such as Surabaya, Cebu, Chiang Mai, and Penang were shadows of their former selves (Rimmer and Dick 2009).

Available evidence suggests that, despite the persistence of primary urban concentrations, Asian countries are experiencing more internally dispersed patterns of growth — which has repercussions for the internal distribution of manufacturing operations. Recent work by the World Bank shows that medium-sized cities in Asia are big enough to offer firms specialized externalities (Huang and Bocchi 2009). This mirrors findings from the United States and Japan, where smaller cities have emerged as specialized centres of production (Henderson 2010).

There are several drivers for this process. First, Asia looks to continue to enjoy high growth rates. The World Bank estimates that by 2025, Asia's contribution to global economic growth is likely to be at least 50 per cent, with eight countries — China, India, Japan, South Korea, Taiwan, Indonesia, Malaysia, and Thailand potentially having economies of over US$1 trillion (Gill and Kharas 2007). These countries also currently account for a growing percentage of exports of both manufactures and commercial services (UNIDO 2009; WTO 2009).

Second, countries in Asia are industrializing in a particular way. Trade liberalization, investments in infrastructure and education, as well as regional trade agreements have led to increasing product fragmentation and specialization in the fabrication of specific product components (Huang and Bocchi 2009). This form of industrialization lends itself to a range of smaller urban centres leveraging localization economies.

Third, countries in Asia are urbanizing rapidly. East Asia's 750 million urban inhabitants will be joined by an expected 550 million additional people by 2025 (Gill and Kharas 2007). By this year, China is expected to have a 59 per cent urbanization rate. Rates for Singapore, Korea, and Malaysia will be above 80 per cent and Indonesia and the Philippines will have rates between 50–55 per cent. India's urbanization rate is predicted to rise from 30 per cent in 2010 to 37 per cent in 2025. The country will experience an increase in the number of urban dwellers from 340 million in 2010 to some 590 million by 2030 (UN Population Division 2009; McKinsey Global Institute 2010).

However, this growth will not all be concentrated in capital cities, as secondary urban centres will increase in number and relative wealth. According to the World Bank's *An East Asian Renaissance*, over the period 2005–15, more than half of all growth in urban areas in East Asia will be in cities with less than 500,000 people (Gill and Kharas 2007). Greater incomes and ensuing investment in education and infrastructure means that many of these urban areas have the potential to become production sites (Boston Consulting Group 2010).

Thus, clusters of firms in high-technology sectors have appeared in secondary urban centres that are removed from capital cities or traditional sites of industry. In China and India, investing in "second tier" cities is an increasingly common phenomenon (Spire Consulting 2010; Hutchinson and Ilavarasan 2008). Regional dispersion policies enacted in Korea have led to important new clusters starting up in provinces away from the capital. In the Philippines, Cebu has built up a considerable agglomeration of electronics firms. In Malaysia, both Penang and Johor have important electronics clusters as well as important investments in the medical device and petrochemical sectors, respectively.

These developments call for an analysis of the current and potential roles of sub-national governments in seeking to foster and attract new industries.

THE ROLE FOR SUB-NATIONAL GOVERNMENTS

The implications of externalities, collective efficiency, and the institutional environment for long-term competitiveness imply a heightened role for state and provincial governments. Indeed, there are solid theoretical and practical reasons for sub-national governments to be active agents in fostering economic development.

The principle of subsidiarity argues that government institutions that are closer to their constituents are better-placed to deliver specific types of services, particularly those that provide localized benefits. This is because the level of government closest to end-users will have more information on their needs and the optimal combination of services and taxes for that jurisdiction (Oates 1999).

Furthermore, sub-national governments are subject to the disciplines of the market-place to a greater degree than their national counterparts. Tiebout argues that, assuming perfect information, citizens and firms compare the tax burdens and service delivery options of their constituency with others. They will then "vote with their feet" for the optimal combination of services and taxes. This competition between sub-national units will improve distributive efficiency, as they will be subject to quasi-market pressures and public goods and services will be parcelled out according to local needs and preferences (1956). This level of competition may make sub-national governments more responsive to the needs of local-level economic concerns than their national counterparts.

Third, the same proximity to end-users as well as the possibility of leveraging social capital with firms means that sub-national governments are in a better position than their national counterparts to shape and mould the institutional context in a way that supports economic activity. By tackling information asymmetries, negative externalities, and collective action failures, sub-national governments can maximize the potential positive effects of agglomeration (Storper and Scott, 2003; Landabaso et al., 2003).

In a move to catalyze growth and stimulate policy innovation, national governments in Asia are beginning to devolve responsibilities to lower levels of government. In some cases, this has been through the introduction of decentralization measures, as in the case of the Philippines and Indonesia. In others, such as India, it has been through a generalized "rolling back" of national-level regulatory frameworks that has allowed sub-national governments to move in. In still other cases, such as China, provincial or state-level governments were given financial incentives for achieving higher growth rates in their jurisdictions (Rudolph and Rudolph 2001; Yao 2009).

However, while greater agency at the sub-national level may be a welcome development, there is very little in the way of comparative work that examines quite how and under what conditions sub-national governments can pursue their own industrialization strategies effectively. Work on state-led development (Evans 1995; Kohli 2004), national innovation systems (Freeman 1995; Lundvall 1992; Nelson 1993), and varieties of capitalism (Hall and Soskice 2001) all have theoretical and practical insights but, for the reasons stated above, are not directly applicable to the sub-national level.

A small body of work within comparative politics on sub-national governments and their industrialization strategies offers much insight. Research on state and provincial governments in China, Mexico, Russia, India, and Malaysia shows that, as with their national counterparts, their effectiveness hinges on issues such as bureaucratic capacity and state-business relations. These characteristics are endogenous to the sub-national governments themselves and persist through regime changes at the national level (Segal and Thun 2001; Remick 2002; Sinha 2005; Hutchinson 2008).

In addition, distinct levels of capacity and varying state-society relations across a country's state or provincial governments mean that central priorities will be implemented differently at the local level, resulting in a "mosaic" effect. According to Gills and Philip "even within the same country, i.e., at the sub-national scale, local and regional differentiation can produce differential outcomes of economic policy pursued nationally" (1996, p. 586).

However, while the effectiveness of sub-national governments is also influenced by factors such as state capacity and state-business relations, they differ in a key way from their national counterparts. They are not entirely autonomous and are located in a subordinate position, assuming an intermediary role between the national state and the local population (Tendler 2002).

That said, there is room for agency in pursuing sub-national priorities — regardless of the regime type at the national level. Rather, state and provincial leaders gauge the impact of their decisions on two levels — national and local — the relative weight of which is dependent on the country's political context. Where sub-national leaders are elected, the weight attached to local priorities will increase. However, in contexts where sub-national leaders are appointed by national-level authorities or party heads, the importance placed on national-level considerations will be augmented correspondingly (Sinha 2005).

Thus, available research on sub-national governments shows that, despite their subordinate position vis-à-vis national governments, they can be agents with influence over their own economic fortunes. National and local priorities must be reconciled, and the degree of autonomy afforded to states and provinces is contingent on the country's political context which changes over time. As with their national counterparts, the ability of sub-national governments to successfully pursue their own economic strategies is influenced by their bureaucratic capacity and relationships with the private sector.

POLICY-MAKING AT THE SUB-NATIONAL LEVEL

In many cases, this new agency at the sub-national level has sparked competition between states and provinces. In the best of cases, attempts to catalyze economic growth have led to investment drives, substantial internal restructuring, and greater responsiveness towards local firms. On the other hand, uncritical attempts to attract new industries can result in duplication of efforts by nearby provinces or a "race to the bottom", where firms are courted exclusively through financial inducements rather than by proposing value-added attributes (Tendler 2002; Gordon and Cheshire 1998).

At the most aggregate level, the "industrial policy model" is characterized by measures aimed at achieving an overall macroeconomic environment conducive to structural transformation, accompanied by targeted policies to support specific sectors — albeit through direct state action or via encouraging the private sector. That said, while sub-national governments may have acquired new visibility and additional responsibilities, they usually do not have a wide range of tools or a large quantity of resources at their disposal.

However, while the scope of the responsibilities of state or provincial governments may differ in each country and their actions may be constrained by national-level institutions and initiatives, the key attribute is the authority to raise revenue — through taxation and other means — and spend on initiatives that fall within their mandated responsibilities (Gray and Dunning 2002).

The first area where sub-national governments can be proactive is prioritizing the efficient and effective delivery of services that are specified in their mandate. Land zoning and development including industrial parks, provision of utilities such as water and electricity, provision and maintenance of basic infrastructure, and processing of permits and licences to set up businesses are often competencies for sub-national governments.

While often overlooked, effective provision of these services can be a key component of a locationally-rooted competitive advantage. For example, a mapping exercise of state-level competitiveness in India showed the total number of permits required to start a business ranged from three in Chhattisgarh to 20 in Orissa (Oxford Analytica 2009). For this reason, Mexico, India, Indonesia, Vietnam, and the Philippines have begun yearly ranking exercises of the business environment in municipalities and states or provinces.[3]

However, beyond providing a conducive local environment for business, there are many other policies that sub-national governments can implement to attract investment, stimulate inter-firm collaboration, foster the acquisition of technological capabilities, and maximize the potentials of agglomeration. A survey of the Regional Advantage, Collective Efficiency, Regional Innovation Systems, and Clusters frameworks allows a basic typology of policies to be constructed (Scott and Storper 2003; Humphrey and Schmitz 1996; European Commission 2006; Andersson et al. 2003).

The first type can be termed "broker" policies, which encourage value-enhancing dialogue and collaboration among firms and supporting organizations. This includes: platforms for planning and exchange to encourage firm linkages through initiatives such as competence mapping or technology roadmaps; encouraging interactions between firms and generators of industry-relevant knowledge such as universities and research institutes; and collating statistics on local firms, national-level grants and loans, as well as international market trends. These activities can be restricted to facilitation processes, but can also be complemented with targeted investments in areas such as industrial parks and business incubators (Andersson et al. 2003). Broker policies can also be helpful in addressing negative externalities such as low levels of trust, labour poaching, or excessive competition on price as opposed to quality (Scott and Storper 2003).

The second is "demand-side" policies. Public procurement policies can be, and frequently have been, a source of demand for a new or fledgling industry. In addition, through encouraging "self-discovery", government action can reduce the risk and/or initial cost of new production technology or attempting to produce a new good for the market (Rodrik 2004). However, while public sector support through such policies has contributed to the emergence of many high-tech industries, this needs to be exercised with caution. First, there is potential for excessive intervention, leading to sub-optimal and overly-dependent new industries, as well as public influence on parameters such as budget, timing, and phase-out. Second, the public sector may not be the most demanding of clients, with technical requirements lagging behind those on the open market (Andersson et al. 2003).

The third type consists of tackling market failures in key areas. The standard is industry-relevant training, in which small and medium enterprises are notoriously unwilling to invest. Other related activities involve: providing public research facilities; investing in basic research; facilitating access to capital; encouraging the development and provision of technology-related services particularly in new sectors; and marketing the region national and internationally (Storper and Scott 2003; European Commission 2006).

These policy options are far from negligible, and when combined with greater proximity to, and social capital with, the private sector can be harnessed to considerable effect. In fact, it can be argued that these attributes make sub-national governments better placed to foster certain types of industrial activities than their national counterparts.

According to Weiss, successful economic transformation is defined as one of two outcomes: engineering a transition from an economy based on agriculture to one based on industries or services, termed *structural transformation*; or creating new production activities, speeding up technological learning, and disseminating innovative practices within a specific sector, termed *sectoral* or *industrial-technological transformation* (Weiss 1998, p. 66).

Doner (2007) argues that these two types of transformation are, in practice, very different, requiring substantially different levels of institutional capacity as well as policies. Structural transformation is understood to mean fostering the development of a more diverse industrial sector. In contrast, sectoral transformation is understood to mean the long-term process of indigenous technological learning. According to Doner, this is comprised of the following three aspects:

(a) Moving from lower value-added to higher value-added activities regarding processes, functions, products, or sub-sectors.
(b) Increasing the quantity and technological complexity of inputs from local firms.
(c) Satisfying the requirements of global value chains insofar as price, quality, and delivery are concerned.

He argues that policies have different degrees of complexity, which are determined by: the number of actors involved in taking decisions; the level of technical knowledge required for implementation; and the size and influence of potential "losers" in any allocation of resources. More complex policies require specific organizational attributes to be carried out effectively. These include the abilities to: maintain constant communication with private sector operators; construct and sustain relationships of trust; and monitor commitments made by the actors involved in the upgrading enterprise (2007, pp. 70–73).

Using this metric, Doner argues that policies aimed at encouraging structural transformation are less complex than those aimed at fostering upgrading. Thus measures such as enforcing property rights, ensuring macroeconomic stability, and directing credit are not overly onerous in terms of technical knowledge or working with large numbers of actors. Even more sophisticated policies such as those associated with trade liberalization can draw on policy measures implemented elsewhere.

Conversely, upgrading involves working with firms to: acquire indigenous technological capabilities; foster inter-firm linkages; and improve internal processes to meet the requirements of global production networks. This involves implementing measures to: establish linkages with foreign firms; secure access to new markets; absorb new technology; work with other firms to create value chains; and pursue constant upgrading of the work-force. In turn, these measures require: working with greater numbers of actors; acquiring a deep reservoir of technological knowledge; and accepting longer-term diffuse benefits for participants.

While many of the policies required to pursue structural transformation lie out of the realm of sub-national governments, much of what the pursuit of sectoral transformation entails lies within the purview of many sub-national governments. Given the importance of local knowledge, proximity, and social capital for fostering sectoral upgrading, sub-national governments may be able to more closely approximate the institutional conditions necessary for upgrading. And, tackling information externalities, collective action dilemmas, and providing industry- or location-specific public goods such as technical training and collective facilities may well be feasible for sub-national states.

Having argued that sub-national governments can and do have agency in fostering the development of industry, and having laid out some of the types of policies that can be implemented to this effect, the next section will set out this book's key aims, sector of interest, and choice of cases.

RESEARCH DESIGN

This book aims to deepen our understanding of how and whether sub-national governments can play a significant role in promoting industrial transformation. In order to do this, it will compare and contrast ten cases of emerging regions largely, but not exclusively, from Asia.

In this book, sub-national government is taken to mean a territorial sub-division of a national government,[4] specifically a meso-level unit such as a state or province. It is taken to be synonymous with region, which is defined as "a meso-level political unit set between the national or federal and local levels of government that might have some cultural or historical homogeneity but which at least ha[s] some statutory powers to intervene and support economic development, particularly innovation" (Cooke 2001, p. 953).

Given the variance in governance structures across nations, this exercise will compare state, provincial, and municipal governments in countries with varying: income levels; technological capabilities; and separation of powers between national and sub-national governments.

The analytical focus of the project will be strengthened by focusing on one sector, enabling the different regions' progress to be compared along the same metric. Thus, cases will be subjected to the same market cycles, industry trends, technological requirements, and production standards. In this case, the focus industry will be the electronics sector, which is taken to comprise the sub-groups of: electronics components; industrial equipment; and consumer electronics (Dicken 2003, p. 400).

Using Weiss' definition of economic transformation, regions will be assessed according to their efforts to foster *structural* and, more particularly, *sectoral* transformation.

The choice of electronics as the sector of interest has two main advantages. First, given its potential for structural change and foreign exchange-earning potential, the electronics sector has traditionally been seen as an attractive export-oriented industry. Consequently, it is quite wide-spread in the region, and offers a range of potential industry centres and policy approaches for study. Second, the electronics sector has traditionally been concentrated in the United States, Europe, and Japan. As a result, all of the cases proposed below are late entrants to this sector, which turns the attention to policy and constructed competitive advantage, as opposed to pre-existing institutional contexts.

This book has a strong comparative policy focus, with an aim to draw out lessons learned for decision-makers at the national and sub-national levels. In order to facilitate effective comparison and contrast, the chapters ahead are centred on the themes and questions that follow.

(a) The electronics industry
This involves analysing the growth and dynamism of the given region's electronics sector. Key aspects include: number of firms/employees; composition by sub-sector; ownership structure; skill and capital-intensity of tasks undertaken by firms; the extent and nature of linkages of firms within locally-present clusters; and evidence of product/process innovation.[5]

(b) Overall national context
This involves examining the overall national economic and policy context within which each sub-national case operates. Central questions include: the relations between the state/provincial and national authorities regarding economic policy and industrial development; the formal and informal responsibilities that state/provincial authorities possess regarding industrial policy and economic development; the revenue sources available to sub-national governments; and whether there is an explicit regional policy in place.

(c) The sub-national context
This requires establishing: how the given state/provincial governments sought to promote the emergence and subsequent development of the electronic sector; what initiatives were pursued to this effect (improving the local business environment, broker policies, demand-side policies, and tackling market failures); what role the private sector played in pursuing collective efficiency; and the relationship between the sub-national state and the local private sector.

In order to explore these questions, this book is grouped into five parts.

In the first part, this introduction is followed by an overview of key trends in the electronics sector and the challenges inherent in negotiating entry into global production and innovation networks for sub-national regions in Asia.

The second, third, and fourth parts group the ten cases so as to first facilitate comparison within regions and, subsequently, between them.

The second part brings together four cases from industrializing countries in Southeast Asia. The province of Cebu, Philippines, and the state of Johor, Malaysia stand out as cases of locally-based economies that have been shaped by considerable sub-national state initiative, emerging as centres of electronics production in their own right. The chapters on Da Nang, Vietnam, and Thailand provide useful cases of municipal and provincial governments in countries with a strong tradition of centralized administration. Da Nang provides an interesting example of a geographically remote industrializing region that capitalized on greater amounts of autonomy to dramatically improve its regulatory framework and infrastructure. The chapter on Thailand puts forward the dynamics between the country's elected provincial representatives and centrally-appointed provincial governors and what this means for sub-national agency.

The third part consists of chapters on Chengdu, People's Republic of China, and Tamil Nadu, India. The PRC has enacted important decentralization measures and has also seen the rise of a number of dynamic regions that are seeking to capture the excess demand from primary centres of industry. Chengdu is one such case, as the city has emerged as a key inland manufacturing centre. Tamil Nadu, for its part, has rapidly emerged as the pre-eminent centre of export-oriented electronics manufacture. Enabled by the 1991 liberalization measures and spurred on by competition from its neighbouring territories, the state government has moved to improve the local business climate, upgrade its skill base, and attract and retain foreign direct investment.

The fourth part of the book brings together cases from industrialized countries. Singapore is, of course, a city-state. However, while its government remit is more extensive than any sub-national government, the extent to which it leverages on its proximity to firms provides much insight into the more hands-on policies open to state and provincial governments. Gumi, Korea, is a case of both national and sub-national governments working in tandem to decentralize industrial activity. The case of Kaohsiung, Taiwan is an interesting counter-example, as it is the classic "secondary" region that struggles to outcompete the capital region. The final chapter on North Brabant, Netherlands, is from further afield and provides an interesting case of a provincial government with considerable influence, but little formal power.

ADDITION TO KNOWLEDGE

From a disciplinary point of view, this book is aimed at regional studies/ economic geography, economics, and comparative politics. Work on analysing political and economic outcomes in multi-levelled systems of governance is just beginning and is, as yet, limited to very few cases. This book aims to form part of this emergent tradition, placing sub-national states at the centre of analysis. Several reasons militate for a sub-national perspective on the relationship between institutions and economic outcomes.

First, in restricting analysis to aggregate data or institutional contexts, much economic and political research suffers from what Rokkan et al. term "whole-nation" bias, which overlooks variations within countries, particularly between "competing economic, political, or cultural centres" (1970, p. 49). However, institutions, taken to mean "humanly devised constraints that shape human interaction", are influenced by local, political, historical, and cultural specificities. Some of these institutional contexts will be more conducive to specific types of economic activity than others, enabling them to grow more successfully (Martin 2000). Studying regions within countries also forces researchers to be more precise in operational and theoretical terms, thus producing "greater gradation in existing concepts" and greater nuance in applying national level concepts and ideas to the sub-national level (Sinha 2012).

Second, a uniquely national perspective obviates the initiative of sub-national actors in shaping their own destinies. An emerging body of research on industrializing countries shows how differences in structure, capacity, and strategy at the sub-national level explain the varying effectiveness of national-level initiatives in different parts of a country. In addition, rather than being mere extensions of national-level governments, sub-national governments have their own interests to advance. In turn, policy approaches are selected based on these, and are affected by local state-society relationships.

Third, a sub-national focus helps address the small "N" problem that has plagued much of the state-led development literature. This is because there are relatively few successful instances of successful government-facilitated industrialization at the national level. However, classifications are influenced, often unduly, by national level aggregates and may hide successful initiatives undertaken by sub-national governments (Snyder 2001). Thus, looking at governments, policy frameworks, and public-private sector interaction at the sub-national level dramatically increases the cases available for study, and may also uncover unexplored dynamics.

However, it must also be stated that the sub-national approach is not without its own methodological challenges. First, the definition of sub-national units such as provinces, states, and municipalities differs across countries and may change over time. Second, states and provinces often do not have "hard" borders. Rather, their boundaries can be "fluid and shifting", with greater or less degrees of permeability and permanence (Moncada 2012). Third, regardless of the degree of decentralization of power, sub-national units are not detached from national governments and are subject to influence from above (Thun 2006).

That said, many of the same issues of impermanence and cross-border influence also affect national governments (Moncada 2012). And, while sub-national governments are undeniably influenced by their national counterparts, influence also flows the other way. Sub-national governments can influence central governments through mechanisms such as lobbying, policy transfer, selective implementation of national-level decrees, and even fiscal behaviour (Sinha 2005; Wibbels 2000).

However, while much research focussed on the national level does not explicitly factor in these issues, it is hoped that a sub-national perspective will be more open to these issues and influences. This perspective will not only deepen our understanding of the interaction between government action and economic outcomes at the sub-national level, but will also shed light on the necessary conditions for effective government agency in general.

Notes

1 A cluster is defined as a "spatial and sectoral concentration of firms". Timothy Bresnahan, Anthony Gambardella, and Annalee Saxenian, "'Old Economy' Inputs for 'New Economy' Outcomes: Cluster Formation in the New Silicon Valleys". *Industrial and Corporate Change*, vol. 10, no. 4 (2001): 836.

2 The Neo-classical model assumes that, apart from an initial investment, acquiring technology has no further costs, constraints, or risks. Thus, technology acquisition is implicit in capital investment and can be acquired by any firm anywhere. Those influenced by the endogenous growth school contend, in contrast, that firms do not dispose of complete information on a particular technology and its alternatives and technical "mastery" of an acquired technology is not effortless and cheap. Rather, effective use of technology has a strong "tacit" element that requires investments in skills, organizational practices, and technical expertise. The process is inherently risky as investments will not necessarily result in increased productivity, and is difficult for firms seeking to enter a new sector, who must learn how to use a particular technology while simultaneously competing against those who have mastered it (Lall 2004).

3 <http://www.doingbusiness.org/reports/subnational-reports> (accessed 24 May 2012).

4 A government is defined as "the formal institutions, offices, processes and personnel through which the day-to-day running of a country, the maintenance of public order and the distribution of resources is managed and maintained" (Axford et al. 2002, p. 330).

5 A useful distinction between types of firm linkages is that of: Marshallian Districts; Hub-and Spoke Districts; Satellite Platforms; and State-Anchored Cities. See Ann R. Markusen 1996, pp. 293–313, pp. 337–40.

References

Amin, A. "An Institutionalist Perspective on Regional Economic Development". *International Journal of Urban and Regional Research* 23 (1999): 365–78.

Amin, A. and N. Thrift. "Neo-Marshallian Nodes in Global Networks". *International Journal of Urban and Regional Research* 12 (1992): 171–86.

Andersson, T., S. Schwaag Serger, J. Sorvik, and E.W. Hansson. *The Cluster Policies Whitebook*. Malmo: International Organization for Knowledge Economy and Enterprise Development, 2003.

Axford, B., G.K. Browning, R. Huggins, and B. Rosamund. *Politics: An Introduction*. Second Edition, London: Routledge, 2002.

Boston Consulting Group. "Winning in Emerging-Market Cities: A Guide to the World's Largest Growth Opportunity". BCG Perspectives, September 2010.

Cooke, P. "Strategies for Regional Innovation Systems: Learning Transfer and Applications". Cardiff: Centre for Advanced Studies, Cardiff University, 2001.

Davis, J. and J.V. Henderson. "Evidence on the Political Economy of the Urbanization Process". *Journal of Urban Economics* 53 (2003): 98–125.

Dicken, P. *Global Shift: Transforming the World Economy*. Fourth Edition. London: Sage Publications, 2003.

Doner, R. *The Politics of Uneven Development: Thailand's Economic Growth in Comparative Perspective*. Cambridge: Cambridge University Press, 2007.

European Commission. "Constructing Regional Advantage: Principles, Perspectives, Policies". Brussels: Directorate-General for Research, European Commission, 2006.

Evans, P. *Embedded Autonomy: States and Industrial Transformation*. Princeton: Princeton University Press, 1995.

Freeman, C. "The 'National System of Innovation' in Historical Perspective". *Cambridge Journal of Economics* 19, no. 1 (1995): 5–24.

Friedman, T.L. *The World is Flat: The Globalized World in the Twenty-First Century*. London: Penguin, 2006.

Gill, Indermit and Homi Kharas. *An East Asian Renaissance: Ideas for Economic Growth*. Washington D.C.: International Bank for Reconstruction and Development, 2007.

Gills, B. and G. Philip. "Editorial: Towards Convergence in Development Policy?: Challenging the 'Washington Consensus' and Restoring the Historicity of Divergent Development Trajectories". *Third World Quarterly* no. 4 (1996): 585–92.

Giuliani, Elisa and Martin Bell. "Industrial Clusters and the Evolution of their Knowledge Networks: Revisiting a Chilean Case". SPRU Electronic Working Paper Series No. 171. Brighton: Science Policy Research Unit, University of Sussex, 2008.

Gordon, Ian R. and Paul C. Cheshire. "Territorial Competition: Some Lessons for Policy". *The Annals of Regional Science* 32, no. 3 (1998): 321–46.

Gray, H.P. and J.H. Dunning. "Towards a Theory of Regional Policy". In *Regions, Globalization, and the Knowledge-Based Economy*, edited by J.H. Dunning. Oxford: Oxford University Press, 2002.

Hall, Peter and David Soskice. *Varieties of Capitalism: The Institutional Foundations of Comparative Advantage*. Oxford: Oxford University Press, 2001.

Hamaguchi, N. "Regional Integration, Agglomeration, and Income Distribution in East Asia". In *Reshaping Economic Geography in East Asia*, edited by Yukon Huang and Alessandro Magnoli Bocchi. Washington D.C.: World Bank, 2009.

Henderson, J.V. "Cities and Development". *Journal of Regional Science* 50, no. 1 (2010).

Hill, H., Budy Resosudarmo and Yogi Vidyattama. "Indonesia's Changing Economic Geography". CCAS Working Paper No. 12. Kyoto: Doshisa University, 2008.

Huang Yukon and Alessandro Magnoli Bocchi. "Lessons from experience: reshaping economic geography in East Asia". In *Reshaping Economic Geography in East Asia,* edited by Yukon Huang and Alessandro Magnoli Bocchi. Washington D.C.: World Bank, 2009.

Humphrey, J and H. Schmitz. "The Triple C Approach to Local Industrial Policy". *World Development* 24, no. 122 (1996): 1859–77.

Hutchinson, F.E. "Developmental States and Economic Growth at the Sub-National Level: The Case of Penang". In *Southeast Asian Affairs 2008*, edited by D. Singh and Tin Maung Maung Than. Singapore: Institute of Southeast Asian Studies, 2008.

Hutchinson, F.E. and P. Vignesawara Ilavarasan. "Open Windows of Opportunity and Local Interdependencies: The IT/ITES Sector and Economic Policy at the Sub-national Level in India". *Economic and Political Weekly* 43, no. 46 (2008): 64–70.

Kohli, A. *State-Directed Development: Political Power and Industrialization in the Global Periphery*. Cambridge: Cambridge University Press, 2004.

Lall, S. "Reinventing Industrial Strategy: The Role of Government Policy in Building Industrial Competitiveness". G-24 Discussion Paper 28. Oxford: Oxford International Development Centre, 2004.

Landabaso, M., Christine Oughton, and Kevin Morgan. "Learning Regions in Europe: Theory, Policy, and Practice through the RIS Experience". In *Systems and Policy for the Global Learning Economy*, edited by David V. Gibson, Chandler Stolp, Pedro Conceicao, and Manuel V. Heitor. London: Praeger, 2003.

Leamer, E. and M. Storper. "The Economic Geography of the Internet Age". NBER Working Paper 8450. Cambridge, MA: National Bureau of Economic Research, 2001.

Lucas, R. "On the Mechanics of Economic Development". *Journal of Monetary Economics* 22, no. 1 (1988): 3–42.

Lundvall, B.-A. *National Systems of Innovation: Towards a Theory of Innovation and Interactive Learning*. London: Pinter Publishers, 1992.

Markusen, A. "Sticky Places in Slippery Space: A Typology of Industrial Districts". *Economic Geography* 72, no. 3 (1996): 293–313.

Marshall, A. *Principles of Economics*, 1st edition. London: Macmillan, 1890.

Martin, R. "Institutional Approaches to Economic Geography". In *A Companion to Economic Geography*, edited by E. Sheppard and T.J. Barnes. Oxford: Blackwell Publishers, 2000.

McKinsey Global Institute. "India's Urban Awakening: Building Inclusive Cities, Sustaining Economic Growth". Boston: Mckinsey Global Institute, 2010.

Moncada, E. and R. Snyder. "Subnational Comparative Research on Democracy: Taking Stock and Looking Forward". *Comparative Democratization* 10, no. 1 (2012).

Montero, A.P. *Shifting States in Global Markets: Subnational Industrial Policy in Contemporary Brazil and Spain.* University Park: Penn State University Press, 2002.

Nelson, R.R. *National Systems of Innovation: A Comparative Analysis.* Oxford: Oxford University Press, 1993.

Newlands, D. "Competition and Cooperation in Industrial Clusters: The Implications for Public Policy". *European Planning Studies* 11, no. 5 (2003): 521–32.

Oates, W.E. "An Essay on Fiscal Federalism," *Journal of Economic Literature*, Vol. XXXVII, September (1999): 1120–49.

OECD. *New Forms of Governance for Economic Development.* Paris: Organization for Economic Cooperation and Development, 2004.

Ohlin B. *Interregional and International Trade.* Cambridge MA: Harvard University Press, 1933.

Oxford Analytica Consulting. *India Deconstructed: Risks and Opportunities at State Level.* Oxford: Oxford Analytica Consulting, 2009.

Paul, D. "Re-scaling IPE: Subnational States and the Regulation of the Global Political Economy". *Review of International Political Economy* 9, no. 3 (2002): 465–89.

Remick, E. "The Significance of Variation in Local States: The Case of Twentieth Century China". *Comparative Politics* 34 (July 2002): 399–419.

Rimmer, P.J. and Howard Dick. *The City in Southeast Asia: Patterns, Processes and Policy.* Singapore: NUS Press, 2009.

Rodrik, D. *Industrial Policy for the Twenty-First Century.* Faculty Research Working Paper RWP 04-047. Cambridge MA: JFK School of Government, Harvard University, 2004.

Rokkan S. with A. Campbell, P. Torsvik, and H. Valen. *Citizens, Elections, Parties: Approaches to the Comparative Study of the Processes of Development.* Oslo: Scandinavian University Books, 1970.

Romer, P. "Increasing Returns and Long-Run Growth". *Journal of Political Economy* 94, no. 5 (1986): 1002–37.

———. "The Origins of Endogenous Growth". *The Journal of Economic Perspectives* 8, no. 1 (1990): 3–22.

Rudolph, L.I. and S.H. Rudolph. "The Iconization of Chandrababu: Sharing Sovereignty in India's Federal Market Economy". *Economic and Political Weekly* 36, no. 18 (2001): 1546–60.

Saxenian, A. *Regional Advantage: Culture and Competition in Silicon Valley and Route 128*. Cambridge MA: Harvard University Press, 1994.

Scott, A.J. and M. Storper. "Regions, Globalization, Development". *Regional Studies* 37, nos. 6–7 (2003): 579–93.

Segal. A. and E. Thun. "Thinking Globally, Acting Locally: Local Governments, Industrial Sectors, and Development in China". *Politics and Society* 29, no. 4 (2001): 557–88.

Schmitz, H. "Small Shoemakers and Fordist Giants: Tales of a Supercluster". *World Development* 23, no. 1 (1995): 9–28.

Scott, A.J. *Regions and the World Economy: The Coming Shape of Global Production, Competition, and Political Order*. Oxford: Oxford University Press, 1998.

Snyder, R. "Scaling Down: The Subnational Comparative Method". *Studies in Comparative International Development* 36, no. 1, Spring (2001): 93–110.

Sinha, A. "Scaling Down and Up". *Comparative Democratization* 10, no. 1 (2012).

———. *The Regional Roots of Developmental Politics in India: A Divided Leviathan*. Bloomington: Indiana University Press, 2005.

Spire Research and Consulting. "The Next Frontier in Asia: How Asia's Second-tier Cities are the World's New Marketing Arena". *Spire-Journal* Q3, 2010.

Storper, M. "Globalization and Knowledge Flows: An Industrial Geographer's Perspective". In *Regions, Globalization, and the Knowledge-Based Economy*, edited by J.H. Dunning. Oxford: Oxford University Press (2002), pp. 42–62.

———. *The Regional World: Territorial Development in a Global Economy*. New York: The Guildford Press, 1997.

Storper, M. and A.J. Scott. "The Wealth of Regions: Market Forces and Policy Imperatives in Local and Global Context". *Futures* 27, no. 5 (2003): 505–26.

Tendler, J. "The Economic Wars Between the States". Paper presented at the OECD/State Government of Ceara Meeting on Foreign Direct Investment and Regional Development, Fortaleza, 12 December 2002.

Thun, E. *Changing Lanes in China: Foreign Direct Investment, Local Governments, and Auto Sector Development*. New York: Cambridge University Press, 2006.

Tiebout, C.M. "A Pure Theory of Local Expenditures". *The Journal of Political Economy* 64, 5 (1956): 416–24.

UNIDO. *Industrial Development Report 2009: Breaking In and Moving Up: New Industrial Challenges for the Bottom Billion and Middle-Income Countries*. Vienna: UNIDO, 2009.

UN Population Division. *World Urbanization Prospects, The 2009 Revision*. New York: UN Department of Economic and Social Affairs, 2009.

Weiss, L. *The Myth of the Powerless State: Governing the Economy in a Global Era.* Cambridge: Polity Press, 1998.

Wibbels, E. "Federalism and the Politics of Macroeconomic Policy and Performance". *American Journal of Political Science* 44, no. 4 (October 2000): 687–702.

Williams, G., Landell-Mills, P., and A. Duncan. "Uneven Growth Within Low-income Countries: Does It Matter and can Governments Do Anything Effective". Brighton: The Policy Practice Ltd., 2005.

World Bank. *World Development Report 2009: Reshaping Economic Geography.* Washington D.C.: World Bank, 2009.

WTO. *International Trade Statistics, 2009.* Geneva: World Trade Organization, 2009.

Yao Yang. "The Political Economy of Government Policies Toward Regional Inequality in China". In *Reshaping Economic Geography in East Asia,* edited by Yukon Huang and Alessandro Magnoli Bocchi. Washington D.C.: World Bank, 2009.

2

HARNESSING ASIAN CAPABILITIES FOR TRANSFORMING THE ELECTRONICS AND IT SECTORS: RECENT TRENDS, CHALLENGES AND A WAY FORWARD

K.J. Joseph

INTRODUCTION

Path-breaking changes in electronics technology in general and microelectronics in particular have given birth to the Information Technology (IT) revolution and its related innovations. In the closing decades of the last century, the IT revolution facilitated not only technological changes, but also far-reaching institutional changes as seen through deepening globalization processes (Soete 2006).

Perhaps there is no other region like East Asia, which took advantage of the microelectronics revolution and rise of export-oriented production, preludes to today's IT revolution and globalization. A considerable body of work indicates that outward-oriented economic policies, accompanied by a catalytic role for foreign firms and targeting of global production networks within a context of activist state policy has been at the core of the East Asian Miracle (Ernst and Kim 2002; Rodrik 1992; Sen 1983). Drawing inspiration from the development experience of Southeast Asian countries, many less developed countries, especially in Asia, are striving to promote industrial transformation by developing their electronics and IT production base and promoting the use of IT. The moot question is "How rocky is the road ahead for these countries?"

As globalization processes and the IT revolution gained momentum, there were significant changes in the organization of electronics production that led to the establishment of Global Production Networks (GPN), which later evolved into Global Innovation Networks (UNCTAD 2005). This was facilitated to a great extent by the formation of the World Trade Organization, which led to the widespread dismantling of barriers to trade and investment. For IT goods, this was further accentuated by the Information Technology Agreement (Joseph and Parayil 2008).

These developments also encouraged Outward Foreign Direct Investment (OFDI) from developing nations, such as China, India and other countries, with a view to further strengthening their technological capabilities and enhancing their market access (Pradhan 2004; Kumar and Chadha 2009). The process of trade and investment liberalization was further accentuated with the establishment of a multitude of regional trading agreements (UNCTAD 2004). One notable case with regard to electronics has been the e-ASEAN Framework Agreement, which involved liberalized trade and investment on the one hand and capacity-building on the other (Joseph 2006).

All these processes resulted in an unprecedented rate of increase in trade and investment between countries in the South in general and Asia in particular (UNCTAD 2005*b*). Given that the world economy is now more open and integrated, the international division of labour is now much less constrained than previously. Moreover, for developing countries, an open world economy facilitates the importation of ideas, technologies, and know-how that can be used to establish production capabilities even in high-technology sectors such as electronics.

However, the purchase of technology is not a panacea. As argued by Freeman (2011), the purchaser always receives a more reduced information set than that possessed by the seller. Therefore technologies cannot simply be taken off the shelf and put into use effortlessly. Without a functioning infrastructural base accompanied by investment in education, training, R&D and other scientific and technical activities, very little can be accomplished by way of assimilation of imported technologies. Hence the key issue, and central concern of this paper, relates to the current positioning of those developing countries aspiring to develop their electronics production base, in terms of their ability to offer a conducive institutional environment and infrastructural facilities for achieving their intended goals.

The remainder of this chapter is organized as follows. Section two presents the analytical framework, which articulates the bearing of a country's trade and investment regimes with its institutional architecture and infrastructural facilities (both human and physical). Section three examines the implications for industrially-aspiring sub-national states of recent developments such as the advent of Global Production and Innovation Networks, as well as the emergence of new players from Asia as sources of outward foreign direct investment. The last section presents a perspective wherein, apart from highlighting the need for building an institutional architecture at the sub-national and sectoral level, a call is made for an e-Asia Framework Agreement that involves not only liberalized trade and investment but also built-in provisos for capacity building (technological, physical and human) at the sub-national level within the framework of South-South Cooperation.

ANALYTICAL BACKGROUND

The successful pursuit of industrial transformation assumes key importance given that available information indicates that the returns to globalization and integration with global production and innovation networks have been unevenly distributed. The East Asian "Tigers" of Singapore, South Korea, and Taiwan have been able to build up their technological capabilities, successfully moving from the status of Original Equipment Manufacturers (OEM) to Original Brand Manufacturers (OBM) (Hobday 1994; 2002). In turn, they have emerged as major sources of OFDI in their own right. Beyond establishing a solid IT hardware production base, they have successfully harnessed information technology for development by ensuring its diffusion to different sectors of their economies and societies.

However, in the cases of Thailand, Malaysia, and Indonesia, their successful entries into various electronics global production networks have not been matched by a commensurate ability to develop a software base or harness IT for Development. Based on a detailed analysis of the electronics industries in Southeast Asia, Ernst (2001) argued that, due to the "sticky" specialization of exportable commodities, simple export-oriented production can no longer guarantee sustained growth and welfare improvement. Moreover, a narrow domestic knowledge base has led to limited industrial upgrading and limited backward and forward linkages. Evidence also indicates that IT-induced prosperity in general and electronics production in particular has been confined to a few locations, leading to an enclave type of development and contributing towards widening regional and personal inequalities (Joseph 2006).

Based on various studies, Ernst (2001) argues the need for industrial upgrading in most Southeast Asian countries. The issue of industrial upgrading is the most pertinent for countries that remain at the low end of the global production network. However, for a large number of countries, the development of an electronics production base, as well as broader IT-based development, remains a distant dream. In these cases, the key is to identify the strategies that will enable them to make an entry into a global production network.

While addressing recent trends and their implications for industrial transformation at the sub-national level, it is important to have an understanding of the significance of the electronics sector in the ongoing IT revolution. Though the relative share of hardware components has declined over time, as Ernst (2001) rightly remarked, both are complementary and need each other. Hence, it is necessary to have a fair understanding on the ways and means by which electronics contributes to development in general and industrial transformation in particular. This may be viewed broadly at two levels: (a) on account of the growth of electronics sector — the hardware and software; and (b) on account of its contribution towards facilitating IT-induced prosperity. The former refers to the contribution in output, employment and export earnings from electronics which has been well-documented in many East Asian countries. The latter refers to the fact that, without hardware, the IT software alone cannot deliver increased efficiency, productivity and competitiveness through facilitating information exchange and reducing transaction costs.

However, in recent literature on IT and Development, the focus of attention has been essentially on IT use and only limited attempts have been made of examining policies towards electronics production. As argued by Mytelka and Ohiorhenuan (2000), the often-suggested strategies place developing countries in a situation of perpetual *attente* — waiting for the transfers of technology from the North and focusing their attention on the need to attract multinational corporations to their shore. The studies on technology diffusion, however, have shown that, along with demand-side factors, supply-side factors are also important determinants of diffusion. Hence, the greater domestic availability of electronics goods acts as a catalyst in the process of diffusion. Therefore, enhancing the diffusion of IT need not imply a neglect of electronics production. To the extent that the present levels of income are important determinants of IT use, there is no reason why sub-national states need to forgo the income earning opportunities offered by the production of electronics goods, which could also be instrumental in industrial transformation at the sub-national level.

Production of IT Goods and Services: Challenges and Potential

Studies have shown that in the U.S., where the macroeconomic benefits of IT revolution are already apparent, the electronics sector accounted for about 8.3 per cent of the GDP and nearly a third of GDP growth between 1995 and 1999 (U.S. Department of Commerce 2000). Electronics production also contributed to lowering inflation rates, since a growing proportion of economic output has been in sectors marked by rapidly falling prices.[1] Benefits from electronics production have not been confined to the U.S. alone. As already noted, the electronics sector has shown to be a major source of economic output, exports and job creation in countries like South Korea, Singapore, Taiwan and others. Therefore, it appears that sub-national states in the developing world could gain a great deal by focusing on the production of electronics, instead of the present lop-sided approach towards IT use.

It has, however, been argued that, given the electronics sector's very high entry barriers, it is not necessarily an easy proposition for aspiring sub-national states in the South to enter global production networks. For example, industry segments such as microprocessors and key types of electronic equipment are almost closed because standards are set by leading IT players, mainly U.S. companies such as Intel, Cisco and others. Other segments of the electronics industry are highly capital-intensive, scale-intensive and require specialized skills that only a few countries and regions can hope to achieve (Kraemer and Dedrick 2001). Moreover early entrants such as Singapore, Hong Kong, South Korea, Taiwan, and Ireland have preempted many of these opportunities by securing strategic positions in many GPNs.

While there is some merit in the above argument, a closer look at the characteristics of the electronics sector reveals that the doors are not that firmly closed for newcomers. The electronics sector is a multi-product industry and is characterized by a wide range of GPNs dedicated to each. In broad terms, the sector may be divided into two categories — equipment and components. Electronics equipment may be separated into consumer electronic products and electronic capital goods, although this distinction is increasingly getting blurred. The former comprises audio and video equipment and other consumer equipment. The latter, in turn, comprises the following broad categories: medical equipment; control instruments and industrial electronic equipment; computers; and communication equipment. Each of these broad categories are comprised of a large number of sub-groups and final products. Similarly, the electronic components sub-sector may be broadly divided into: active components; passive components; and electro-mechanical components. In each of these broad categories there are again a large number of products. Thus, there are a wide range of products that come under the electronics sector and they vary in terms of technological sophistication, dynamism and investment requirements (Joseph 2006; 1997). In addition, the demand for electronic goods is likely to increase as the rate of IT diffusion increases both in the developed and developing world. Therefore, the key issue in the context of recent developments is to locate the factors that facilitate the ability of aspiring sub-national states to profitably enter electronic hardware production networks.

Facilitators of Electronics and IT Production

Trade and Investment

Analytically, it can be argued that trade policy reforms play a dual role. They are instrumental in promoting both the use and production of IT by operating from both the demand and supply sides. From the demand side, as Kraemer and Dedrick (2001) argue, one of the best ways to promote IT use is to avoid or reduce barriers to use. Needless to say, any government policy that makes electronic goods more expensive, especially in countries where affordability is a crucial issue, will discourage its use and reduce the possible benefits from IT. Thus trade policy reforms, in the form of lowering taxes and tariffs and dismantling non-tariff barriers, can have the effect of promoting demand and use. Trade policy reforms also ease domestic supply constraints and create a more competitive environment, leading to lower prices and better quality products and, thus, promoting the use of IT.

The influence of trade policies on the development of electronic goods production is due to the nature of the production process itself. In an assembly-oriented industry like electronics, production essentially involves assembling a number of components and sub-assemblies based on a design. The production of needed components and sub-assemblies may be highly skill-, capital- and/or scale-intensive such that no country has the capacity to produce all the needed components and other accessories. Hence, there is a need for segmenting the production process across different locations. This is what led to the emergence of global production networks.

Thus, in global production networks, each component or sub-assembly task is made or carried out in different locations, according to their respective comparative advantage. This essentially means that production in any country will call for significant imports and the bulk of the output will have to be exported to other countries rather than sold in the domestic market. Hence, if production and therefore investment in electronics is to take place in a specific country, the trade regime needs to enable the free flow of inputs into and outputs out of the economy. This explains, at least to some extent, why India — which had an electronics industry with production levels higher than South Korea in 1971 — lagged behind as it followed a restrictive trade regime (Joseph 1992).

In the case of electronics production, the link between trade and investment notwithstanding, it has been shown that local capabilities are critical for attracting investment and promoting production. In a context where low-cost labour is taken for granted, the ability of developing countries/regions to participate in global production networks is governed by their ability to provide certain specialized capabilities that the MNCs need in order to complement their own core competence (Lall 2001; Ernst and Lundvall 2000). Countries and regions that cannot provide such capabilities are kept out of the circuit of global production networks, regardless of their trade regime.

Cantwell (1995) also argued that MNCs, in recent years, have followed knowledge-based asset-seeking strategies to reinforce their competitive strengths. Hence, from the perspective of developing the electronics industry at the sub-national level, a liberal trade regime constitutes a necessary, but not sufficient, condition. More importantly, to eliminate the risk of getting trapped at the low end of the value chain and to facilitate movement along the continuum of Original Equipment Manufacturer (OEM) to Original Brand Manufacturer (OBM) and finally to Original Design Manufacturer (ODM) (Hobday 1994), there is the need for building up an innovation system while simultaneously pursuing a liberal trade and investment regime.

In a similar vein, a survey by Saggi (2002) concludes that the absorptive capacity of the host country is crucial for obtaining significant benefits from FDI. Without adequate human capital or investment in R&D, spillovers from FDI are unfeasible. This calls for complementing liberalized trade and FDI policies with appropriate policy measures and institutional interventions with respect to education, R&D, and human capital such that learning capabilities are enhanced in all parts of the economy. This is the central concern of studies on innovation systems. In this context, as argued by Bresnahan et al. (2001), the initiatives needed to enable an entry into electronics GPNs may be more arduous compared to those needed to upgrade the position of those already in a production network.

Innovation System

It is by now recognized that an economy's ability to bring about industrial transformation, especially by harnessing a knowledge- and skill-intensive sector such as electronics in a sustained manner depends, to a great extent, on its National Innovation System (NIS). While the historical roots of the concept of NIS can be traced back to the work of List (1841), the modern version of this concept was introduced by Lundvall (1985) in a booklet on user-producer interaction and product innovation.

Freeman (1987), while analysing the economic performance of Japan, brought the concept to an international audience. He defined a National Innovation System as "the network of institutions in the public and private sectors whose activities and interactions initiate, import, modify and diffuse new technologies" (p. 1). This definition highlights the processes and outcomes of innovation.

Since then there has been burgeoning literature (Lundvall 1992; Nelson 1993; Freeman 1995; Edquist 1997) focusing on different dimensions of the innovation system.[2] Based on the evolutionary approach to innovation, Nelson and Winter (1977; 1982), Nelson (1981; 1995) Carlsson and Stankiewiez (1995), and Carlsson et al. (2002) have advanced the technological systems approach focusing mainly on the generation, diffusion and utilization of technology.

The NSI framework was further enriched by studies on regional systems of innovation (DeBresson, 1989; DeBression and Amesse 1991) and sectoral systems of innovation (Breschi and Malerba 1997; Malerba 2002). Thus the innovation system may be supranational, national, regional or sectoral. These approaches complement rather than exclude each other and selection of the system of innovation should be sectorally- or spatially-delimited depending on the context and object of study (Edquist 1997).

As regards industrial transformation at the sub-national level, Regional Innovation Systems (RIS) appears to be an appropriate conceptual framework. An RIS is defined as a "constellation of industrial clusters surrounded by innovation supporting organizations" (Asheim and Coenen 2005). Viewed in this manner, it goes beyond industrial clusters, which simply refer to the geographic concentration of firms in the same or related industries (Porter 1998; Pietrobelli and Rabellotti 2004). Drawing insights from geography, the concept of RIS was developed on the basis of, and inspired by, successful regions and clusters such as Silicon Valley (Cohen and Fields 1998; Saxenian 1994), Baden Württemberg (Staber 1996) and the Third Italy (Beccatini 1990; Piore and Sabel 1984). Hence most of the literature on regional innovation systems reflects the traits and characteristics of the developed world.

An RIS according to Andersson and Karlsson (2004), is comprised of two components.

- The regional production structure — comprised of individual firms and their networks.
- The regional supportive infrastructure — comprised of all institutions that support economic activity and innovation such as government agencies, research institutes, technology centres, and so forth.

Thus, across the different interpretations, the RIS approach stresses the systemic dimensions of the innovation process and the dynamic interaction between the different components of the system — individuals, organization and institutions (viewing innovation as an evolutionary, path dependent and interactive process — and not a linear one).

To the extent that the focus of present discussion is on a specific sector, insights from sectoral innovation systems can also be of great utility. A sectoral system of innovation and production is a set of new and established products for specific uses and the set of agents carrying out market and non-market interactions for the creation, production and sale of those products. A sectoral system has a knowledge base, technologies, inputs and an existing, emergent and potential demand. The agents comprising the sectoral system are organizations and individuals (e.g., consumers, entrepreneurs, scientists). Organizations may be firms (e.g., users, producers and input suppliers) and non-firm organizations (e.g., universities, financial institutions, government agencies, trade unions, or technical associations), including sub-units of larger organizations (e.g., R&D or production departments) and groups of organizations (e.g., industry associations). Agents are characterized by specific learning processes, competencies, beliefs, objectives, organizational structures and behaviors. They interact through processes of communication, exchange, cooperation, competition and command, and their interactions are shaped by institutions (rules and regulations). Over time, a sectoral system undergoes processes of change and transformation through the co-evolution of its various elements (Malerba 2002).

Empirical evidence across countries in the South also indicates that the elements of a sectoral system that were instrumental varied from sector to sector and country to country. While the crucial factor behind technological progress in sectors like electronics in Taiwan has been the learning and capabilities of domestic firms under the weak patent regime (Amsden and Chu 2003), the role of the government has been highlighted in the case of telecommunications and aircraft in Brazil (Mani 2004; Dahlman and Frischtak 1993; Viotti 2000) and software in India (Joseph 2002; 2006).

In several sectors, Mazzoleni and Nelson (2006) have shown that universities and public research laboratories performed advanced research and trained human capital, which were important as the experience of several countries indicate. The catch-up process of countries in different sectoral systems has also been affected by the specific types of networks. In some sectoral systems like electronics, as argued by Lundvall (1992), vertical networks with suppliers have provided new inputs and shared relevant information for production and innovation, and led to learning and capability development by domestic firms. In the context of global production networks, studies have also shown that specialization in different stages of the global value chain has been another way to catch up (Gereffi et al. 2005; Ernst 2002; Morrison, Pietrobelli and Rabellotti 2006). While the large and growing domestic demand has been relevant to catch-up for most sectors in countries like China, exporting for the world market has played a major role in catch-up in small or medium size countries. These differences, as argued by Malerba (2004), need to be seen against the fact that sectors are not homogenous and are characterized by different technologies, actors, networks and institutions.

These are important insights in understanding sectoral dynamics in terms of their innovation and production processes. However, from the perspective of developing countries, one also needs to reckon with the new international environment in which they operate. With the removal of trade barriers, domestic firms, regardless of the sector in which they operate, are exposed to international competition. Thus, the infant industry protection and government subsidies widespread in most of the earlier catch-up episodes have a very limited role at best today. The unprecedented exposure to international competition in turn has had influence on their innovative behaviour and competitive strategies of local firms. This has been manifested in the increasing incidence of joint ventures and takeover of local firms by foreign firms.

Similarly, the strong intellectual property rights regime being imposed on the developing countries of today entails an environment significantly different from the one in which the East Asian Tigers developed. Today, there is little scope for reverse engineering and duplicative imitation-based innovation strategies widespread in the earlier regime.

In addition, the role of university-industry interaction that was significant in prior episodes of catch-up is likely to be more limited due to recent cuts in social sector expenditure. In this new environment, the observation that countries that are technologically backward have the potential for generating growth more rapidly than more advanced countries (Abramovitz 1986) may not be as applicable as it was earlier. Therefore, under the new disposition, the basic building blocks of the sectoral system, as articulated by Malerba, while remaining intact, might exert their influence in a distinct way from the earlier catch-up episodes. Hence, any attempt at developing sectoral/regional systems of innovation in developing countries needs to take note of the existing context and devise appropriate strategies to address them.

RECENT TRENDS AND CHALLENGES

There is hardly any country in the developing world today that has not initiated policy measures and institutional interventions to harness IT for Development. But the development of an electronics production base seems to have not received the attention that it deserves, presumably because various agencies, including multilaterals, consider the promotion of IT use as their key agenda. In this context, the experience of ASEAN newcomers appears instructive. Induced by the development experience of the first ASEAN member countries wherein electronics production played a significant role in catching-up, the policy framework of these countries underscores the need for developing an electronics production base as a means of pursuing industrial transformation. Despite their best efforts, the production bases for electronics production in these countries remain, with the plausible exception of Vietnam, rudimentary. This needs to be seen within a context of successful trade and investment liberalization, but rather more modest success in developing vibrant innovation systems (Joseph 2006).

Industrial transformation at the sub-national level has also become an issue of immense policy relevance for emerging countries like India and China. Of late, China and India have joined the globalization bandwagon, not only through their active engagement in the production and export of IT and electronics, but also by harnessing the power of new technology for addressing their varied developmental problems. Moreover, both China and India have emerged as attractive locations for FDI in R&D and are active participants in Global Innovation Networks. In the sphere of OFDI, these countries today are no less significant players. At the same time, it has been shown that their impressive economic performance has hardly been inclusive and that there has been widening interpersonal and interregional disparity in development. This, in turn, casts doubt on the effectiveness of the hitherto followed strategies in achieving industrial transformation at the sub-national level and bringing about more regionally inclusive development outcomes.

ITA and Global Networks in Production and Innovation

Although outsourcing and sub-contracting have a long history in the electronics sector, outsourcing gained momentum during the 1990s, largely due to the emergence of contract manufacturers. While analysing the development of the electronics industry and its competitive dynamics along with patterns of industrial organization, Ernst (2002*a*, 2002*b*) observed that globalization in the electronics industry is characterized by the international dispersal of the value chain to highly concentrated locations. This has been an outcome of an organizational innovation that resulted in the formation of global production networks induced by MNCs' increased outsourcing requirements in the context of heightened international competition. In their effort to achieve cost minimization, they search for low cost foreign capabilities that are complementary to their own competencies. The creation of GPNs reflects increasing pressures to exploit complementarities that result from the interactive nature of knowledge creation (Antonelli 1998). A peculiar character of GPNs is that manufacturing is de-coupled from product development and is dispersed across firm and national boundaries.

GPNs, while being global, are characterized by a heavy concentration between and within countries. To the extent that GPNs call for free inflow of inputs and outputs from the country, the countries that followed a liberal trade regime were those who secured entry into GPNs. There has also been concentration in a few specialized local clusters because the specialized capabilities that the MNCs seek are confined to select locations. By the same logic, the concentration of dispersal increases as we move towards more complex capital-intensive equipment and components.

However, most of these countries are found specialized in the mass production of a few products mainly for the export market. This has led to a kind of sticky specialization, with limited backward and forward linkages especially for materials and production equipment. This gives rise to a very high level of import dependence and limited value addition. In the case of Thailand, Mephokee (2003) noted that Thai IT firms play a small subcontracting role by supplying minor components for foreign firms in the IT industry. However, the production of these components is largely import-dependent, due to four reasons. First, the production technology belongs to foreign parent companies. Secondly, there are no domestic components, because the production technology is not available in Thailand. Thirdly, the quality of domestic components cannot meet the requirements of foreign companies. Finally, it is easier to deal with foreign suppliers with whom a long-term relationship has already been established. Thus, the study concludes that Thai firms have little room to play in their local IT industry.

A striking feature, which can partly be attributed to the strategy being adopted, is the mismatch between local production and consumption both at the component and equipment level. To illustrate, in the case of telecommunication equipment, Thailand exports almost 70 per cent of its production and at the same time imports more than 70 per cent of its domestic demand (Joseph 2006). The case with semiconductor devices appears to be similar. The narrow production base with export orientation also has the effect making the industry highly susceptible to international market fluctuations. In such a context, the need for upgrading of East Asia's electronics industry has been underlined. The key issue is to what extent recent developments like the Information Technology Agreement (ITA) of WTO and the emergence of Global Innovation Networks facilitate the much-needed upgrading of Asia's electronics industry.

Since the Information Technology Agreement of WTO came into force in 1997, tariffs and other duties and charges on the goods covered by the Agreement have been abolished. This, in turn, has made some countries that were conventionally considered to be unattractive for GPNs due to their restrictive trade regimes to emerge as attractive locations. Thus China to a great extent and India to some extent have emerged as priority investment targets for leading global electronics companies. Having abundant stocks of low-cost manpower and a liberal trade and investment regime, these countries pose a serious challenge to established manufacturing locations with a similar competitive advantage.

Between 1995 and 2004, the share of R&D spent by Western European multinationals outside their home countries increased from 26 per cent to 44 per cent, by Japanese multinationals from 5 per cent to 11 per cent, and by North American multinationals from 23 per cent to 32 per cent. Since then, there has been a substantial increase in investment by these multinationals in developing economies like Brazil, India, and China. According to a survey of the world's largest R&D spenders (UNCTAD 2005), China has been considered as the third most important offshore R&D location (after the USA and UK), followed by India (6th) and Singapore (9th). A more recent survey by the Economic Intelligence Unit in 2006 showed that India and China are the second and the third most important offshore R&D locations (after the USA and ahead of the UK). Leading global corporations thus consider India, the USA and China to be the next overseas locations for future R&D.

The unprecedented growth in global innovation networks on the one hand and increasing participation of developing countries on the other, has attracted significant scholarly attention. Drawing from the received wisdom, it is possible to locate a host of "centripetal forces"[3] which induce the firms to centralize R&D activities in headquarters and "centrifugal forces"[4] that work towards the dispersal of R&D activities across different locations beyond the home country. The unprecedented increase in the pace at which GINs are being formed (UNCTAD 2005; Hall 2011), however, suggests the presence of certain factors that reduce uncertainty as well as the costs of coordination and transaction and thus undermine the power of centripetal forces. There are two important factors for this rebalancing and resultant increase in the mobility of knowledge as argued by Albuquerque et al. (2011). The first relates to the improvement in the information communication infrastructure and its extensions around the world. The second refers to policies to liberalize trade and investment, which helped firms exploit the benefit of technological change. To this, we may add the emergence of new locations that are perceived as capable of providing complementary capabilities, especially human capital, at lower cost. For instance, it has been predicted that by 2010 China would have more science and engineering doctorates than United States (Freeman 2005; National Science Board 2008).

OFDI from the South: China and India

As is evident from Table 2.1, leading Asian countries have been major investors in Asia. Rajan et al. (2011) have shown that, during the period 1990–2004, Japan has been the largest investor into emerging Asia, accounting for 17–18 per cent of the total inflows and showing an increase from 13–14 from 1990–94. According to UNCTAD (2008), OFDI from developing countries increased from US$6 billion between 1989 and 1991 to US$225 billion in 2007, which indicated an increase in their share in global outflows from 2.7 per cent to nearly 13 per cent. But a recent development with implications for industrial upgrading and industrial transformation has been the emergence of selected developing countries, especially from Asia, as major sources of OFDI.

While OFDI from some of the major economies in Asia slowed down in early 2009 in the wake of the global financial crisis, OFDI from China maintained an upward trend and that from India showed only a marginal decline (UNCTAD 2009). Despite the global financial crisis, FDI from China reached US$53.8 billion in 2008, an increase of over 100 per cent from US$26.5 billion in 2007, and its outflows continued to grow in 2009

Table 2.1
Top Bilateral FDI Flows between Asian Countries, 1997–2005#
(Millions of US$)

Donor	Host	Average		In percent to Asia	
		1997–2000	2001–5	1997–2000	2001–5
Hong Kong	China	17,750.8	17,819.1	16.0	16.8
China	Hong Kong	7,266.9	5,459.4	6.5	5.2
Japan	China	3,276.2	5,194.5	3.0	4.9
Taiwan	China	2,774.8	3,361.3	2.5	3.2
Singapore	China	2,706.3	2,136.7	2.4	2.0
Japan	Thailand	1,347.0	2,324.9	1.2	2.2
Japan	Hong Kong	1,417.6	2,044.6	1.3	1.9
Singapore	Hong Kong	2,835.3	353.1	2.6	0.3
Japan	Singapore	1,281.5	1,276.6	1.2	1.2
Singapore	Malaysia	844.1	1,133.8	0.8	1.1
Singapore	Thailand	441.7	1,381.9	0.4	1.3
Japan	Korea	607.8	717.3	0.5	0.7
Taiwan	Hong Kong	268.9	446.6	0.2	0.4
Japan	Philippines	232.9	377.5	0.2	0.4
Malaysia	China	290.8	316.7	0.3	0.3
Hong Kong	Malaysia	272.3	296.5	0.2	0.3
Hong Kong	Thailand	360.1	160.8	0.3	0.2
Japan	India	249.3	244.7	0.2	0.2

This data is based on FDI inflow data to the host economy.
Source: Rajan, R.S. with Gopalan S. and Hattari, R. (2011).

(Hong 2011). China currently ranks 13th in the world as a source of FDI and third among all developing and transition economies. The FDI outflow from India was US$18.8 billion in 2008, slightly less than the US$21.4 billion seen in 2007. As a result, the share of China and India in total East, South and Southeast Asian outflows increased from 23 per cent in 2007 to 37 per cent in 2008 (Hong 2011). It is important to note that about 75 per cent of China's OFDI was directed towards Asia. Though Chinese OFDI in Asia is highly concentrated at present in Hong Kong (88 per cent), with the signing of China-ASEAN Free Trade Agreement in 2010 and ASEAN China Investment Agreement in 2009, Chinese investment is likely to become regionally more dispersed. Similarly, India also has signed the Free Trade Agreement with ASEAN and the discussion on services and investment is in progress.

The Challenges

Given the opportunities provided by GINs, GPNs and FDI from the South for industrial upgrading, the often-followed strategy has been to attract more FDI through various incentives. This has resulted in intense competition not only between developing countries but also between different regions within them to provide incentives. The final outcome of such wasteful incentive competition is bound to be detrimental to the interests of developing countries. In a context wherein incentives and cheap labour are taken for granted by MNCs, the policies of regional governments aspiring to enter into GPNs need to satisfy very demanding requirements.

In this context, drawing from Bresnahan et al. (2001) it can be argued that the factors that enabled national entities to enter into global production networks could be different from those that enable industrial upgrading of those who have already entered them. Agglomeration economies, external effects and social increasing returns from any source may arise naturally once the regions have successfully entered GPNs.

But, the most difficult and risky part is to negotiate entry in the first place. At this stage, factors such as linkages with a sizable and growing demand, along with technological capabilities in the firm-level supply of manpower at all levels including production, R&D, managerial, adequate infrastructure, venture financing and, above all, an enabling policy environment accompanied by an uncorrupt bureaucracy are crucial.

For countries with limited markets, government procurement policies, especially at the early stage can be of great help. All these fall into what we have discussed as vibrant innovation systems which include: high-quality but low-cost infrastructure and information communication systems; streamlined administrative procedures that facilitate smooth supply chain management and quick adjustments to change in markets and technology; and an efficient support industry and services with certified procedures that guarantee world class quality standards and short time-to-market cycles (Ernst 2008). There is also the need to evolve an interlinked system of research centres, universities, firms and other organizations that can tap into the growing stock of global knowledge, assimilate and adapt it to local needs and create new knowledge. While we underline the need for qualified skilled manpower there could be different sources for skill for different regions. Here the role of universities is important. Yet it could also be accomplished through training imparted by industry (like the large IT firms in India), or repatriates from other countries as shown by the experience of Taiwan.

Segal and Thun (2001) have shown that although national institutional characteristics provide the overall framework for growth and regulate the overall process, variation in developmental outcomes is the result of the specifics, characteristics and abilities of local institutions. Even at the local level, different industrial sectors have different developmental needs, and a policy that works for one sector will not necessarily work for another. To the extent that the needs of one sector vary from another and the institutional structures that are required to meet firm-level needs are local more often than national, there is a need for focusing on local institutions and firms within a particular sector. Hence, the focus of policy-makers needs to be on judiciously building an appropriate institutional architecture at the regional level (regional innovation system) complemented by a focus on sectors that are part of their comparative advantage (sectoral innovation system).

However, as noted earlier, given the changed environment in which developing countries operate there are serious constraints today as compared to yesteryear. Here the emergence of India and China could be a blessing in disguise. The huge potential markets in these countries for electronic products and services provide new trade and investment opportunities for Asian firms. In addition, Asian electronics firms could also exploit the low-cost and high-skilled manpower available in these countries. Here we need to take note of the presence of accumulated capabilities by select developing countries in the South in the sphere of electronics. Also, the international context for building innovation systems is less friendly than ever before while innovation systems are no longer limited by national boundaries. Hence, it is pertinent to explore ways by which Southern capabilities can be harnessed for building up an appropriate institutional architecture to bring about industrial transformation.

TOWARDS A PERSPECTIVE

During the 1970s and 80s, South-South cooperation was much debated among developing countries. The issue seems to have taken a back seat during the last decades as developing countries were increasingly experimenting with trade and investment liberalization under globalization. But today, due to the increasing disenchantment among developing countries with globalization, the topic is re-gaining momentum.[5]

In a sense, the potential for Electronics and IT-based industrial transformation of the developing world through South-South cooperation is due to the fact that while the Western world held a monopoly over earlier GPTs, in the case of IT, the capabilities are more diffused with substantial capabilities in Asia. While Japan and developing Southeast Asian countries hold a leading position in the production of IT goods and services, and China recently entered the league of IT goods production, in the field of IT software and services India has emerged as a major player in the world market (Heeks 1996; Kumar 2001; Arora et al. 2001; Joseph and Harilal 2001; Joseph 2009). In addition, there are a number of IT-based solutions developed by the South that address South-specific issues like last mile connectivity and affordability (Joseph and Parayil 2008). More importantly, Asia is the home of the highly-skilled manpower much-needed for IT in general and electronics in particular. The challenge therefore, is to harness the synergy between IT hardware, software and human capabilities of Asian countries to help industrial transformation by promoting both the production and use of IT.

While South-South Cooperation in the field of information technology is in its incipience, there are a number of regional and bilateral arrangements for harnessing IT for Development.[6] Perhaps, the most notable one is the e-ASEAN Framework Agreement. The e-ASEAN initiative has to be seen against the background of the economic and digital divide between the new ASEAN (Cambodia, Lao PDR, Myanmar and Vietnam) and old ASEAN member countries (Brunei Darussalam, Indonesia, Malaysia, Philippines, Singapore and Thailand). The e-ASEAN initiative, among others, is an integral part of the Initiative for ASEAN Integration. Unlike the Information Technology Agreement of WTO, which is essentially a tariff-cutting mechanism agreed upon mostly by developed countries, the e-ASEAN Agreement aims at tariff cutting along with facilitating capacity-building. Thus the underlying strategy is one of "ASEAN helping ASEAN".

Here it is pertinent to highlight the limits of such a regional arrangement as compared to the benefits of broad-based cooperation. To be more specific, we may examine to what extent the old ASEAN could help in capacity building, both physical and human, in the new ASEAN countries. This in turn depends on the capabilities of the old ASEAN in the field of IT, which could be viewed in terms of capabilities in IT hardware and IT software and human resources. In the case of IT hardware capability, the countries of ASEAN-6 such as Singapore, Malaysia Thailand are known for their IT manufacturing and export base. But as already noted, in most of these countries, IT hardware investment and production is dominated by the MNCs with a limited role for domestic firms.

The issue is more acute in the case of IT manpower, because the old ASEAN countries are also faced with an excess demand situation, both in terms of quality and quantity, with respect to IT manpower. For example, an estimate for Thailand, despite its concerted efforts to build up human capital, has shown that, by 2006, the excess demand for IT manpower would be of the order of 26,000 (Suksiriserekul 2003). Even in the case of Singapore, which is highly developed, there is an acute shortage of IT manpower.[7] Thus in achieving the declared objective of bridging the development divide between the old and new ASEAN members by harnessing IT, cooperation among the ASEAN countries may be complemented with more broad-based cooperation involving other countries in Asia.

Thus, the scope for cooperation among countries in Asia is obvious. But, what is at present missing is an institutional arrangement for promoting it as well as research backed by theory and empirics to sustain it. Here lies the need for an e-Asia Framework Agreement aiming at the establishment of a more regionally-diversified production base for IT and electronics and promotion of IT use across different sectors of the economy and segments of society. Towards achieving this objective, the Agreement, in tune with the Information Technology Agreement of the WTO should focus on liberalizing the trade in IT goods and services. At the same time, drawing from the e-ASEAN Framework Agreement, the e-Asia Framework Agreement should be instrumental in the creation of an institutional architecture at the sub-national and sectoral level *inter alia* by harnessing Asia's capabilities in both hardware and software. Given the paramount importance of human capital in developing IT production and promoting IT use, a special focus may be given to developing the IT manpower base. In general, the Agreement should facilitate an integrated development of the IT sector by promoting both production and use instead of the ongoing lop-sided approach wherein developing countries are often considered as passive adopters of technology.

Notes

1 The report argues that actual inflation fell by 0.5 percentage points a year from 1994 to 1998 due the effect of declining prices of IT (electronic) goods. Also the electronic sector, including telecommunications, employed 7.4 million

workers in 1998 and this accounted for 6.1 per cent of the total employment with an annual wage rate more than 1.5 times that for all private employees. A cynic may argue that to sustain such growth in employment, output and wages in developed countries, the diffusion — and not production — of IT in developing countries needs to grow at higher rate.

2 Here the readers are referred to the large number of papers presented in the GLOBELICS conferences held in Rio in 2003, Beijing in 2004 and Pretoria in 2004. The papers are available at <www.globelics.com>.

3 The centripetal forces included the need to protect firm-specific technology to avoid R&D leakage (Rugman 1981) due to the tacit nature of technological knowledge, need for closer coordination in decision making in the face of uncertainty of innovation (Patel and Pavitt 1991) and to take advantage of scale economies in R&D and the high cost of co-ordination and control (e.g., Vernon 1974).

4 The centrifugal forces include demand-oriented factors that emanate from the need to be nearer to the export market to exploit regions' differential advantage in production and in R&D (Cantwell 1995). Of the supply-side factors operating as centrifugal forces, the most important one appears to be the access to scientific and technological skills, including scientific infrastructure, that are available in the host countries on more advantageous terms than in the home market (Ernst 2008).

5 The High-level Conference on South-South Cooperation held at Marrakech on 18, December 2003 at the instance of G-77 had a special Roundtable on IT for Development. The Roundtable underlined the role of ITs in enhancing the capacity of enterprises of developing countries and called for concrete actions to help countries improve the use of IT, including through e-commerce, e-finance, e-governance and e-tourism. The Roundtable also highlighted the need for South-South Cooperation in ITS. See for details: <http://www.g77.org/marrakech/RT-IT.htm>.

6 Kumar and Chadha (2009) made the case for IT cooperation among SAARC countries. In all the bilateral cooperation between India and other countries, IT is an important element.

7 In an interview in New Delhi in 2004, Singapore's Infocom Development Authority (IDA) Chairman Lam Chuan Leong said that Singapore had an Infocom resource pool of 93,000, which is significantly short of projected demand. They, therefore, would like to outsource 25,000 professionals from India over the next five years to meet the growing demand.

References

Abramovitz, M. "Catching up, Forging ahead and Falling Behind". *Journal Economic History* 46, no. 2 (1986): 385–406.

Albuquerque Eduardo da Motta, G. Britto, O.S. Camargo and G. Kruss. "Global Interactions Between Firms and Universities: Global Innovation Networks as First Steps Towards a Global Innovation System". Discussion Paper No. 419. Belo Horizonte: CEDEPLAR/FACE/UFMG, 2011.

Amsden A. and W. Chu. *Beyond Late Development*. Cambridge MA: MIT Press, 2003.

Andersson, M. and C. Karlsson. "Regional Innovation Systems in Small & Medium-Sized Regions: A Critical Review and Assessment". CESIS Working Paper Series, no. 10. Stockholm: Centre of Excellence for Science and Innovation Studies, 2004.

Antonelli, C. *The Microdynamics of Technological Change*. London: Routledge & Kegan Paul, 1998.

Asheim, B. and L. Coenen. "Knowledge Bases and Regional Innovation Systems: Comparing Nordic Clusters". *Research Policy* 34, no. 8 (2005): 1173–90.

Arora, A., Arunachalam V.S., Asundi J. and Ronald F. "The Indian Software Services Industry". *Research Policy* 30, no. 3 (2001): 1267–87.

Beccatini, G. "The Marshallian Industrial District as a Socio-economic Notion". In *Industrial Districts and Inter-firm Cooperation in Italy*, edited by F. Pyke, G. Becattini and W. Sengenberger. Geneva: International Institute for Labor Statistics, 1990.

Breschi, S. and F. Malerba. "Sectoral Innovation Systems: Technological Regimes, Schumpeterian Dynamics and Spatial Boundaries". In *Systems of Innovation: Technologies Institutions and Organizations*, edited by C. Edquist. London: Pinter, 1997.

Bresnahan Timothy, Alfonso Gambardella and Annalee Saxenian. "Old Economy Inputs for A New Economy Outcomes: Cluster Formation in the New Slicon Valleys". *Industrial and Corporate Change* 10, no. 1 (2001): 835–60.

Cantwell, J.A. "The Globalization of Technology: What Remains of the Product Cycle Model?" *Cambridge Journal of Economics* 19 (1995): 155–74.

Carlsson, B. and R. Stankiewiez. "On the Nature, Function and Composition of Technological Systems". In *Technological Systems and Economic Performance, The case of Factory Automation*, edited by B. Carlsson. Boston: Kluwer Academic Publishers, 1995.

Carlsson, B., S. Jacobsson, M. Holmen and A. Rickne. "Innovation Systems: Analytical and Methodological Issues". *Research Policy* 31 (2002): 233–45.

Carlos, A., Primo Braga, John A. Dally, and Bimal Sareen. "The Future of Information and Communication Technology for Development". Paper presented in the IT Development Forum. Petersberg, Germany, 2003.

Chen, Shin-Horng. "Global Production Networks and Information Technology: The Case of Taiwan". *Industry and Innovation* 9, no. 3 (2002): 249–65.

Cimoli, M., A. Hofman and N. Mulder. *Innovation and Economic Development: The Impact of Information and Communication Technologies in Latin America*. Cheltenham: Edward Elgar, 2010.

Cohen, S.S. and G. Fields. "Social Capital and Capital Gains, or Virtual Bowling in Silicon Valley". Berkeley Roundtable on the International Economy. University of California, Berkeley, 1998.

Dahlman C. and C. Frischtak. "National Systems Supporting Technical Advance in Industry: The Brazilian Experience". In *National Innovation Systems: A Comparative Analysis*, edited by R. Nelson. New York & Oxford: Oxford University Press, 1993.

DeBresson, C. "Breeding Innovation Clusters: A Source of Dynamic Development". *World Development* 17, no. 1 (1989): 1–6.

DeBresson, C, and F. Amesse. "Network of Innovators: A Review and Introduction to the Issue". *Research Policy* 20 (1991): 363–79.

DOI. "Creating a Development Dynamic: Final Report of the Digital Opportunity Initiative". Washington DC: UNDP, 2001, available at <http://www.opt-init.org/framework/DOI-Final-Report.pdf>.

Edquist, C. *Systems of Innovation: Technologies Institutions and Organizations.* London: Pinter, 1997.

Ernst, D. "From Digital Divides to Industrial Upgrading: Information and Communication Technology and Asian Economic Development". East-West Center Working Paper, Economics Series No. 36. Honolulu: East-West Center, 2001.

———. "Global Production Networks and the Changing Geography of Innovation Systems; Implications for Developing Countries". *Economics of Innovation and New Technology* 11, no. 6 (2002*a*): 497–523.

———. "Electronics Industry". In *The IEBM Handbook of Economics*, edited by Lanzonick W. London: Thomson, 2002*b*.

———. "Innovation Offshoring and Outsourcing: What are the Implications for Industrial Policy?". *International Journal of Technological Learning, Innovation and Development* 1, no. 3 (2008): 309–29.

Ernst, D. and B.A. Lundvall. "Information Technology in the Learning Economy: Challenges for Developing Countries". East-West Center Working Paper, Economics Series, No. 8. Honolulu: East-West Center, 2000.

Ernst, D. and L. Kim. "Global Production Networks, Knowledge Diffusion and Local Capability Formation". *Research Policy* 31, no. 8–9 (2002): 1417–29.

Freeman, C. *Technology Policy and Economic Performance: Lessons from Japan.* London: Pinter, 1987.

———. "The National Innovation Systems in Historical Perspective". *Cambridge Journal of Economics* 19, no. 1 (1995): 5–24.

Freeman, Christopher. "Technology Inequality and Economic Growth". *Innovation and Development* 1, no. 1 (2011): 11–25.

Freeman, R.B. "Does Globalisation of the Scientific/Engineering Workforce Threaten U.S. Economic Leadership?". NBER Working Paper 11457. Cambridge, MA: National Bureau of Economic Research, 2005.

Gereffi G., Humphrey J., Sturgeon T. "The Governance of Global Value Chain". *Review of International Political Economy* 12, no. 1 (2005): 78–100.

Hobday, M. "Export-led Technology Development in the Four Dragons: The Case of Electronics". *Development and Change* 25, no. 2 (1994): 333–61.

———. "Innovation and Stages of Development: Questioning the Lessons from East and South East Asia". Paper prepared for SOM/TEG- Conference at the University of Groningen, The Netherlands: Empirical Implications of Technology-Based Growth Theories, August 2002.

Hall, B.H. "The Internationalization of R&D". UNU-MERIT Working Paper 2011–49. Maastricht: United Nations University, 2011.

Heeks, R. *India's Software Industry: State Policy, Liberalization, and Industrial Development.* New Delhi: Sage Publications, 1996.

Hong Zhao. "The Expansion of Outward FDI: A Comparative Study of China and India". *China: An International Journal* 9, no. 1 (2011): 1–25.

IMF. "*World Economic Outlook: The Information Technology Revolution*". Washington DC: International Monetary Fund, 2001.

India, Ministry of Information Technology. "Actions Taken Report of the National Task Force on Information Technology and Software Development", IT Action Plan: Part I. New Delhi: Ministry of Information Technology, 2000.

Indjikian R. and D.S. Siegel. "The Impact of Investment in IT on Economic Performance: Implications for Developing Countries", *World Development* 33, no. 5 (2005): 681–700.

Joseph, K.J. "Growth Performance of Indian Electronics: A Comparative Analysis with South Korea". *Indian Industrialization*, edited by Arun Gosh et al. Oxford: Oxford University Press, 1992.

———. *Industry under Economic Liberalization: The Case of Indian Electronics.* New Delhi: Thousand Oaks; London: Sage Publications, 1997.

———. "Growth of IT and IT for Development: Realities of the Myths of the Indian Experience". WIDER Discussion Paper No. 2002/78. Helsinki: United Nations University, 2002.

———. *Information Technology, Innovation System and Trade Regime in Developing Countries: India and the ASEAN.* New York: Palgrave Macmillan, 2006.

———. "Sectoral Innovation Systems in Developing Countries: The Case of IT in India". In *Handbook of Innovation Systems and Developing Countries: Building Capabilities in a Global Setting*, edited by Lundvall, B.A., K.J. Joseph, C. Chaminade, and J. Vang. Cheltenham: Edward Elgar, 2009.

Joseph, K.J. and G. Parayil. "Can Trade Liberalization Bridge Digital Divide: Assessing the Information Technology Agreement of WTO". *Economic and Political Weekly* 48, no. 1 (2008): 46–53.

———. "India-ASEAN Cooperation in Information Communication Technologies (ITs): Issues and Prospects". In *ASEAN-India Economic Relations: The Road*

Ahead, edited by N. Kumar, R. Sen and M. Asher. Singapore: Institute of Southeast Asian Studies, 2005.

Joseph, K.J. and K.N. Harilal. "Structure and Growth of India's IT Exports: Implications of an Export-Oriented Growth Strategy". *Economic and Political Weekly* 36, no. 34 (2001): 3263–70.

Kaushik, P.D. and N. Singh. "Information Technology and Broad-Based Development: Preliminary Lessons from North India". *World Development* 32, no. 4 (2004): 591–607.

Kraemer, K.L. and J. Dedrick. "Information Technology and Economic Development: Results and Policy Implications of Cross-Country Studies". *Information Technology, Productivity and Economic Growth*, edited by M. Pohjola. Oxford: Oxford University Press, 2001.

Kumar, N. "Indian Software Industry Development: International and National Perspective". *Economic and Political Weekly* 36, no. 45 (2001): 4278–90.

Kumar, N. and A. Chadha. "India's Outward Foreign Direct Investments in Steel Industry in a Chinese Comparative Perspective". *Industrial and Corporate Change* 18, no. 2 (2009): 249–267.

Kumar, N. and K.J. Joseph. "National Innovation Systems and India's IT Capability: Are there any lessons for ASEAN Newcomers?". In *Asia's Innovation System in Transition*, edited by Bengt Ake Lundvall, P. Intarakumnerd and Jan Vang. Cheltenham: Edward Elgar, 2006.

Lall, S. *Competitiveness, Technology and Skills.* Cheltenham: Edward Elgar: 2001.

Lundvall, B.A. *Product Innovation and User-Producer Interaction.* Aalborg: Aalborg University Press, 1985.

Lundvall, B.A. *National Systems of Innovation: Towards a Theory of Innovation and Interactive Learning.* London: Pinter Publishers, 1992.

Malerba, F. *Sectoral Systems of Innovation: Concepts, Issues and Analyses of Six Major Sectors in Europe.* Cambridge: Cambridge University Press, 2004.

———. "Sectoral Systems of Innovation: A Framework for Linking Innovation to the Knowledge Base, Structure, and Dynamics of Sectors". *Economics of Innovation and New Technology* 14, nos. 1–2 (2005): 63–82.

———. "Sectoral Systems of Innovation and Production". *Research Policy* 31 (2002): 247–64.

Mani, S. "Coping with Globalization: An Analysis of Innovation Capability in Brazilian Telecommunication Equipment Industry". UNU-INTECH Working Paper 2004–3. Maastricht: United Nations University, 2004.

Mazzoleni R. and R. Nelson. "The Roles of Research at Universities and Public Labs in Economic Catch-up". LEM Working Paper Series No. 2006-01. Turin: Universita di Torino, 2006.

Mephokee, C. "Thai Labour Market in Transition: Toward a Knowledge-Based Economy". In *Human Resource Development Toward a Knowledge-Based Economy: The Case of Thailand*, edited by M. Makishima and Suksiriserkul, Somchai. Tokyo: Institute of Developing Economies, 2003.

Morrison A., C. Pietrobelli C. and R. Rabellotti. "Global Value Chains and Technological Capabilities: A Framework to Study Industrial Innovation in Developing Countries". CESPRI Working Paper No. 192. Milan: Bocconi University, 2006.

Mytelka, L.K. and John F.E. Ohiorhenuan. "Knowledge-Based Industrial Development and South-South Cooperation". *Cooperation South*, no. 1 (2000): 74–82.

National Science Board, *Science and Engineering Indicators 2008*. Arlington, VA: National Science Foundation, 2008.

NECTEC. "IT for Poverty Reduction: Examples of Programmes/Projects in Thailand". Bangkok, NECTEC, 2000.

Nelson, R.R. "Recent Evolutionary Theorizing about Economic Change". *Journal of Economic Literature* 33 (1995): 48–90.

———. *National Innovation Systems: A Comparative Analysis*. Oxford: Oxford University Press, 1993.

———. "Research on Productivity and Productivity Differentials, Dead Ends and New Departures". *Journal of Economic Literature* 19 (1981): 1029–64.

Nelson, R.R. and S.G. Winter. *An Evolutionary Theory of Economic Change*. Cambridge, MA: The Belking Press of Harvard University Press, 1982.

———. "In Search of a Useful Theory of Innovation". *Research Policy* 16, no. 1 (1977): 36–76.

OECD. "Background Report to the Conference on Internationalization of R&D". Brussels, March 2005.

Patel, P, and K. Pavitt. "Large Firms in the Production of the World's Technology: An Important Case of 'Non-Globalisation'". *Journal of International Business Studies* 22 (1991): 1–21.

Pietrobelli, C. and R. Rabellotti. "Upgrading in Clusters and Value Chains in Latin America: The Role of Policies". *Sustainable Department Best Practices Series*. New York: Inter-American Development Bank 97, 2004.

Piore, M. and C. Sabel. *The Second Industrial Divide*. New York: Basic Books, 1984.

Pohjola, M. "Information Technology and Economic Growth: A Cross-Country Analysis". In *Information Technology, Productivity and Economic Growth,* edited by M. Pohjola. New York: Oxford University Press, 2001.

Porter, M. E. "Clusters and the New Economics of Competition". *Harvard Business Review* 76, no. 6 (1998): 77–90.

Pradhan Jaya Prakash. "The Determinants of Outward Foreign Direct Investment: A Firm-level Analysis of Indian Manufacturing". *Oxford Journal of Development Studies* 42, no. 4 (2004): 619–39.

Rajan R.S., S. Gopalan and R. Hattari. *Crisis, Capital Flows and FDI in Emerging Asia*. New Delhi: Oxford University Press, 2011.

Rodrik, D. "The Limits of Trade Policy Reform in Developing Countries". *Journal of Economic Perspectives* 6, no. 1 (1992): 87–105.

Rugman, A.M. "Research and Development by Multinational and Domestic Firms in Canada". *Canadian Public Policy* 7, no. 4 (1981): 604–16.

Saggi, K. "Trade Foreign Direct Investment and International Technology Transfer: A Survey". *The World Bank Research Observer* 17, no. 2 (2002): 191–235.

Saxenian, A.-L. *Regional Advantage: Culture and Competition in Silicon Valley and Route 128*. Cambridge: Harvard University Press, 1994.

Schware, R. "Software Entry Strategies for Developing Countries". *World Development* 20, no. 2 (1992): 143–64.

Segal, Adam and Eric Thun. "Thinking globally, acting locally; local governments industrial sectors, and development in China". *Politics & Society* 29, no. 4 (2001): 557–88.

Sen, A K. "Development: Which Way Now?". *The Economic Journal* 93 (1983): 745–62.

Soete, L. "A Knowledge Economy Paradigm and its Consequences". UNU-MERIT Working Paper Series No. 2006-001. Maastricht: UNU/MERIT, 2006.

Staber, U. "Accounting for variations in the performance of industrial districts: the case of Baden-Wuttemberg". *International Journal of Urban and Regional Research* 20 (1996): 299–316.

Stoneman, P. *The Economics of Technological Diffusion*. Oxford: Blackwell Publishers, 2002.

Suksiriserekul, S. "IT Manpower Development". In *Human Resource Development Toward a Knowledge-Based Economy: The Case of Thailand*, edited by M. Makishima and Suksiriserkul, Somchai. Tokyo: Institute of Developing Economies, 2003.

UNCTAD. *World Investment Report: Transnational Corporations and the Infrastructure Challenge*. Geneva: United Nations Conference on Trade and Development, 2008.

———. *World Investment Report: Transnational Corporations, Agricultural Production, and Development*. Geneva: United Nations Conference on Trade and Development, 2009.

———. *World Investment Report: Transnational Corporations and the Internationalisation of R&D*. Geneva: United Nations Conference on Trade and Development, 2005.

———. "Bilateralism and Regionalism in the Aftermath of CANCUN: Re-establishing the Primacy of Multilateralism". UNCTAD, TD(x1)/BP/12, Geneva 2004.

———. *E-Commerce and Development Report 2003*. Geneva: United Nations Conference on Trade and Development, 2003.

———. *E-Commerce and Development Report 2002*. Geneva: United Nations Conference on Trade and Development, 2002.

U.S. Department of Commerce. *Digital Economy 2000 Report*. Washington, D.C.: US Department of Commerce, 2000.

———. *Country Commercial Guide*. Washington, D.C.: US Department of Commerce, 2000, available at <http://www.usatrade.gov/website/ccg.nsf/ShowCCG?OpenForm&Country=THAILAND>.

Vernon, R. "The Location of Economic Activity". In *Economic Analysis and The Multinational Enterprise*, edited by J. Dunning. London: George Allen and Unwin, 1974.

Viotti, E.B. "Passive and Active National Learning Systems: A Framework to Understand Technical Change in Late Industrializing Countries and Some Evidence from a Comparative Study of Brazil and South Korea". Conference Paper for the 4th International Conference on Technology Policy and Innovation. Curitiba, Brazil, 28–31 August 2000.

Wong, P.K. "The Contribution of Information Technology to the Rapid Economic Growth of Singapore". In *Information Technology, Productivity and Economic Growth*, edited by M. Pohjola. New York: Oxford University Press, 2001.

Wong, P.K. and A.A. Singh. "IT Industry Development and Diffusion in Southeast Asia". In *Information Technology in Asia: New Development Paradigms*, edited by C.S. Yue and J.J. Lim. Singapore: Institute for Southeast Asian Studies, 2002.

World Bank. *Information Communication Technology: A World Bank Group Strategy.* Washington D.C.: The World Bank, 2002.

SECTION II

Cases from Industrializing Southeast Asia

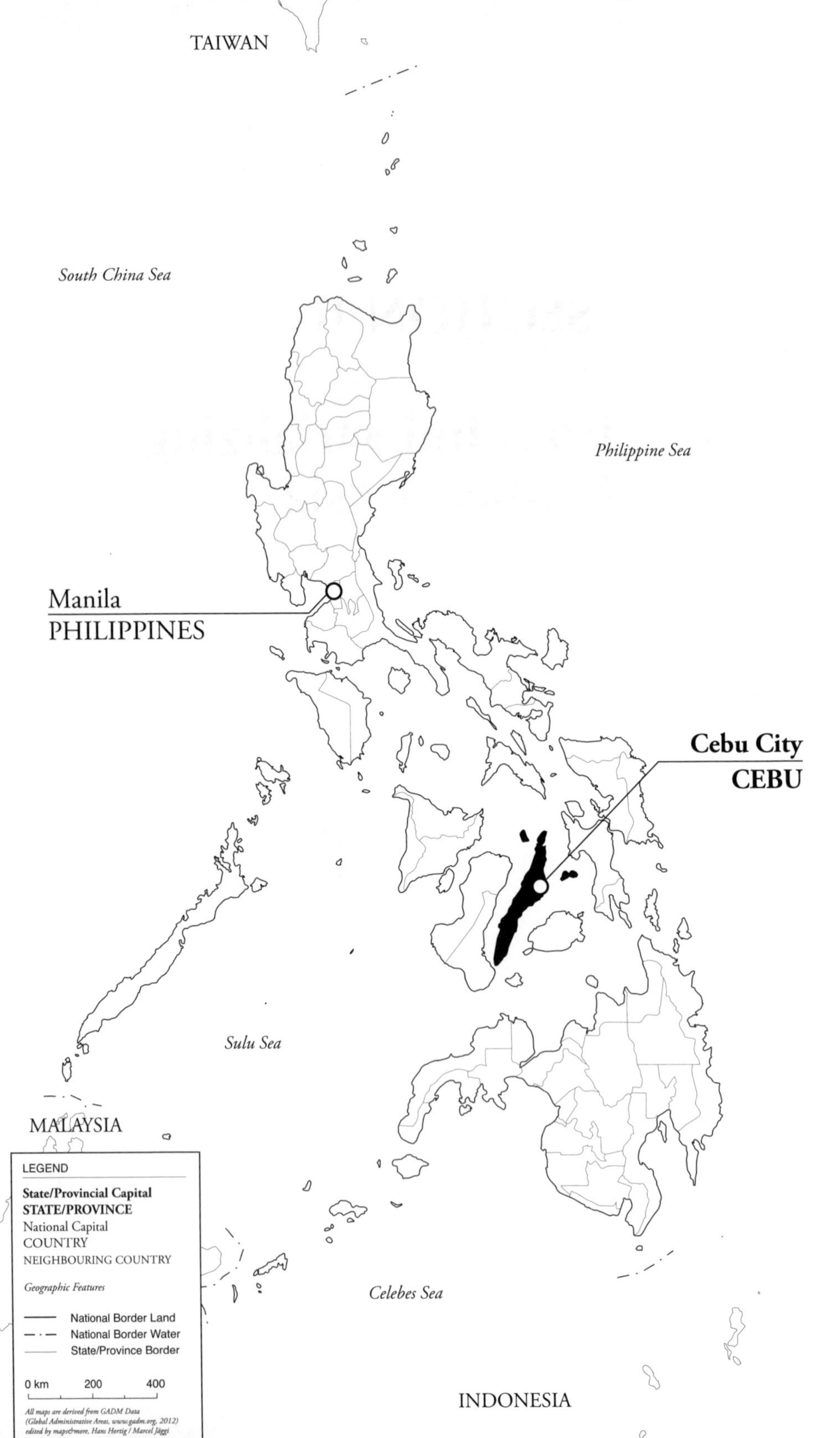
TAIWAN
South China Sea
Philippine Sea
Manila
PHILIPPINES
Cebu City
CEBU
Sulu Sea
MALAYSIA
Celebes Sea
INDONESIA
LEGEND
State/Provincial Capital
STATE/PROVINCE
National Capital
COUNTRY
NEIGHBOURING COUNTRY
Geographic Features
National Border Land
National Border Water
State/Province Border
0 km
200
400
All maps are derived from GADM Data
(Global Administrative Areas, www.gadm.org, 2012)
edited by maps&more, Hans Hortig / Marcel Jäggi

3

A RELATIONAL VIEW ON REGIONAL DEVELOPMENT: THE CASE OF THE ELECTRONICS SECTOR IN CEBU, PHILIPPINES

B.J. van Helvoirt

INTRODUCTION

Over the past years, the pattern and nature of interactions between economic agents has been increasingly recognized as an important factor for shaping the geography of regional economic performance. This chapter uses insights from this "relational turn" in economic geography and applies them to analyse regional performance in the Philippines. The chapter draws on empirical evidence from a study of an island province in the Central Visayas area of the country. The province of Cebu provides an interesting research area for an analysis of the influence of politico-economic relational networks on regional economic development, as it has experienced a strong emergence in the global marketplace that was influenced by networks of politico-economic relations. As this study shows, the politico-economic relational constellation of Cebu has a profound impact on the region's economic development trajectory.

SETTING THE SCENE: CEBU'S SHIFTING POSITION IN THE GLOBAL MARKETPLACE

The island province of Cebu is centrally located in the Central Visayas region in the heart of the Philippine archipelago. Its largest conurbation, Metro Cebu, is the main urban centre of the southern Philippines. It has emerged as a significant centre of manufacturing and service industries in the last two decades, attracting both foreign and domestic investment and producing for world markets. Outside the Manila area, it is the regional economy that is most associated with globalization and the "global shift" of industries to low-cost locations. As a result, Metro Cebu has emerged as a second node in the Philippines (after Metro Manila) with direct global connections — shipping, aviation, and telecommunications.

Cebu's economic profile is marked by three features that clearly distinguish it from other provinces in the Southern Philippines. The first is its size. As it is host to two-thirds of the Central Visayas' total population, the regional economy generates approximately 70 per cent of its exports (Figure 3.1). This leading regional position can be related to Cebu's other two characteristics: its economic diversity and strong insertion in the global economy. Cebu's urban economy has a diversified economic base with manufacturing and service industries such as education, tourism and leisure, shipping, and IT-related industries as key economic pillars. With its international sea- and airport, many of these activities are oriented towards the global market.

Cebu's outward orientation has deep roots. While pre-Spanish Cebu was not a major settlement, it did have the largest port in the Visayas that traded with Chinese merchants. However, it was not until 1860 — when the Spanish opened the port of Cebu to world trade, following the abolition of Manila's monopoly — that the province started to ship exports of cash crops such as sugar, hemp, and tobacco that were grown throughout the Visayas. British and American trading houses established a presence in Cebu, representing (with Spain) the principal markets for local exports (Fenner 1985). The American era (1898–1946) further boosted trade and spurred Cebu's industrialization, with sugar and coconut oil mills being erected with American capital. But also local (Chinese-) Filipino entrepreneurs embarked on industrial activities, giving rise to a local bourgeoisie.

After independence in 1946, Cebu's position as a domestic entrepôt and shipping centre was further strengthened when the country embarked upon a "Filipino First" import substitution policy. Cebu became a logical base for the distribution and sales networks of major Filipino manufacturers, and the main hub of domestic shipping lines. Only after the removal from power of Marcos in 1986 did the Philippines join the export-oriented growth model successfully pioneered by other Asian countries. It was at this moment in time that Cebu experienced an enduring period of rapid economic growth, spurred by its successful repositioning in the global marketplace. An effective marketing campaign attracted foreign tourists to the newly constructed beach resorts and FDI to the Mactan Export Processing Zones (MEPZ), spurring a new type of industry: the production of electronics by multinational manufacturers.

The non-violent people's uprising against the Marcos dictatorship which led to his ousting generated much positive attention for the Philippines, also among foreign investors. This attention was cleverly embarked on by the local government of Cebu via an international marketing strategy, which spurred an unprecedented influx of FDI and export-driven growth. While the Mactan Export Processing Zone (MEPZ I) had been established on Mactan Island in 1979 (it was built by Marcos, but remained largely idle due to a lack of interest from foreign investors who were reluctant to invest in the Philippines at that time), it did not take off until 1986. From 1986, the number of foreign companies grew from eight to 36 in 1991, including prominent electronics manufacturers like General Motors, Timex and NEC (Churchill 1993).

Between 1986 and 1990, capitalization of new businesses grew by 1,287 per cent, while the Board of Investment (BOI) registered a rise of investments by an incredible 3,730 per cent. In this short period, Cebu evolved into a net exporter: even after the devastating typhoon Ruping had left the island largely in ruins a year before, a 19 per cent growth of manufactured (mainly electronics) exports heavily contributed to an overall export growth of 5.8 per cent in 1991. As can be seen in Table 3.1, the growth in firms on the MEPZ I also continued afterwards, spurring the local government to set up MEPZ II. Figure 3.1 clearly illustrates the pivotal importance of these two EPZs for the export-oriented economy of Cebu and the larger Central Visayas region (see Table 3.5 in the Appendix for an overview of all SEZs in Cebu).

Japanese electronics manufacturing firms, which form the single largest pool of electronics manufacturing FDI in Cebu, generally have a strong direct control over their production facilities in Cebu, with many firms having (close to) 100 per cent foreign equity (Table 3.4 in the Appendix).

Table 3.1
Mactan Export Processing Zone I & II Firm and Employment Development, 1995–2004

	1995	1996	1997	1998	1999	2000	2001	2002	2003	2004
No. of firms										
MEPZ I	83	97	101	103	104	109	108	108	107	110
MEPZ II			14	19	25	32	37	40	39	48
No. of workers										
MEPZ I	29,243	32,811	36,074	35,920	37,118	39,487	40,189	37,350	39,000	42,095
MEPZ II			1,473	2,845	2,940	4,676	7,305	6,809	6,809	8,968

Sources: Department of Trade and Industry (DTI) Central Visayas, from Van Helvoirt (2009).

Another aspect is that many Japanese firms outsource their jobs to other Japanese firms. In the case of semi-conductor firms, often only the packaging is outsourced to local Filipino firms. Nonetheless, there is a growing number of local suppliers for manufacturers of electronics such as hard disk drives, for instance plastic molded parts or metal pressed parts. Moreover, Japanese firms do send their Filipino workers, such as the engineers, supervisors and operators to Japan for training. This encourages the transfer of technological know-how from the MNCs to the local economy (Zosa 2004).

Aside from the development of EPZs, a real estate investment-driven "boosterist"[1] (Short 1996) agenda undertaken by local politicians and business people stimulated the construction of modern infrastructure, including the provincial road network, the construction of modern business parks and a sufficient supply of water and electricity for industrial use. Important also was the fact that Cebu bypassed Manila and established direct connections with foreign economic centres in terms of logistics (aviation, shipping) as well as business contacts. The ensuing "Ceboom" repositioned

Figure 3.1
Mactan Export Processing Zone I & II Exports and Their Share of Total Central Visayan Exports, 1995–2004

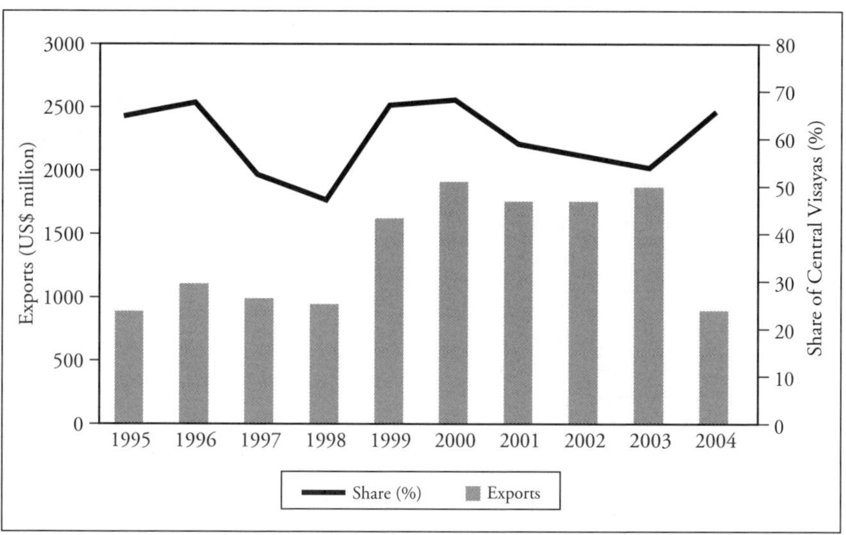

Sources: Department of Trade and Industry (DTI) and Cebu Investment Promotions Centre (CIPC), from Van Helvoirt (2009).

the province as a significant centre for electronics manufacturing and trading by MNCs from Japan, Korea and the U.S. in particular — supported by a myriad of local (sub)contractors and suppliers. With an export value of US$946 million that comprised over one third of the region's total export value in 2000 (Zosa 2004), the electronics industry has been a leading economic driver of the Cebuano economy (Table 3.2).

Cebu's principal electronics export products are semiconductors and a diverse range of electronic components and equipments. The Japanese electronics manufacturer Mitsumi, based in the town of Danao, is the single largest employer of the entire Visayas region, employing approximately 10,000 workers. The type of industrial activity in Cebu is largely restricted to production of electronics components and assembling of (semi)finished electrical equipment. However, in recent years some MNCs have also set up R&D facilities in Cebu, such as Lexmark's Research and Development Corporation, which serves as the base of Lexmark's R&D operations in the Asia-Pacific region. In 2004, three of the five largest single export companies in Cebu were in electronics: Cebu Mitsumi, Inc. (connectors), Lexmark International (printer circuits and cartridges), and Teradyne Philippines (semiconductors).

Table 3.2
Major Economic Export Activities of Cebu, 2000

Rank	Products	Value (US$M)	Percentage of total Cebu export value
1	Electronics	946.24	33.65%
2	Other industrial goods	360.22	12.81%
3	Electrical equipment	234.24	8.33%
4	Furniture	219.62	7.81%
5	Garments	157.47	5.60%
6	Steel/metal products	100.95	3.59%
7	Marine products	71.71	2.55%
8	Gifts, toys and housewares	46.40	1.65%
9	Vehicles/machinery parts	34.59	1.23%
10	Mineral products	28.12	1.00%

Source: Zosa 2004.

As Zosa (2004) illustrates, Cebu's economic competitiveness mainly manifests itself through its: quality human resources; dynamic export sector and tourism industry; proximity to international entry and exit points; infrastructure; cost of doing business; quality of life; and responsiveness of the local government to business needs. When rated among 26 Asian cities (JETRO 2003), Cebu is the most competitive in terms of monthly wage of mid-level managers and supervisors and the second most competitive in terms of monthly housing rent for foreigners. On the other hand, Cebu, as a destination of Japanese investments, is hampered by the relatively high wages of both unskilled and skilled labour, an unstable wage rate environment and moderately high costs of telecommunication, water, gas, and container transport.

A RELATIONAL VIEW ON REGIONAL DEVELOPMENT

This chapter starts from the assumptions of the "new regionalism" literature (Coe et al. 2004), which emphasizes the role of local institutional arrangements for economic activity. Regions have regained the interest of economic geographers as prime platforms of activities and transactions in the global economy (Scott and Storper 2003). The "new regionalism" literature places particular emphasis on endogenous institutions and local interactions among firms. Focus is placed on localized relational assets that are conceptualized as "untraded interdependencies" (Storper 1997), "institutional thickness" (Amin and Thrift 1994) or "associational economies" (Cooke and Morgan 1998). These assets, which are embedded in local social, cultural and institutional arrangements, are claimed to have a direct impact on a region's competitive potential via networks of interdependency, reciprocity and trust among firms (Amin 1999).

As such, this line of literature stresses the importance of regionally-confined relational networks for regional development in a global context. This relational reasoning on the regional economy corresponds with the "cultural turn" within political economy that has sparked the new cultural political economy (CPE). This recent strand of literature analyses politico-economies from a sociological perspective, namely how they are embedded in, and shaped by, place-specific societies and cultures (Sayer 2001). Rather than viewing politico-economies as national systems that are guided by formal regulations and procedures that can be reconstructed elsewhere, cultural political economy is actor-oriented and argues that the behaviour of these non-universal actors is influenced by their specific socio-cultural environments.

New regionalism and decentralization have given birth to the rise of meso-level governments functioning in a context of multilevel governance, which refers to a strong interdependence between various governmental and nongovernmental actors in policy-making (Hooghe and Marks 2001). In other words: the local state has become an actor in a multi-scalar relational network with other (non)governmental actors. In this network, two relational dimensions are particularly important: the horizontal public-private interplay between the local state and the local business community; and the vertical core-periphery relations along the hierarchical structure of the national government.

Regarding the public-private interplay, a key indicator to identify the role of the nation-state in business systems is the "strength" of the state (Whitley 1999). This strength can be defined as the dominance and autonomy of the state versus the private sector and its interest groups. Evans (1995) describes embedded autonomy as the key characteristic that distinguishes developmental states from "less developmental" predatory and intermediate states, in the sense that they are broadly embedded in civil society but without losing their autonomy from powerful interest groups.

As depicted by van Helvoirt (2009), there is a significant body of literature that describes the influence of local and regional political entities on regional economic development, focusing specifically on the Philippines. This extensive body of work suggests that political embeddedness and autonomy are undisputedly important analytical components for this case study. Authors such as Anderson (1988), Hutchcroft (2001; 2000; 1998; 1994), Kerkvliet (2002; 1995), Sidel (1999; 1998) and McCoy (1994), among others, have carried out an extensive body of work that not only analyses the Philippine political system from a relational perspective, but also zooms in on the regional and local playing field. In these studies, the existence of highly personalistic public-private relations (brokerage, patronage) that have a strong influence on policy-making and economic development is often depicted and analysed. Hutchcroft (1998) stresses that the Philippine political economy can be identified by a weak separation between the "official" and the "private" sphere, with most power residing in the private sector. Such dense interconnectivity between the public and private sector — often crystallised in dyadic personalised relationships

— manifests itself most prominently on the local and regional level in the Philippines, as Sidel (1999) and McCoy (1994) extensively describe. The dual position of local political leaders as key public and private actors, or their close relationship with powerful local entrepreneurial groups — central issues in Sidel's work — and how this affects public policy making are also unavoidable focus points for this case study on regional economic development.

The second relational tier of the local state is vertical in its nature and runs within the hierarchical structure of the Philippine state. While decentralization and new regionalism have given rise to a hollowing-out of the national state (Jessop 2003) and this is especially true for a geo-politically dispersed country like the Philippines (see Hill 2002), the national government remains an influential actor for regional economies. To a large extent the policy stage for cities and provinces is set by national government regulations. More importantly, the increase of local state authority and responsibility is accompanied by a rising need to retrieve financial resources from the national budget (De Dios 2007). Savitch and Kantor argue that tight vertical networks between the centre (national government) and periphery (local governments) are important for the latter. Formal and informal collaboration with national ministries can provide provincial and city governments with valuable resources and political leverage in bargaining with the private sector (Savitch and Kantor 2002). As various extensive studies on regional development in the Philippines have indicated (see e.g., Balisacan and Hill 2007; 2003), central-local relations within the Philippine Government are also of vital importance when analysing the strength of the local state and its role in the process of regional economic development.

Using the theoretical concepts outlined above, focus of this study will be on the key politico-economic relational networks of the island province of Cebu. Before showcasing the findings of the empirical analysis, the analytical framework and methodology of the case study are briefly discussed in section three. The empirical analysis of the study will start with section four, which focuses on the horizontal type of networking between the local private and public sectors. Consequently, section five will analyse the vertical relational networking of the local state with the national government. Then, section six will couple the two networks to define the local politico-economic system of Cebu. Finally, the regional development outcomes that derive from the politico-economic system are discussed in section seven.

GEOGRAPHICAL CONTEXT, ANALYTICAL FRAMEWORK AND METHODOLOGY OF THE STUDY

While there are other explanations for the emergence of the regional Cebuano economy — resource endowment and geographical features play a role — this chapter argues that the island's economic emergence in the global marketplace has been amplified via two politico-economic relational networks: (1) local public-private coordination (horizontal networking); and (2) national-local ties within the Philippine state (vertical networking). As will be further argued in this study, the combination of these two networks has been influential in shaping a unique Cebuano development trajectory that is marked by distinct outcomes in terms of wealth creation and distribution.

Different qualitative research methods have been used to investigate the two relational networks mentioned. Primary data was gathered by face-to-face interviews with key public and private stakeholders in the Cebuano political economy. Furthermore, local academics and journalists were also interviewed to include the perspectives from well-informed — but not directly involved — local informants. These interviews were performed by the author during various visits to Cebu over the period 2006–2008. Additionally, extensive desk research was carried out by the author in the same period, investigating a wide variety of sources (e.g., regional and local development plans, newspaper articles, studies from local and foreign academics).

The outline of the empirical remainder of this chapter is as follows: the first empirical section will analyse the horizontal type of public-private networking in the region. Subsequently, the second empirical section will discuss Cebu's vertical type of political relational networking. Finally, the chapter will merge the findings and formulate conclusions about Cebu's relational network and its coherent impact on regional development in terms of wealth creation and distribution.

HORIZONTAL NETWORKING: PUBLIC-PRIVATE COORDINATION

In many ways, the "economic miracle" or "*Ceboom*" that happened in Cebu during the late 1980s and early 1990s was shaped by local initiatives. As will be illustrated below, dynamic political leadership and strong private sector involvement have been the key ingredients for a decisive public-private Cebuano growth alliance. This alliance has radically transformed the local economy and repositioned Cebu in the global economy in a manner that parallels the more general model of development pursued in Southeast Asia (most notably, Singapore) during the eighties and nineties: export- and FDI-driven industrialization through strong government action on pro-growth economic policies (World Bank 1993). A noteworthy aspect of Cebu's growth is that it occurs on the local level: different from the national "city-state" of Singapore, Cebu has been able to pursue a unique model of regional economic growth within the national politico-economic context of the Philippine state.

After Marcos was forced to step down in 1986, this enabled many of his political rivals to revive their public activities. This was especially true for one of his most prominent enemies that was hit hard during the years of Martial Law: the Osmeña family from Cebu. The peaceful removal of Marcos not only cleared the path for the Osmeñas to return to the political stage, it also created a global wave of positive attention on the Philippines that had long suffered from a terrible reputation under the former dictator. In Cebu, these two movements seem to have been critical ingredients for its economic growth.

A central role in the post-Marcos development of Cebu is played by the local Osmeña family, and in particular Emilio "Lito" Osmeña, who served as Governor of Cebu from 1988 to 1992 and subsequently became the economic advisor of President Ramos for national government flagship projects from 1992 to 1998.[2] Lito was accompanied by his cousin Tomas "Tommy" Osmeña, who was a two-term mayor of Cebu City — the capital city of the province, which is also an economic centre of major importance — from 1988 to 1995, and supporter of his pro-growth agenda.[3] Their pro-business — and growth-oriented administration initially focused on finding innovative solutions to cut through or circumvent red tape. By promising minimum regulation and intervention, the provincial government started to attract new investors. Moreover, these investors were welcomed with streamlined application procedures and a relative freedom to exercise their business operations (Law 1997; Mojares 1994).

While these local pro-business policies have certainly been important in attracting new investors, it would have not become a success without a proper physical business environment. To push their ambitious and costly pro-growth development agenda, the Osmeña tandem broke with the traditional ways of fundraising by the local government. Impatient with waiting for funds to trickle down from the national government in Manila for the much-needed capacity upgrade of key infrastructural facilities, the local leaders actively searched for new funding partners and schemes.

The first partners were found by looking abroad: through an extensive public-private lobby from Cebu and guarantees of the national Philippine Government, the Japanese Overseas Economic Cooperation Fund and the Asian Development Bank provided extensive loans. Additionally, some of the most innovative funding schemes were coordinated in close partnership with major national property corporations that had close ties with international sources of capital and technology (Sajor 2003).[4] In 1988, the provincial government offered to sell a 45-hectares golf course in Cebu City via public bidding. The winning bid from a consortium of big investors exceeded the set floor price threefold, leaving the provincial government with a soon-to-be developed new business park close to the city centre and P551 million to spend on its infrastructural plans. A second major land deal of the Osmeña administration was the selling of the former Lahug airport site on the equity market in 1991 through the issuance of convertible bonds with a total value of P300 million. These bonds were exchangeable for shares of Cebu Property Ventures and Development Corporation (CPVDC), a joint venture of the provincial government and Ayala Land Incorporation (a major Philippine real estate development corporation). Consequently, this site was developed into the Asiatown IT Park, a Philippine Economic Zone Authority-registered economic zone with Japanese and American call centres and electronics R&D companies. As shown in Table 3.5 in the Appendix, this IT Park is just one of the many economic zones that have been created to transform Cebu into an attractive location for (foreign) firms active in export-oriented (electronics) manufacturing, IT and business process outsourcing (van Helvoirt 2009).

Aside from joint ventures and long-term personal relations with major business dynasties, the Osmeña administration also institutionalized coordination with the local private sector in various forms. One of the key activities that were undertaken by the local public and private sector together was the marketing of Cebu as an investment destination, using the catch phrase of "Cebu: an island in the Pacific". Using their close ties with the local business community, the Osmeñas personally led overseas trade missions comprised of the leading local business associations, being the Cebu Filipino-Chinese Chamber of Commerce and Industry and the Cebu Chamber of Commerce and Industry. During these missions, the public-private Cebuano delegation networked with chambers of commerce in the flourishing NICs of East Asia (Sajor 2003). Furthermore, the (foreign) manufacturing firms located on the MEPZs had also formed their own local business association (since 1976): the Mactan Export Processing Zones Chamber of Exporters and Manufacturers, which allowed the export-oriented firms on the Mactan zones to address any issues or requests to the local government in a structured and organised manner. Additionally, a local marketing vehicle was founded in 1994 under the name of the Cebu Investment Promotion Centre. This public-private initiative is a joint undertaking of the Province of Cebu, the Local Government Units of Metropolitan Cebu and at least 50 of the largest corporations operating in Cebu, with the primary function of promoting Cebu to foreign direct investors (van Helvoirt 2009).

The major transformations described above have laid the foundations for an economic vibrancy and public-private interactivity that still marks Cebu today. While the 1997 Asian Crisis put a stop to the period of fast-paced economic growth, the strong public-private interaction has remained alive until this day. This is also true for the economic primacy of Cebu, which has succeeded in keeping its leading position as the primary economic centre of the southern Philippines.

VERTICAL NETWORKING: NATIONAL-LOCAL POLITICAL RELATIONS

Cebu has always been in the limelight of national politics. Its strategic economic position, combined with its powerful political dynasties, backed by a large pool of voters has always made Cebu a "first-class" province. However, the position of Cebu vis-à-vis Manila has a two-faced character. Cebuanos have always contested the concept of a "Filipino national identity" (Banlaoi 2004). The island has a strong history of self-awareness and "Cebuano pride" that discards intervention and top-down control from "imperial Manila". On the other hand, while many Cebuanos want to believe they truly are "an island in the Pacific" that operates independently from the presidential Malacañang Palace, the development of Cebu as an economic powerhouse in the southern Philippines could not have been realized without strong support from national government leaders. This contradictive duality of national-local political relations between Manila and Cebu has been particularly strong during the "Ceboom" era. As discussed below, local political leadership played an essential role in this through their close and rather independent connections with national powerbrokers that allowed them to pursue an "autonomous" local growth agenda.

While based in Cebu, the Osmeñas are a political dynasty of national standing. Ever since the family patriarch Don Sergio Osmeña climbed his way up to the presidential seat in 1944, the Osmeñas have kept their contacts and influence in national politics. While the Marcos era forced them to retreat from the political arena, they quickly regained their prominence under President Corazon Aquino and used the new momentum to push their local growth agenda at the national level (van Helvoirt 2009). While Lito only served for one term as Governor of Cebu, he remained important for the development of the province afterwards. Though he lost his bid for vice president to Joseph Estrada, President Fidel Ramos kept Osmeña in his administration by giving him the prominent position of Presidential Chief Economic Adviser. In this role, Osmeña was in charge of flagship projects and had a powerful voice in the National Economic and Development Authority and Department of Trade and Industry. This assured a continued and expanded national government dedication to the development of Cebu as the prime economic centre of the southern Philippines. As Sajor (2003) argues, this influential position in the Ramos administration enabled Osmeña to attract significant national government support and resources for Cebu-based development projects. As such, he was able to get generous subsidies from the national government to upgrade the island's telephone lines, electric power facilities, transport terminals, traffic management, garbage collection, police force and airport terminal. These were all essential assets and valuable selling points in Cebu's quest to attract (foreign) investors for its local export-oriented economy. Moreover, the close ties with national government leaders allowed Cebu to gain easier access to international development project financiers that made the island less dependent upon financing from the national government that occurs via the normal routing within the regulatory framework of the Local Government Code.[5]

The 1991 enactment of the Code is the most significant boost to the existing local government system in the Philippines, as it has devolved significant powers from the provincial offices of National Government Agencies to Local Government Units (LGUs).[6] First of all, LGUs gained responsibility for the delivery of various aspects of basic services (most importantly, health, agriculture, environment and natural resources, social services and public works funded by local funds). Second, certain regulatory and licensing powers were transferred to local governments. These include: the reclassification of agricultural lands; enforcement of environmental laws; enforcement of the National Building Code; and processing and approval of subdivision plans. Third, the direct involvement of civil society — in particular, non-governmental organizations and people's organizations — in the process of local governance was enabled via representation in local government bodies and participation in local decision-making.

Fourth, and most importantly, the financial resources available to LGUs were increased by (a) broadening their local taxing powers, providing them with a specific share of the national wealth exploited in their area (e.g., via mining, fishery and forestry); and (b) increasing their automatic share from national taxes, i.e., the Internal Revenue Allotment shares, from a low 11 per cent to as much as 40 per cent. Of the funds automatically transferred from the national government to LGUs, 20 per cent goes to communities (*barangays*), 34 per cent to municipalities, 23 per cent to cities, and 23 per cent to the provinces. Furthermore, LGUs were allowed and encouraged to be more entrepreneurial by providing them with more opportunities to: enter into joint ventures with the private sector; engage in Build-Operate-Transfer[7] arrangements; float bonds; and obtain loans from local private institutions. All of these measures were initiated to encourage LGUs to be more "business-like" and competitive in their operations (Atienza 2006). As depicted above, the Cebu local government found creative methods to capitalize on the opportunities provided by the Local Government Code legislation (van Helvoirt 2009).

Opposed to the close harmonious relations between Manila and Cebu is the push for more autonomy by local officials. Governor Lito Osmeña's local growth-oriented agenda was partly based on his Manila-defiant entrepreneurial attitude, making him "a Cebuano first, a businessman second, and a politician third" (Churchill 1993, p. 10). Although his vision of an independent "Republic of Cebu" was never really pursued, Osmeña did foster a Cebuano belief in self-reliance and the persistent pursuance of a local growth agenda, even if this caused conflicts with neighbouring provinces in the region. Importantly, this local autonomy can also be attributed to the fact that Cebu's political leaders have generated access to international funds (as described in the previous section) that makes it less dependent upon financial contributions from the national state.

DEFINING CEBU'S POLITICO-ECONOMIC CONSTELLATION

Having analysed the two sets of politico-economic networks of the regional economy, this section will merge these two sets into one constellation. This coupling of the two relational networks will be used to come to a full description of the regional politico-economy.

The urban economy of Metro Cebu is diverse and complex, and closely connected to several levels of scale, including "the global". This has made a serious impact on the development trajectory of Cebu. Successful local political business tycoons redefined the Cebu economy as a huge real estate operation, reclaiming land and opening it up to outside investors in business parks that catered to the need of national and global business. The way in which this political economic restructuring was undertaken is reminiscent of the urban regime literature by Stone et al. This involves: forging growth alliances by mobilizing local and national political support; public sector investment in infrastructure and support services; and attracting sections of international capital from countries such as the U.S., Japan, Korea to fill the reclaimed business areas. The strong connectivity of the elites with the private sector — local, national and even international — has helped to transform Cebu into the second-largest business centre of the country. Furthermore, public-private coordination at the local level has been strong, with the Cebu Chamber of Commerce as the leading representative of the local business community. Aside the well-entrenched position of local political elites in the business scene, their close ties with national political leaders has been a second strategic relational asset that has influenced the development of the Cebuano economy. These strong central-local ties, based on the electoral and economic weight of Cebu in the Philippines, have been instrumental in gaining the necessary financial and political support from Malacañang for major infrastructural projects such as the expansion and modernization of the international sea- and airports.

In Cebu, as depicted in Figure 3.1, the relational networks have a strong multi-scalar dimension. Strong global integration has connected local stakeholders directly with national and international agents. Two forms of connectedness seem particularly important here: (1) the position of Cebu in the global marketplace (linking it to global capital and MNCs); and (2) the strong connectedness of local government leaders with private and other public stakeholders. These two forms of international connectedness are not separate dynamisms, but are bonded patterns in a regional politico-economy that can be identified as an urban regime. Following Stone (1989),

Figure 3.2
Cebu's Relational Politico-Economy

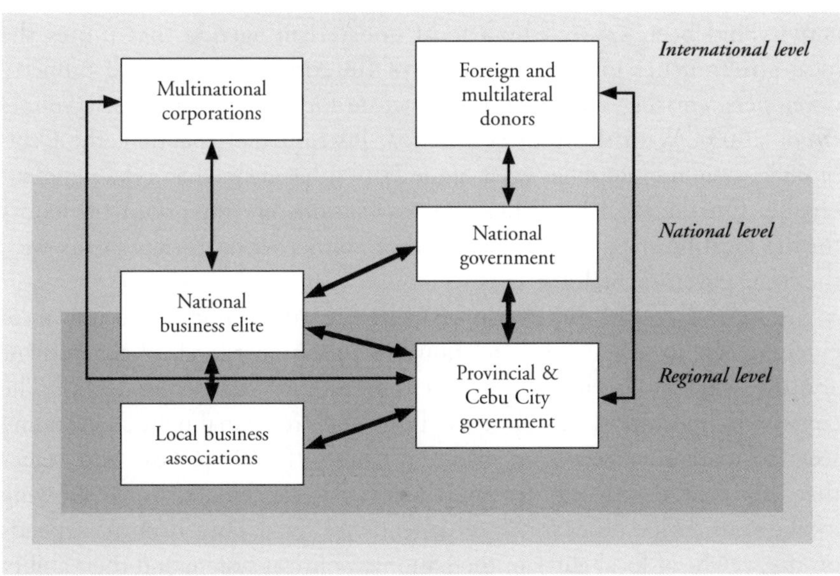

Source: van Helvoirt 2009.

a decisive dynamism that makes an urban regime is the interplay between public and nongovernmental (often private) actors. Regime analysis views power as fragmented and regimes as collaborative arrangements through which local governments and private actors assemble the capacity to govern (Mossberger and Stoker 2001). Different from elite theories, regime theory recognizes that it is unlikely that any single group is able to exercise comprehensive control in a complex, globalised environment like Metro Cebu (Stoker and Mossberger 1994). The political elite — i.e., the Osmeñas — of Cebu play an instrumental role in the positioning of Metro Cebu as a modest "global city" (Sassen 2001). Nevertheless, it is only capable of doing so by closely coordinating with other key stakeholders from the private sector — local, national and international — and the national government. As such, it is this multi-actor coalition, on a multi-scalar level, that is essential to Cebu's relational politico-economy.

While the multi-scalar public-private coalition forms the core of the Cebuano politico-economy, there are some features that further shape its specific identity. First, the coalition seems to privilege the participation of certain business interests over others, as it is particularly concerned with promoting economic growth via projects that are aimed at converting land use. The growth of Cebu as a regional centre for international business activity has been spurred by a local boosterism agenda that unites the local government's interests with that of (inter)national land and property developers, creating a suitable infrastructure for attracting foreign investors (Sajor 2003). With this project-oriented, instrumental coalition, the Cebu business system exemplifies what Stone (1993) has defined as a development regime. On the local level, business associations are important mediators for the local business community, as they connect local entrepreneurs with the local government leaders.

A second crucial dimension of Cebu's urban regime is the way local elites are able to manage their relationship with higher levels of government and the wider political environment (Stoker and Mossberger 1994). The contrasting specifics of this dimension in Cebu's relational politico-economy seem to resemble a comment made by Keating (1991, p. 66), who argues that "the central state can be oppressive, or it can be a resource allowing localities to escape other forms of dependence. [...] This, in turn, depends on the weight of local elites in the national political system and their ability to forge coalitions to extract resources on their own terms". The unique position of Cebu in the political economy of the Philippines enables its political leaders to be a somewhat less compliant "autonomous" province that does not entirely depend on top-down dole-outs by the national government, but can actually make bottom-up claims for national support as well. Using this support, local elites have combined economic and political positions of command to take a lead in forging the decisive connections. This echoes the statement by Segal and Thun (2001, p. 558) who argue that "local governments do not simply try to reproduce and catch up with development efforts initiated by the central government, but are often the actual architects of growth, designing and implementing development policies that are conducive to local institutional frameworks and specific development needs".

REGIONAL DEVELOPMENT OUTCOMES

Having outlined the regional politico-economic constellation of Cebu, this section will elaborate on the regional development outcomes it yields. Two dimensions of regional development are discussed below; i.e., the creation and distribution of wealth.

Undeniably, Cebu's urban development regime has been successful in generating economic growth — in terms of FDI and export-oriented trade — via the creation of attractive business locations and infrastructures. Additionally, it has also encouraged cooperation and partnership between the local government and the private sector "imbued with entrepreneurial spirit" (Etemadi 2001, p. 33). These influential dynamics have shaped the contemporary modest "world city" (Friedman 1997) of Cebu. However, as elaborated in the former paragraph and more extensively by Sajor (2001; 2003), this public-private coalition is largely focused on stimulating an FDI and real estate development-driven economic growth; a pro-business strategy that weakens the redistribution function of the local state. As such, the strong public-private coalition has a contradictory development impact in the sense that it does spur rapid economic growth. However, the opportunities to benefit from this growth are not accessible to all actors in the local economy, but remain largely restricted to the elite members of the urban regime.

While resembling a growth alliance as described in the American context by Stone (1989), an important deviation from the model is with the extent of dynastic control over the levers of public policy and the longevity of control over public resources. While growth alliances are typically diverse and temporary (by virtue of their combination of different interest groups), the patrimonialism of the Philippine political economy has made for a more protracted programme in Cebu, which shows clear risks of lock-in in public-private land and real estate development policy where this no longer is very effective in fostering local development. In more general terms, the two development impacts that originate from Cebu's global-urban development regime resemble Van Naerssen's (2004) observation — referring to earlier work of Berner (2002) — that the exposure of Cebu to global forces has intensified local mechanisms of inclusion and exclusion. Clearly, the strong economic growth Cebu has experienced in the past decades has increased the opportunities for local entrepreneurs to "go out and explore" their chances on the lucrative global market. It has also raised the level of well-being for the many Cebuanos who work in the city's global-oriented service and manufacturing sectors. Still, as depicted by Andriesse et al. (2011), the mechanism of local exclusion within these social groups can be observed as well, for instance in the local subcontractors and labourers who are laid off when production on the export processing zones slows down.

APPENDIX

Table 3.3
The Philippine Electronics Industry by Major Products

Components and devices (semiconductors)
Consumer electronics (audio and video equipment)
Office equipment (photocopy machines, electronic calculators)
Control and instrumentation
Electronic data processing (PCs, disk drives, motherboards)
Telecommunications (mobile telephones, radios, satellite receivers)
Communications and radar (radar detectors, CCTV)
Medical and industrial (spiro analysers, smoke detectors)
Automotive electronics (brake systems, wiring harness, car radios)

Source: Agarwalla 2005.

Table 3.4
Selected Examples of Japanese Electronics Firms in Cebu

Name company	Equity	Major Product(s)
Cebu Microelectronics Corp.	99.92%	Assembly of super precision electric discharging mechatronics (EDM) tools, etc.
Goji Industry Corp.	100%	Manufacture of CD-ROM mechanism and magnetic audio heads
Halsangz Plating Cebu Corp.	99.95%	Electroplating of electronic parts
Intec Cebu Inc.	100%	Manufacture of printed circuit board (PCB) for CD and CD-ROM players
Mactan Showa Electric Wire Inc.	100%	Production of stud wires for semiconductors
Muramoto Audio-Visual Phils. Inc.	100%	Floppy disk drives for computers, CD-ROM
NEC Technologies Phils Inc.	99.99%	Transmission and telecommunication equipment and system
Phil. Makoto Corp.	99.92%	Magnetic eraser heads for industrial semiconductor
Taiyo Yuden (Phils) Inc.	99.96%	Electronic components
Cebu Mitsumi Inc.	100%	Electronic Products: magnetic tape head, floppy disk drive, CD-ROM

Source: Zosa 2004.

Table 3.5

Special Economic Zones in Cebu Province, by Location and Major Activity

Name	Location	Major economic activity
Arcenas IT Estate Building	Cebu City	IT related activities
Asia Town IT Park	Cebu City	Business Process Outsourcing
Bigfoot IT Park	Lapu-Lapu City	IT related activities
Cebu IT Tower	Cebu City	IT related activities
Cebu Light Industrial Park	Lapu-Lapu City	Export-oriented manufacturing
Cebu South Road Properties	Cebu City	None (under development)
HDWF-WTCI IT Tower	Cebu City	IT related activities
HVG Arcade IT Park	Mandaue City	Business Process Outsourcing
Innove IT Plaza	Cebu City	IT related activities
JY Square IT Centre	Cebu City	IT related activities
KRC IT Zone	Mandaue City	IT related activities
Lexmark Plaza	Cebu City	IT related activities
Mactan Economic Zone	Lapu-Lapu City	Export-oriented manufacturing
Mactan Economic Zone II	Lapu-Lapu City	Export-oriented manufacturing
Mango Square	Cebu City	IT related activities
MRI Ecozone	Danao City	Export-oriented manufacturing
New Cebu Township	Naga	Export-oriented manufacturing
Oakridge IT Centre	Mandaue City	IT related activities
Pioneer House Cebu	Cebu City	IT related activities
Polambato-Bogo Economic Zone	Bogo City	Export-oriented manufacturing
Synergis IT Centre	Cebu City	IT related activities
Taft IT Park	Mandaue City	IT related activities
West Cebu Industrial Park	Balamban	Heavy industrial manufacturing

Source: Philippine Economic Zone Authority (PEZA), from van Helvoirt (2009).

Notes

1 Short (1996, p. 210; quoted in Sajor 2001, p. 7) defined contemporary pro-growth boosterism as the aggressive promotion of a pro-business agenda for a particular city against the background of competing cities and alternative investment opportunities. Its leading players — usually influential politicians and local government leaders, corporate chairpersons, realtors, local banks and chambers of commerce — all seek to define the city as an economic growth machine. These players have a strong consensus regarding stimulating investment and economic growth while limiting the re-distributional function of the state.

2 As Sajor (2003) argues, this prominent position enabled Lito Osmeña to attract significant national government support and resources for Cebu-based development projects. As such, he was able to get generous subsidies from the national government to upgrade the island's telephone lines, electric power facilities, transport terminals, traffic management, garbage collection, police force and airport terminal.

3 Additionally, the Cebu City Council was dominated by Bando Osmeña — Pundok Kauswagan, Tomas Osmeña's local party: it held eight of the sixteen seats in 1988, fourteen in 1992 and all sixteen in 1995 and again in 1998 under Mayor Alvin Garcia, a former vice-mayor under Osmeña (Etemadi 2000, p. 38). Moreover, John Henry "Sonny" Osmeña, Lito's elder brother, was Senator from 1987 to 1995 and Raul del Mar, a cousin of Tomas Osmeña, was Congressman of the 1st district of Cebu City from 1987 to 1998.

4 Two of the biggest Manila-based firms that operated in Cebu during the "Ceboom" are Ayala Land Incorporation and Aboitiz Group of Companies. Based on total sales, these two national conglomerates are ranked as the 3rd and 12th largest companies in the Philippines. Ayala has international partnerships with Hong Kong-based, Dutch and French land development and construction corporations. Aboitiz has substantial partnerships with Japanese corporations (Sajor 2003).

5 Under the Local Government Code of 1991 not only do Local Government Units (LGUs) have certain responsibilities (delivery of basic public services), but it also provides LGUs with local taxing power and a automatic share of national taxes via the Internal Revenue Allotment (IRA). While these fiscal decentralization measures have enhanced local state autonomy, there are still some major drawbacks. As both De Dios (2007) and Atienza (2006) argue, for many LGUs the local tax base does not generate sufficient income to finance their expanded service tasks, making them still very much dependent upon the national government for their resources. Additionally, the skewed IRA distribution formula among various levels of local government has resulted in "cities gaining a windfall while municipalities and provinces, which bear 92.5 per cent of all devolved services, find it harder to cover the cost of devolution" (Atienza 2006, p. 435).

6 Before the implementation of the code, the functions assigned to LGUs were limited to the levying and collection of local taxes, the issuance and enforcement of regulations governing local business activities, and the administration of certain services and facilities (e.g., garbage collection, public cemeteries, public markets and abattoirs (Manasan 2007, pp. 278–79)).

7 Build-Operate-Transfer (BOT) is a form of project financing, wherein a private entity receives a franchise from the local government to finance, design, construct and operate a public facility (e.g., water supply or power plant) for a certain period, after which ownership is transferred to the local government. During the time that the private sector entity operates the facility, it is allowed to charge facility users a fee to regain its investment costs, as well as operating and maintenance expenses.

References

Agarwalla, G. "Philippine Electronics Equipment Production and Manufacturing". World Bank Working Paper. Washington D.C.: World Bank, 2005.

Amin, A. "An Institutionalist Perspective on Regional Economic Development". *International Journal of Urban and Regional Research* 23, no. 2 (1999): 365–78.

Amin, A. and N. Thrift. "Living in the Global". In *Globalisation, Institutions and Regional Development in Europe*, edited by A. Amin and N. Thrift. Oxford: Oxford University Press, 1994.

Anderson, B. "Cacique Democracy in The Philippines: Origins and Dreams". *New Left Review* 169 (1988): 3–31.

Andriesse, E., N. Beerepoot, B. Van Helvoirt, and A. Van Westen. "Business Systems, Value Chains and Inclusive Regional Development in South-East Asia." In *Value Chains, Social Inclusion and Economic Development*, edited by A.H.J. Helmsing and S. Vellema. London: Routledge, 2011.

Atienza, M.A.L. "Local Governments and Devolution in the Philippines". In *Philippine Politics and Governance: An Introduction*, edited by N.M. Morada and T.S.E. Tadem. Quezon City: University of the Philippines Diliman, 2006.

Balisacan, A.M. and H. Hill, *The Dynamics of Regional Development: The Philippines in East Asia*. Cheltenham: Edward Elgar, 2007.

———. *The Philippine Economy: Development, Policies, and Challenges*. Quezon City: Ateneo de Manila University Press, 2003.

Banlaoi, R.C. "Globalisation and Nation-Building in the Philippines: State Predicaments in Managing Society in the Midst of Diversity". In *Growth and Governance in Asia*, edited by Y. Sato. Honolulu: Asia-Pacific Centre for Security Studies, 2004.

Berner, E. "Global Citadels and Ghettos: The Dynamics of Inclusion and Exclusion in Metro Cebu". *Philippine Quarterly of Culture and Society* 29 (2001): 211–25.

Churchill, P.R. "Cebu: Aberration or Model for Growth?" *Philippine Quarterly of Culture and Society* 21, no. 1 (1993): 3–16.

Coe, N.M., M. Hess, H.W.C. Yeung, P. Dicken, and J. Henderson. "'Globalizing' Regional Development: A Global Production Networks Perspective". *Transactions of the Institute of British Geographers* 29, no. 4 (2004): 468–84.

Cooke, P.N. and K. Morgan, *The Associational Economy*. Oxford: Oxford University Press, 1998.

De Dios, E.S. "Local Politics and Local Economy." In *The Dynamics of Regional Development: The Philippines in East Asia*, edited by A.M. Balisacan and H. Hill. Cheltenham: Edward Elgar, 2007.

Etemadi, F.U. "Civil Society Participation in City Governance in Cebu City". *Environment and Urbanisation* 12, no. 57 (2000): 57–72.

———. *Towards Inclusive Urban Governance in Cebu*. Partnership and Poverty Working Paper 25. Birmingham: International Development Department, School of Public Policy, University of Birmingham, 2001.

Evans, P.B. *Embedded Autonomy: States and Industrial Transformation*. Princeton: Princeton University Press, 1995.

Fenner, B.L. *Cebu Under the Spanish Flag, 1521–1896: An Economic-Social History*. Cebu City, Philippines: San Carlos Publications, 1985.

Friedman. J. *World City Futures: The Role of Urban and Regional Policies in the Asia-Pacific Region*. Hong Kong: Hong Kong Institute of Asia-Pacific Studies, Chinese University of Hong Kong, 1997.

Henderson, J.V., Z. Shalizi, and A.J. Venables. "Geography and Development". *Journal of Economic Geography* 1, no. 1 (2001): 81–105.

Hill, H. "Spatial Disparities in Developing East Asia: A Survey". *Asian-Pacific Economic Literature* 16, no. 1 (2002): 10–35.

Hooghe, L. and Marks, G. "Unraveling the Central State, But How? Types of Multi-level Governance". *The American Political Science Review* 97, no. 2 (2001): 233–43.

Hutchcroft, P.D. "Booty Capitalism: Business-Government Relations in the Philippines". In *Business and Government in Industrializing Asia*, edited by A.J. MacIntyre. Ithaca: Cornell University Press, 1994.

———. *Booty Capitalism: Politics of Banking in the Philippines*. Ithaca: Cornell University Press, 1998.

———. "Colonial Masters, National Politicos, and Provincial Lords: Central Authority and Local Autonomy in the American Philippines, 1900–1913". *The Journal of Asian Studies* 59, no. 2 (2000): 277–306.

———. "Centralisation and Decentralisation in Administration and Politics: Assessing Territorial Dimensions of Authority and Power." *Governance* 14, no. 1 (2001): 23–53.

Jessop, B. *State Theory: Putting the Capitalist State in Its Place*. Cambridge: Polity Press, 2003.

JETRO. *The 13th Survey of Investment-Related Cost Comparison in Major Cities and Regions in Asia*. Tokyo: Japan External Trade Organization, 2003.

Keating, M. *Comparative Urban Politics*. Aldershot: Edward Elgar, 1991.

Kerkvliet, B.J.T. *Everyday Politics in the Philippines: Class and Status Relations in a Central Luzon Village*. Lanham: Rowman & Littlefield, 2002.

———. "Toward a More Comprehensive Analysis of Philippine Politics: Beyond the Patron-Client, Factional Framework". *Journal of Southeast Asian Studies* 26, no. 2 (1995): 401–19.

Law, L. "Cebu and *Ceboom*: The Political Place of Globalisation in a Philippine City." In *Pacific Rim Development: Integration and Globalisation in the Asia-Pacific Economy*, edited by P.J. Rimmer. Sydney: Allen & Unwin Australia Pty Ltd., 1997.

Manasan, R.G. "Decentralisation and the Financing of Regional Development". In *The Dynamics of Regional Development: The Philippines in East Asia*, edited by A.M. Balisacan and H. Hill. Cheltenham: Edward Elgar Publishing, 2007.

McCoy, A.W., ed. *An Anarchy of Families: State and Family in the Philippines*. Madison: University of Wisconsin Centre for Southeast Asian Studies, 1993.

Mojares, R.B. "The Dream Goes On and On: Three Generations of the Osmeñas, 1906–1990". In *An Anarchy of Families: State and Family in the Philippines*, edited by A.W. McCoy. Madison: Ateneo de Manila University Press, 1993.

Mossberger, K. and G. Stoker. "The Evolution of Urban Regime Theory: The Challenge of Conceptualization". *Urban Affairs Review* 36, no. 6 (2001): 810–35.

Sajor, E.E. "Globalization and the Urban Property Boom in Metro Cebu, Philippines". *Development and Change* 34, no. 4 (2003): 713–42.

———. *The Real Estate Boom in the 1990s and Land Use Allocation for Socialized Housing in Metro Cebu, Philippines*. UMP-Asia Occasional Paper no. 52. Patumthani, Thailand: Urban Management Programme: UNDP/UNHCS/World Bank, 2001.

Sassen, S. *The Global City: New York, London, Tokyo*. Princeton: Princeton University Press, 2001.

Savitch, H.V. and P. Kantor. *Cities in the International Marketplace*. Princeton: Princeton University Press, 2002.

Sayer, A. "For a Critical Cultural Political Geography". *Antipode* 33, no. 4 (2001): 687–708.

Scott, A.J. and M. Storper. "Regions, Globalization, Development". *Regional Studies* 37, nos. 6–7 (2003): 579–93.

Segal, A. and E. Thun. "Thinking Globally, Acting Locally: Local Governments, Industrial Sectors, and Development in China". *Politics & Society* 29, no. 4 (2001): 557–88.

Short, J.R. *The Urban Order: An Introduction into Cities, Culture and Power*. Cambridge, MA: Blackwell Publishers, 1996.

Sidel, J.T. *Capital, Coercion and Crime: Bossism in the Philippines*. Stanford: Stanford University Press, 1999.

———. "The Underside of Progress: Land, Labour, and Violence in Two Philippine Growth Zones, 1985–1995". *Bulletin of Concerned Asian Scholars* 30, no. 1 (1998): 3–12.

Stoker, G. and K. Mossberger. "Urban Regime Theory in Comparative Perspective". *Environment and Planning C: Government and Policy* 12, no. 1 (1994): 195–212.

Stone, C.N. *Regime Politics: Governing Atlanta, 1946–1988*. Lawrence: University Press of Kansas, 1989.

———. "Urban Regimes and the Capacity to Govern: A Political Economy Approach". *Journal of Urban Affairs* 15, no. 1 (1993): 1–28.

Storper, M. *The Regional World: Territorial Development in a Global Economy*. New York: Guilford Press, 1997.

Van Helvoirt, B.J. "Regions, Institutions and Development in a Global Economy. Divergent Regional Business Systems in the Philippines". Ph.D. Dissertation, International Development Studies, Faculty of Geosciences, Utrecht University, 2009.

Van Naerssen, T. "Cebu City in the Global Arena: Its Governance and Urban Development Policy". *Philippine Quarterly of Culture and Society* 32, nos. 3–4 (2004): 179–202.

Whitley, R.D. *Divergent Capitalisms: The Social Structuring and Change of Business Systems*. Oxford: Oxford University Press, 1999.

World Bank, *The East Asian Miracle: Economic Growth and Public Policy*. New York: Oxford University Press, 1993.

Zosa, V. *Philippine-Japan Economic Linkages: A Case Study of Cebu*. Philippine Institute for Development Studies Discussion Paper no. 2004–33. Manila: Philippine Institute for Development Studies, 2004.

CAMBODIA
VIETNAM
PHILIPPINES
Gulf of Thailand
South China Sea
THAILAND
Sulu Sea
Straits of Malacca
BRUNEI
Putrajaya
MALAYSIA
Johor Bahru
JOHOR STATE
(MALAYSIA)
SINGAPORE
INDONESIA
LEGEND
State/Provincial Capital
STATE/PROVINCE
National Capital
COUNTRY
NEIGHBOURING COUNTRY
Geographic Features
National Border Land
National Border Water
State/Province Border
0 km
200
400
All maps are derived from GADM Data
(Global Administrative Areas, www.gadm.org, 2012)
N

4

ONE PRIORITY AMONG MANY? THE STATE GOVERNMENT AND ELECTRONICS SECTOR IN JOHOR, MALAYSIA

Francis E. Hutchinson

INTRODUCTION

Malaysia is a quintessential Little Tiger, one of a group of Southeast Asian countries that managed a transition from a predominantly agriculture-based economy to one with a vibrant manufacturing sector. Once known for its spices, tin, and rubber, Malaysia now has a diversified secondary sector comprising textiles, chemicals, steel, transport equipment, and other products. The electrical and electronics (E&E) sector has been at the centre of Malaysia's structural shift, accounting for the greater part of manufacturing employment, investment, and export earnings (Yusoff and Bhattasali 2008).

The country's E&E sector is essentially concentrated in three firm clusters. In the north, Penang houses market-leading semiconductor and industrial equipment multinationals; a number of contract electronics manufacturers; and many supplier firms, including some local firms with world-class capabilities in the process automation and technology provision sectors. The Klang Valley, comprised of Kuala Lumpur and the surrounding state of Selangor, has a large number of firms in the consumer electronics, semiconductor, and domestic appliance sub-sectors. Johor, in the south of the country, houses the third concentration of firms which includes a large number of contract electronics manufacturers, as well as consumer electronics and computer peripheral producers (Yoon 2013; Rasiah 2002; JSICID 2011).

Available comparative data shows that multinational firms carry out tasks of different intensity in the three locations, and local firms have differing levels of technological capabilities and network strength (Best and Rasiah 2003; Rasiah 2008). Thus, while providing important elements of an overall assessment, a uniquely national-level perspective does not explain these intra-national differences in firm dynamics and technological capabilities.

Given that Malaysia is a federation, with a central government and thirteen state counterparts, an examination of state agency and policies at the sub-national level may shed light on the differential development of each of these firm groupings. While Malaysia's constitution assigns the bulk of responsibilities and revenue sources to the federal government, state governments do carry out important tasks relative to economic policy in general and industrialization in particular.

Research carried out on Penang has documented how its state government and economic development corporation played a role in promoting the development of its electronics sector through implementing targeted investment drives; providing specialized infrastructure; making strategic investments to diminish risk; addressing market failures regarding skilled workers; and seeking to foster inter-firm linkages (Singh 2011; Hutchinson 2008; Rasiah 1999).

The role of the state governments of Selangor and Johor in fostering the emergence and development of their electronics sectors is less well-researched. That said, establishing the role played by the Selangor state government is complicated by its proximity to the Kuala Lumpur federal territory, which has benefited from a disproportionate amount of federal funding for infrastructure, high technology parks, and strategic investments (Best 2007).[1] In addition, Kuala Lumpur houses the headquarters of all federal government agencies, including business support agencies, credit providers, and venture capital firms.

In contrast, Johor, like Penang, is far removed from the Klang Valley, offering an interesting case for research on the issue of sub-national agency. Up until the launching of the South Johor Economic Region in 2006, the state had not been singled out for priority investment by the federal government. And, as with its Penang equivalent, the state government has formulated its own policies for economic development. Johor also boasts a proactive state economic development agency that has moved to provide industrial parks, invest in new sectors, promote local entrepreneurship, and address market failures. Its achievements include employing 65,000 people through a network of 300 companies, including eight listed on the KL stock exchange; and fostering more than 60 local firms.

This chapter will therefore focus on Johor, seeking to establish how state government organizations and policies have, within the overall national policy context, shaped the emergence and subsequent development of its electronics sector. Following this introduction, this chapter has five parts. Part two will put forward the theoretical framework used to analyse the Johor case. Part three will use this framework to analyse the electronics sector in the state, notably its firm structure; the degree to which local technological capabilities have been acquired; and the existence, if any, of inter-firm networks. Part four will analyse how and to what extent state government organizations and initiatives have provided an enabling and supportive environment for the electronics sector. The fifth and final section will put forward the paper's principal conclusions.

Theoretical Framework

One comprehensive approach to analysing a region's ability to provide an environment conducive to innovation and manufacturing competitiveness is Regional Innovations Systems (RIS). For the purposes of this chapter, a region is defined as "a meso-level political unit set between the national or federal and local levels of government that might have some cultural or historical homogeneity, but which at least ha[s] some statutory powers to intervene and support economic development, particularly innovation" (Cooke 2001, p. 953).

The RIS approach focuses on innovation which is defined as the "commercialization of new knowledge in respect of products, processes, and organization". In addition, this definition stipulates that the knowledge developed needs to be used in the market; and innovation is not just the development of entirely new products, but also includes (often small) improvements in existing products or processes (Cooke 2001, p. 953).

In essence, the RIS approach focuses on the ways in which firms relate to each other on one hand, and with the surrounding institutional context on the other. Its starting point is that firms do not possess all capabilities for effective creation or absorption of knowledge in-house. Rather, much of their competitive advantage hinges on being able to effectively access knowledge generated outside, through contact with: other firms; organizations such as universities, business consultants, or research institutes; or collective facilities such as open laboratories.

The wider definition of innovation means the RIS approach does not centre solely on organizations such as research institutes and universities. These organizations are fundamental for certain types of knowledge, but the RIS approach goes beyond them to assess all parts of a region's economic and institutional infrastructure and how it creates, absorbs, and then circulates knowledge. A context that has more opportunities for firms to network among themselves as well as connect with surrounding organizations — which are, in turn, linked, dynamic, and attuned to firm needs — will be more conducive to innovation.

An RIS is made up of two discrete aspects:

- The regional production structure — this includes firms and any networks that bind them together.
- The regional supportive infrastructure — this encompasses all institutions that support economic activity and innovation, such as government agencies, research institutes, universities, technology centres, credit providers and venture capitalists, and business associations (Andersson and Karlson 2004, p. 10).

While a given region may have a mass of firms and an array of organizations, there needs to be interaction and communication within each of the two aspects as well as between them.

The RIS approach differs from more market-oriented approaches such as Porter's cluster theory, in that it starts from the basis that many of the more productive aspects of agglomeration such as the generalized diffusion of knowledge, collective efficiency, and collaboration do not occur automatically. On the contrary, a variety of market failures may impede firms from joining networks, benefiting from the diffusion of knowledge, or engaging in upgrading (see Introduction, this volume).

The rationale behind establishing and strengthening an RIS is thus to increase the innovative potential of firms by correcting market failures, overcoming collective action dilemmas, and increasing the opportunities for collaboration and inter-firm learning. However, while it is tempting to invest in infrastructure, institutions, and incentives, policy-makers need to bear in mind the ability of existing organizations to successfully absorb additional funds (Oughton et al. 2002).

Thus, the role of a regional government can be seen to be one of a facilitator. More than providing a vast array of services, the regional government can act as an "intelligent cell", compiling information, establishing development priorities, making use of existing national institutions and polices, promoting communication and, where necessary, tackling market failures. While the creation of a functioning RIS is a long-term effort, it is very hard for competitors to emulate as, while physical infrastructure and financial capital can be provided relatively easily, social capital takes longer to accumulate.

The next two sections will analyse Johor's Regional Innovation System with regard to the E&E sector. The first will look at its production structure, and the subsequent section will look at its supportive infrastructure, namely the incentives, infrastructure, and institutional context within which firms carry out their operations.

The Electronics Industry in Johor: The Regional Production Structure

After a brief introduction to Johor and its economy, this section will analyse the state's electronics sector, paying particular information to: its size, structure, ownership, and orientation; the level of technological capabilities and the nature of their acquisition; and the extent of networking between firms. In order to place these characteristics within the Malaysian context, references will be made to the electronics sector in Penang where relevant.

Johor's 19,200 sq. kilometres of flat, fertile land has meant the state has long been associated with primary commodities. Its state's initial settlement in the 1840s was driven by the cultivation of pepper and gambier (Trocki 2007). In the early twentieth century, rubber replaced these commodities and was, in turn, superseded by palm oil. Today, Johor is the largest producer of palm oil in Peninsular Malaysia, and also is an important exporter of other agricultural commodities (MPIC 2010).

Despite formidable levels of income accruing from agriculture, the Johor state government began to foster manufacturing in the 1970s — notably in steel and shipping — in an attempt to diversify the economy. Since the 1980s, the manufacturing sector has benefited from sustained levels of foreign direct investment, particularly in the electronics sector.

Over the past three decades, Johor's economy has undergone a structural transformation. In 1983, the state's primary, secondary, and tertiary sectors accounted for 33 per cent, 28 per cent, and 39 per cent of GRP, respectively. In 2008, the primary sector accounted for 8.3 per cent of GRP, and the secondary and tertiary sectors accounted for 44.1 per cent and 47.6 per cent of Johor's GRP, respectively (JSEPU 2009).

The state's manufacturing sector consists of about 4,700 firms employing some 330,000 people. It is diversified, with plastic products, furniture, food processing, and petroleum products constituting — along with electronics

— the most important sub-sectors. The electronics sector — comprising components producers, office and computing machinery, and consumer electronics — accounts for approximately 20 per cent of total manufacturing employment.[2]

The sector is largely foreign-owned, with the largest number of firms having headquarters in Japan (41 per cent), followed by Singapore (35 per cent), Malaysia (18 per cent), and the United States (6 per cent). The sector is largely externally-oriented, with 78 per cent of firms exporting for more than five years. The most common destinations by order of importance are North America, Singapore, Japan, and the European Union. However, there are different patterns, with foreign-owned firms exporting at least 80 per cent of their production, and no local or joint-venture firms exporting more than 50 per cent of their production — thus indicating some linkages between local and foreign firms (RMA 2006).

Table 4.1 has a breakdown of the electronics sectors in Johor and Penang for 2008. In Johor, the electronics sector accounts for some 65,000 employees and 127 firms. The components sub-sector is the largest, accounting for 30,000 workers in more than 70 firms. This is followed by: office and computing machinery with 19 firms and 14,000 employees; television and radio receivers with 29 firms and 11,600 employees; and television and radio transmitters with 6 firms and some 8,300 employees. In each sub-sector, the bulk of the employment is provided by a number of large firms.

Penang's electronics sector is significantly larger in employment terms, with more than 100,000 workers in about 100 firms. As with Johor, the components sub-sector is the largest, in this case comprising some 69,000 workers in 74 firms. The next is office and computing machinery with 23,000 employees in 15 firms. The consumer electronics sub-sectors are significantly smaller. As with Johor, the bulk of employment is concentrated in a number of large firms.

In Johor, employment in the electronics sector has undergone a notable reduction since 2000, when the sector employed some 85,000 people (Table 4.2). The sub-sector most affected was television and radio receivers, whose ranks shrank from 44 firms and almost 39,000 workers in 2000 to 29 firms and 11,000 workers in 2008 — reflecting its declining competitiveness in high-volume, low-mix activities. In contrast, the components, television transmitter, and computer manufacturing sub-sectors have expanded slightly. Thus, Johor seems to be seeing an expansion of the more sophisticated sub-sectors as well as more newer and smaller firms.

Table 4.1

Breakdown of the Electronics Sectors in Johor and Penang by MISC Category, Number of Firms, Employment, and Average Size, 2008

Johor **2008**

	Code	Size	Firms	Employment	Average Employees
Office, accounting, computing machinery	300	Small	3	78	31
		Large	16	13,941	899
Electronic valves, tubes, components	321	Micro	3	7	2
		Small	31	1,990	64
		Medium	12	1,026	87
		Large	28	28,332	1027
Television and radio transmitters	322	All	6	8,287	1381
Television and radio receivers, sound/ video systems	323	Small	22	770	35
		Large	7	10,834	1548
Total for Electronics			127	65,264	514

Penang

	Code	Size	Firms	Employment	Average Employees
Office, accounting, computing machinery	300	Small	3	93	29
		Large	12	23,078	1956
Electronic valves, tubes, components	321	Small	26	836	32
		Medium	11	1,427	127
		Large	37	66,887	1833
Television and radio transmitters	322	All	4	3,899	975
Television and radio receivers, sound/ video systems	323	Small			
		Large	10	7,988	799
Total Electronics			103	104,206	1012

Source: Raw data supplied by the Department of Statistics of Malaysia.

Table 4.2
Firm Numbers and Employment by Sub-sector in Johor and Penang (2000, 2008)

	Johor				Penang			
	2000		2008		2000		2008	
	Firms	Emp.	Firms	Emp.	Firms	Emp.	Firms	Emp.
Computing Machinery	16	12,400	19	14,019	22	29,818	15	23,171
Components	54	27,695	74	31,355	73	56,767	74	69,149
TV transmitters, apparatus	7	6,087	6	8,287	7	4,452	4	3,899
TV radio receivers, sound recording	44	38,877	29	11,604	28	22,957	10	7,988
Total	121	85,059	128	65,265	130	113,994	103	104,207

Source: Raw data supplied by the Department of Statistics of Malaysia.

The electronics sector in Penang has also experienced a contraction in employment, largely concentrated in the television and radio receiver sub-sector. However, the components sub-sector has seen a substantial expansion in employment numbers. Penang is thus witnessing a contraction of the less technologically sophisticated sub-sectors and the expansion of its components sub-sector.

Both states have many firms in supporting sectors such as metal-working services, manufacture of general and special purpose machinery, and plastics products. These sub-sectors are sizeably bigger in Johor than Penang, amounting to 906 firms and 65,800 people in the first state and 485 firms and 36,000 people in the second. However, Johor has a more diverse manufacturing sector, and many of these supporting firms cater to the steel, oil and gas, and petrochemical sectors.

What can be said about the relative sophistication of electronics firms presently in Johor? Relative to Penang, Johor has a larger presence of consumer electronics firms as opposed to the more sophisticated and challenging electronics components and computing machinery sub-sectors. In addition, available evidence suggests that firms in Johor in these sub-sectors concentrate on less technologically sophisticated items.

With regard to electronics components, Penang has a group of 10 semiconductor producers (Intel, AMD, Fairchild, and National Semiconductor inter alia) while Johor has one established semiconductor firm, STMicroelectronics. STMicroelectronics has its Asia-Pacific headquarters as well as wafer fabrication and design centre in Singapore, with semi-conductor assembly operations in Johor. Other components manufacturers in Johor tend to specialize in passive components such as resistors, inductors, capacitors, or magnets. As a result, sophisticated tasks in the semiconductor sub-sector such as R&D or design are not carried out by flagship firms or specialized supplier firms such as the case of Altera in Penang.

Like Penang, Johor has a large body of contract electronics manufacturers (CEMs) that provide integrated manufacturing, logistics, and often design services to flagship firms. Available evidence suggests that these firms concentrate the more sophisticated tasks in regional headquarters in Singapore — and to a lesser extent Penang — operating feeder plants in Johor.

Thus, Flextronics, one of the leading CEMs specializing in electronic displays, printed circuit board (PCB) assembly, and hard disk drive cards, has facilities in Penang and Johor, but is headquartered in Singapore. The same applies for Celestica, the Canadian CEM, which has two facilities in Johor — one for PCB assembly and testing and another for cartridge assembly, filling, and packaging — that are linked into its facility in Singapore (Celestica

2011). Venture Corporation and Wearnes Electronics are structured in a similar fashion, with headquarters in Singapore and production facilities in Johor (Venture 2011; Reuters News, 6 May 2011).

The disk drive manufacturers in Johor have a similar structure to the CEMs. Western Digital has a large manufacturing plant in Johor, with R&D centres in Penang and Singapore (*Business Times*, 11 August 2010). Seagate manufactures aluminium substrates for the assembly of hard disk drives in Johor, but has a technology centre for slider manufacturing in Penang and a hard drive design centre in Singapore (Seagate 2011).

Regarding other segments of the office and computer machinery subsector, Johor houses a significant group of manufacturers of computer peripherals. Firms such as Hewlett Packard, Brother, Epson, and Tektronix produce printers or inkjet cartridges. Other products include CD-ROMs, keyboards, and optical mice and are produced by firms such as PCA, Fujitsu, and Optosensors (JSICID 2011).

The other significant firm grouping in Johor consists of producers of consumer electronics such as CD and DVD players and recorders, camcorders, televisions, projectors, and remote controls by firms such as Panasonic, Mitsubishi, Sharp, Pioneer, Shinwa Technology, and Hitachi (JSICID 2011).

The presence of flagship firms has, in turn, attracted firms in supporting industries, particularly automation and engineering technology. Classic Advantage is headquartered in Singapore, and has had facilities in Johor since 1994 where it manufactures precision engineering and plastic injection moulding equipment. First Engineering makes precision moulds and plastic components, and is also headquartered in Singapore with facilities in Johor, India and China (LTHCorp 2011; First Engineering 2011). Sunningdale Tech, also headquartered in Singapore, is following a similar trajectory (SDaletech 2011).

A sizeable number of small-scale supply firms specializing in machining and equipment supply have also relocated from Singapore to Johor. Taking advantage of the well-developed logistics sector connecting the two territories, this "offshoring" model has been popular with smaller firms, who relocate their manufacturing operations to Johor, while retaining a sales office in Singapore.[3]

Regarding the emergence of local firms, the absence of large numbers of semiconductor firms means that, unlike Penang, there is no local group of automated production equipment producers, specialist designers, or producers of semiconductor test equipment.[4]

What can be said about technological capabilities and the degree of networking between firms in Johor and how they compare to elsewhere in

Malaysia? A survey of 113 electronics firms carried out in Penang and Johor in 2004 gives some insight into the differential level of capabilities.

Regarding foreign-owned and local firms in Johor and Penang, Figures 4.1 and 4.2 depict the relative level of skills intensity (skilled and professional

Figure 4.1
Capabilities in Foreign Firms in Johor and Penang

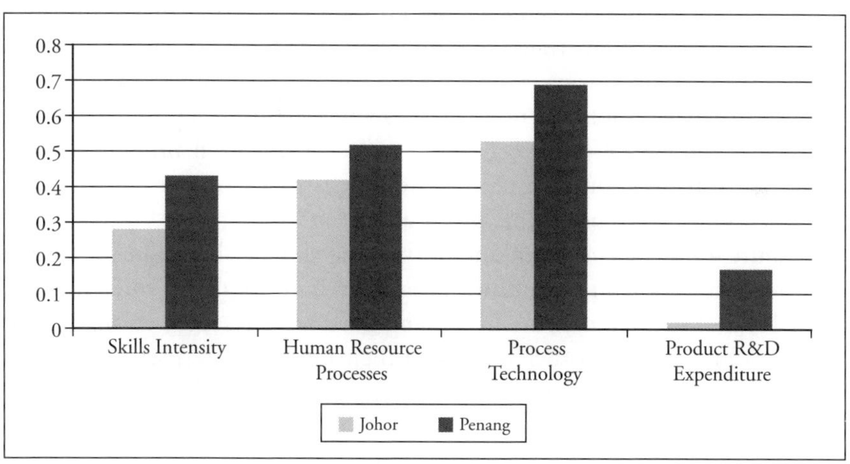

Figure 4.2
Capabilities in Local Firms in Johor and Penang

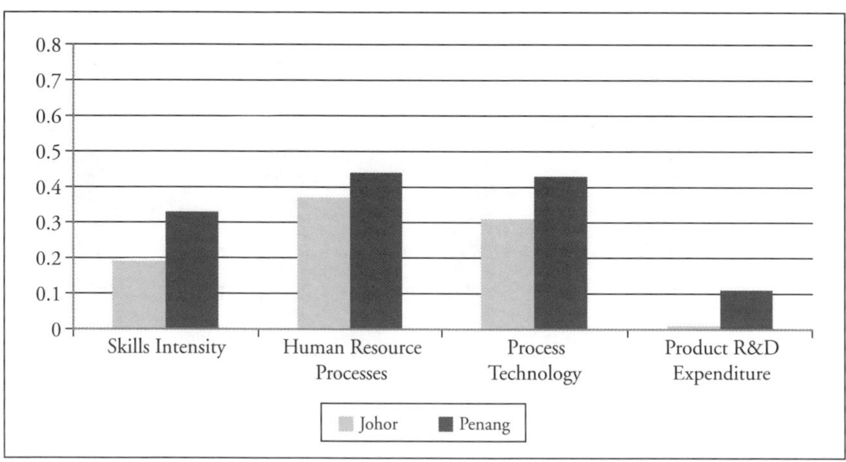

Source: Rasiah 2008.

personnel in the work-force); human resource processes (training expenditure, HR practices and facilities); process technology (age of equipment, quality and process control); and expenditure in product R&D (percentage of sales).

At the most aggregate level, foreign firms in both states have higher levels of skills intensity, HR processes, process technology, and investment in product R&D than local firms. However, foreign firms in Penang score consistently higher across all categories than do their Johor-based counterparts. Indeed, while R&D expenditure of firms in Penang is low, in Johor it is almost non-existent — supporting the assertion that firms there are almost exclusively dedicated to production. The pattern is largely repeated with regards to local firms. However, the gap in capabilities between local firms in Penang and Johor is more marked with regards to skills intensity and process technology. While R&D levels are lower for local firms in Penang than their foreign counterparts, they do invest some resources in this area, whereas Johor-based local firms invest almost nothing.

Figure 4.3 depicts the ranking of the local supplier base and the role of industry associations in Johor and Penang by both local and foreign firms. Both local and foreign firms in Penang gave higher scores to their local suppliers and industry associations than their Johor counterparts. That said, in neither case were the scores high.

Figure 4.3
Network Cohesion in Johor and Penang

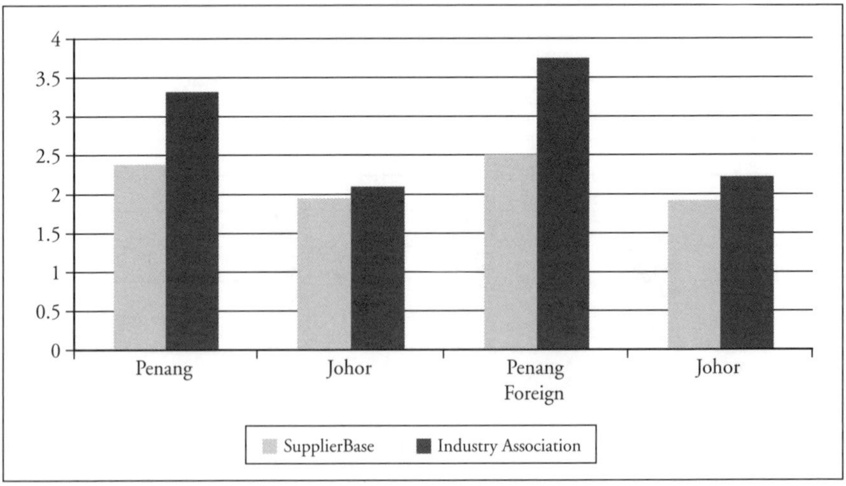

Source: Rasiah 2008.

More recent research supports the lack of inter-firm links in Johor and the limited development of new products. Figure 4.4 depicts the different ways in which firms in Johor learn new process and production systems. Most firms acquire new knowledge through their own efforts (50 per cent) or by purchasing it from technology suppliers (39 per cent). Adapting or reverse engineering competitors' products is another popular method (32 per cent). Only a minority of firms develop additional capabilities collectively (17 per cent) or through an intermediary (5 per cent). A related survey question found that more than 70 per cent of firms in Johor were not involved in product development.

This review of the Johor electronics sector indicates that, relative to Penang, it specializes in less technologically-demanding sub-sectors. Production in Johor is led by a large number of foreign-owned firms that support operations elsewhere. The groupings' external orientation means that many strategic decisions are made offshore, and opportunities for collaboration and interchange at the local level are limited. For this same reason, there is little development of a culture rooted in the local context or of a regional identity. This tendency may also be aggravated by the prevalence of CEMs and consumer electronics firms, whose business model entails fewer opportunities for downstream operations.

Figure 4.4
Source of New Processes and Production Systems in Johor

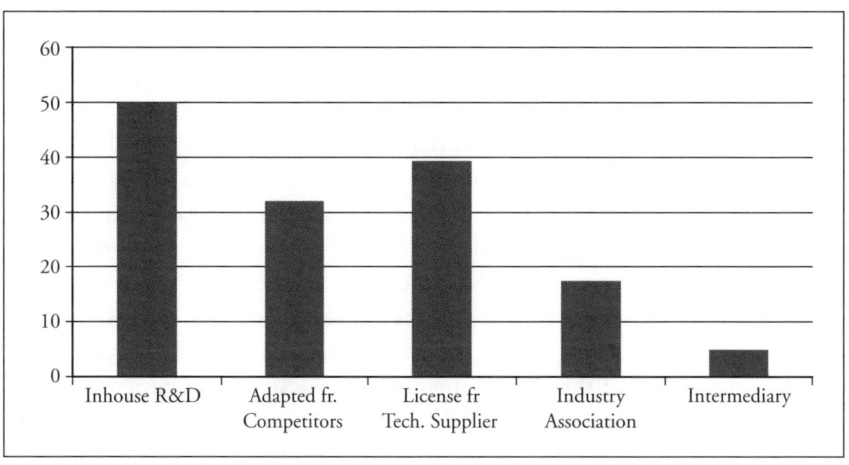

Source: RMA 2006.

The Regional Supportive Structure

Having set out Johor's regional production system, this section will look at its supportive infrastructure, namely those organizations in the state that support economic activity and innovation. The key focus will be on the state government, but the role played by federal government agencies, business associations, and the private sector will be brought in where appropriate.

State Government Organizations in Johor

Malaysia is a centralized federation, with responsibilities and revenue sources geared strongly towards the federal government. Beyond responsibilities for external affairs, defence, and finance, the Constitution attributes to the federal government aspects such as education, labour, and most types of infrastructure. State governments are responsible for land management; agriculture and forestry; and local government and services (maintenance, lighting, markets). The responsibilities for water, housing, as well as town and country planning are listed as shared responsibilities (Ninth Schedule, Constitution of Malaysia). Furthermore, state governments have come to assume informal responsibilities for tasks such as economic planning, promotion and investor liaison, and training.

Insofar as financing, the Constitution attributes the bulk of revenue sources to the federal government, which receives some 90 per cent of total government revenue. State governments are dependent on smaller, less flexible revenue sources such as land, forestry and mines, entertainment, in addition to a series of federal grants (Tenth Schedule, Constitution of Malaysia; Anuar 2000). However, state governments can create subsidiaries with the aim of generating profits to cross-subsidize other activities.

With regards to Johor, its state government has an illustrious tradition. The Johor Civil Service (JCS) traces its origins back to the 1850s, and counts the founder of the country's biggest political party as well as a Prime Minister among its alumni. This civil service is closed, requiring its members to be Muslims from Johor. At present, it has some 170 members, who helm all key positions in the state government.[5]

A number of organizations are responsible for the formulation of economic and industrial policy. The first is the State Legislative Assembly, which is comprised of the elected members of the state government, helmed by a Chief Minister chosen by the majority party. Since the first elections in 1959, the ruling coalition at the national level has also been in power at the state level. Thus, at a macro level, the priorities pursued by the Johor state government have not differed significantly from those of the federal government.

The State Secretary is the highest-ranking administrative officer in the state government, and oversees the work of 10 departments which carry out state government responsibilities. The agency responsible for policy formulation and development planning is the economic planning unit (JSEPU). Headed by a senior JCS official, this unit monitors international and national trends, formulates sectoral plans, and proposes privatization and revenue-generating initiatives. The Unit also produces plans to coordinate the activities of state government departments.

The state government also has a promotional arm, the Johor State Investment Centre (JSIC), which comes under the purview of JSEPU. Also headed by a senior JCS official, the Centre is charged with marketing Johor overseas; facilitating land, licenses, and permits for investors; and gathering information on local-level issues affecting the business community.

The Johor state government also has a small stable of statutory bodies. The most significant of these is JCorp, formerly called the Johor State Economic Development Corporation. Established in 1968, JCorp began operations concurrently with the launching of the New Economic Policy, which was geared to eradicating poverty and restructuring society in the wake of the ethnic unrest that shook Malaysia in 1969. JCorp's organizational objectives mirror these goals, with its corporate mission being to "spearhead" the movement and positioning of the *bumiputra* — or Malay community — in the national economy. JCorp's employees must be Malay, although this requirement is not extended to employees in the majority of its subsidiaries. Its corporate strategy has been to promote industrialization by developing large industrial estates; making strategic investments; and accumulating assets (Johari 1997).

During the first 25 years of its existence, JCorp grew rapidly due to profitable oil palm plantations, sustained land purchases, and a highly leveraged growth strategy. By 1996, the Corporation was at the heart of a conglomerate with 19 divisions, assets worth RM7.4 billion, a turnover of RM2.83 billion, and four firms listed on the Kuala Lumpur stock exchange. Its investments ranged from oil palm plantations to healthcare and from heavy industry to paper production and publishing (Johor Corporation 1997; 1998).

However, JCorp was hit by the 1997/98 financial crisis, losing RM680 million in 1997 and a further RM630 million the following year. In 1998, its total debt reached RM10 billion, and it had to liquidate 35 subsidiaries and request federal government help to restructure its debt. Following this, it closed or divested a number of concerns and re-focussed its energies on core business concerns such as agriculture, property development, and healthcare (Johor Corporation 1999; *The Edge*, 7 January 2011).[6]

At present, JCorp is still a major player in the state's economy with a stable of more than 280 firms, including 8 companies listed on the national stock exchange, and 65,000 employees in Malaysia and overseas. However, while JCorp used to be seen as the implementation arm of the state government, its role is increasingly being curtailed. Since 2004, investor liaison functions have been delegated to JSIC and, at present, there are discussions about divesting some of the conglomerate's assets to pay off an estimated RM6.6 billion in debt (JCorp 2011).

The other key organization in terms of Johor's economic development is the Iskandar Regional Development Authority (IRDA). Established in 2007, IRDA is a federal statutory body charged with promoting the development of the Iskandar Malaysia region, a 2,200 sq km swathe of land fronting Singapore. Iskandar Malaysia is one of five growth corridors that the federal government has created to promote development outside the country's capital. Unique among these corridors, the rest of which are overseen solely by the federal government, Iskandar Malaysia is jointly chaired by the Prime Minister and the Chief Minister of Johor, and IRDA is comprised of both federal and state officials (IRDA 2007).

Key Policy Frameworks

Given Johor's agricultural prowess, the first decades of independence were given over to land development. Both the federal and state governments opened up vast tracts of forest to small-holders. In addition, industrialization was conceived of as a state-led enterprise. Thus, both levels of government invested resources in a range of sectors, including tin processing, steel, ship-building, and palm oil refining. Furthermore, the federal government established a port in Johor in the late 1970s (Guinness 1992).

In the 1980s, commodity prices fell and the state government began to look at industrialization as a method of generating employment and modernizing the economy. Furthermore, the attractiveness of the state was boosted due to the appreciation of the Singaporean dollar and the currencies of the other NICs in the wake of the Plaza Accords in 1985 (Rasiah 2009). Following this, the state benefited from large amounts of manufacturing investment from Singapore, Taiwan, and Japan (Grunsven et al. 1995).

In 1989, the state released the Economic Plan for Johor (1990–2005). The Plan recommended commercializing agriculture and accelerating industrial development. Regarding industry, it recommended focussing on resource-based industries and the E&E sector. In addition, it advocated "economic twinning" with Singapore; encouraging industrial relocation from Japan and other newly-industrialized countries to Johor; and promoting the state as a growth-pole for the southern part of the country. Furthermore, it advocated attracting more knowledge- and capital-intensive industries in view of predicted labour shortages (JSEPU 1989).

During the early 1990s, Johor's economy grew at 10 per cent p.a., higher even than the national average of 8.7 per cent. This rapid growth as well as changing thinking at the national and state levels led to a new framework in 1996, the Johor Operational Master Plan, which was meant to run until 2010. The Plan also sought to promote structural transformation, although the services sector was added as a target area. This was to be achieved through regional cooperation with Singapore and Indonesia, as well as through promoting Johor Bahru as a "modern technopolis". The Plan recommended fostering "modern high-value" services, such as food processing, fabricated metal, electrical and electronics, and the chemicals sub-sector — which were to be facilitated by strategic investments in infrastructure and skills provision (RMA 1996).

In 1997, the Johor state government released the Industrial Master Plan, which stressed an integrated approach to firm groupings, in particular to promote linkages and higher value-added activities. As with its predecessors, the Plan put forward a range of target activities, in this case seven manufacturing and eight service sub-sectors. Regarding E&E, the Plan placed particular emphasis on the development of leading edge capabilities in specific areas. More than previously, the Industrial Master Plan placed emphasis on strengthening the local institutional context within which firms work (MIER 1997).

This policy framework remained in place until 2006, when Khazanah, the federal government's investment arm, and the Johor State Government released the Comprehensive Development Plan for Iskandar Malaysia. Aimed to run until 2025, the plan aims to turn the southern portion of the state into a "strong successful conurbation of international standing". The thrust is to convert the whole region into an integrated whole and bring it more closely into Singapore's orbit, in order to attract investment and capitalize on spillovers.[7]

The Plan recognizes the state's traditional manufacturing activities such as food and agro-processing, petrochemical, oleochemical, and E&E. However, it puts forward six new target areas which receive additional incentives. These are all service activities running from logistics to finance, and education to creative industries (Khazanah Nasional 2006). With this new policy, the emphasis moves squarely away from manufacturing to services as the preferred driver of the economy. Although not stated publicly, the E&E sector is now perceived of as "a mature industry and too volatile".[8]

Thus, over the past 25 years, the Johor state government's policy frameworks have evolved considerably, from a predominantly agricultural focus to one with industry and then services at their centre. Policy-makers have been aware of the short-comings of the state's industrialization model as all plans observed the following characteristics of the E&E sector: high volume, low mix, and relatively low-value added activities; weak linkages between firms; an under-development of the local SME base; a lack of investment in R&D by firms; and a chronic shortage of skilled workers (JSEPU 1989; MIER 1997; RMA 1996).

Having set out the evolution of the Johor state government's policy frameworks, the next sections will look at how these organizations and policies influenced the environment within which the E&E sector developed.

Enabling Environment for Business: The key element of Johor's competitive advantage has always been land, particularly in relation to costs and availability in Singapore. Upon its inception, JCorp was given primary responsibility for developing the state's land bank and generating revenue. Its business model with regard to potential investors was based on offering land in attractive locations accompanied by stream-lined business processes, rather than using low cost as an incentive. The profits from this were used to cross-subsidize a range of strategic investments in new sectors (Ali 1993).

This approach is seen most clearly in the development of its biggest industrial park, Pasir Gudang. In the early 1970s, seeing the long-term potential of Johor Port, JCorp invested substantial resources in securing land around it — which also fronted Singapore. They then pushed for the area to be turned into a free trade zone under their authority, which became the only one of its kind within a port complex in the country (Guinness 1992).

Pasir Gudang's attractiveness was bolstered by making JCorp the local government authority. In effect, the CEO of JCorp became the mayor, with responsibility for all local government prerogatives — enabling approvals for building plans and fitness, fire safety, and trade licences to be issued within 24 hours. Through their responsibility for supplying services such as lighting and maintenance as well as enforcing by-laws, JCorp was also able to offer investors a more protected environment. The Corporation also constructed low-cost housing for workers in the Pasir Gudang area, which today has a population of more than 100,000 residents.

Based on this value proposition, Pasir Gudang filled up over the course of the 1980s — largely with E&E firms, as well as heavy industry and food processing operations. A new park, Tanjung Langsat, was built in 1993, more clearly geared to the palm oil and oil and gas sectors, and JCorp constructed a port with special facilities for handling bulk cargo and chemicals within the park. JCorp also built an access road between Tanjung Langsat and Pasir Gudang to enable firms in each park to access the other's facilities.

JCorp currently has 30 industrial parks in Johor's major population centres, which house more than 1,000 firms. Today, the Corporation is seen to cater more to heavy industry, particularly oil, gas, and petrochemicals. Following previous state government planning, the Corporation acquired sizeable land holdings in the eastern part of Johor. However, the Iskandar region has planned infrastructural investments in the west and central part of the state, leaving JCorp ill-placed to capitalize on these facilities.[9] In addition, since 2009, JCorp no longer has local authority responsibility for Pasir Gudang.[10]

While Johor has been billed as a land-rich state, particularly vis-à-vis Singapore, there are indications that the amount of land available for industry is reaching its limit. This is particularly the case when the conversion of land must be weighed up against its potential for cash crop production. The region's proximity to Singapore also means that land prices have gone up, and in 2009, Johor had the third most expensive land for industrial use in the country (MIDA 2009; RMA 1996).[11]

State governments have traditionally been responsible for water provision as well as local services. However, since the 1990s, the federal government has promoted the privatization of many government services, which have subsequently been nationalized or placed under federal regulatory control. This policy has had different results at the state level, reflecting distinct local opinions regarding the desired role of the federal government.

With regard to solid waste management, the private firm SWM Environment Services has handled solid waste management for the southern part of the country, including Johor, from 1997. As of this year, this operator has been awarded a 22 year concession — under the regulatory control of the federal government. However, the states of Penang, Perak, and Selangor have refused to join this scheme, preferring to retain direct control over this service (*The Edge*, 19 September 2011).

With regard to water, the responsibility for water supply and services was under exclusive control of state governments until 2005 — after which it became a concurrent responsibility. Johor was a pioneer in privatizing water treatment in the early 1990s. In 2000, the state government privatized the entirety of its water supply and distribution to a third party, SAJ Holdings Bhd, under a 30-year concession. Water rates were raised by 40 per cent in 2001 and 30 per cent in 2003 (Santiago 2005). In 2009, SAJ Holdings had a debt of RM3.2 billion and water rates in Johor were the highest in the country — more than three times the cost of water in the cheapest state (MIDA 2009). The Ministry of Finance, through a subsidiary, paid SAJ Holding RM4 billion for its infrastructure and assumed its debt in 2009, in return for the right to assume responsibility for all water sector infrastructure. SAJ Holdings was awarded a 30-year lease to provide water supply services from that date (*New Straits Times*, 12 March 2009). Thus, Johor has opted to relinquish control over its water system and has seen its competitive advantage erode in this aspect.

The Iskandar Malaysia initiative has brought an element of order to the state. Under its plan, southern Johor is divided up into five zones — each with priority sectors, specified incentives and infrastructure, and preferred real estate developers. Thus, land use, infrastructure needs, population centres, and amenities are mapped out. And, prior to rolling out Iskandar Malaysia, the federal government made concessions for a second bridge to Singapore in 1998 and a third port, Tanjung Pelepas, in 1999.

With regard to the E&E sector, three zones have been designated. Two of these encompass the centre of Johor Bahru and the Pasir Gudang/Tanjung Langsat area. The third is in the northern part of Iskandar and centres on the Technology University (Iskandar Malaysia 2008*b*). That said, most manufacturing activities will not qualify for many new incentives. Only companies investing in the target service sector activities and one special zone, Medini, will qualify for benefits such as as exemptions from equity requirements, foreign exchange rules, and unrestricted hiring of overseas skilled workers. For manufacturing activities, the normal incentives such as tax exemptions and investment allowances remain (Khor 2011).

Thus, at first blush, Iskandar Malaysia does not promise a great deal that is new for the E&E sector. Notable large investments in the region have focussed on education, health, property development, oil and gas, and steel manufacture — rather than E&E, which has continued to receive roughly the same level of investment as before.[12] However, it is likely that E&E firms will benefit indirectly from planned infrastructural investments. In addition, the region has received investments from six international universities, including one with a large engineering programme (*New Straits Times*, 24 January 2011).

That said, plans to make the region function as an integrated whole can also disrupt business for the more established residents. Thus, original plans under Iskandar called for all containerized shipping to be routed through Port Tanjung Pelepas in the western part of the state, and for Pasir Gudang's port to specialize in bulk shipping. Objections from firms in Pasir Gudang have been vocal, and this plan has been delayed.[13]

Furthermore, it is not clear that the addition of another layer of bureaucracy will make much difference for investors. While IRDA has a service centre that pledges to process applications quickly, its mandate only encompasses planning, promoting, and facilitation tasks. Ultimate authority still rests with state and local authorities, which IRDA must liaise with to obtain approvals.

In addition, the development of the Iskandar region has not been without local controversy. The state branch of the country's largest political party and some opposition parties have expressed concern about land rights for the Malay community; the rolling back of equity requirements in strategic sectors; and the idea of a passport free zone. National and state level leaders have had to reassure constituents and investors that these issues will be handled effectively (Hew 2008).

Broker Policies: Beyond the provision of land and utilities, the state government has invested resources in planning exercises to understand economic trends as well as coordinate efforts. However, while the Economic Planning Unit is responsible for this, all important plans have been sub-contracted to professional consulting firms outside the state.

Curiously, the largest federal university in the state, the Technology University of Malaysia, has not been involved in any significant capacity. In addition, while technically very competent and — at times — visionary, these plans were not a result of a consultative process. Their focus is very state-led, with emphasis on public investments in physical infrastructure, human capital, and marketing. As a result, the Plans have not been disseminated outside the state government. In particular, the key business associations in the state were not involved in either the planning process or their follow-up, mirroring the pattern taken by the federal government at the national level.

JCorp, when it was charged with investor liaison and, subsequently, the Johor State Investment Centre have undertaken initiatives to market the state and provide initial contacts between firms through producing directories and disseminating information to investors. With Iskandar Malaysia, the level of industry-related knowledge is likely to increase further.

However, the state government has not undertaken competence mapping or technology roadmap exercises to pinpoint gaps in capabilities or missing links in value chains. Consequently state government officials are not able to target priority activities for investment or bolster local capabilities to complete value chains.

Furthermore, JCorp's remit for the sale of industrial land is uniquely profit-based, and no allowance is made for strategic placing of investment activities with the aim of creating eventual synergies. For example, Pasir Gudang, is billed as a location for manufacturing of medium intensity and has oil & gas, steel making, and E&E firms alongside others that make food products. In addition, while state government plans have explicitly recommended cross-subsidizing plots for small firms, JCorp's parks do not have land sited for this purpose (RMA 1996).[14]

However, the state government does have an SME support unit, which acts as a broker between local entrepreneurs, the federal government, and a member of the State Legislative Assembly. Established in 2001, the Johor Corporation for Entrepreneurs facilitates applications for grants and loans, and provides technical and marketing support. However, it does not target the E&E sector or its supporting industries. At present, it works with some 30 firms in rural areas, particularly those involved food processing.[15]

JCorp has tried to create networks of entrepreneurs. Thus, it established the Johor Business and Industry Club in 1991 for *bumiputera* to meet and exchange ideas, accompanied by a business centre with information on how to start up firms. In 2001, this was complemented by the Malaysian Islamic Chamber of Commerce. While these initiatives are promising, neither the Johor state government nor JCorp has established mechanisms for reaching out to the approximately 80 per cent of the local manufacturing sector that is not Malay (RMA 2006).

Of course, the government is not the only mechanism through which communication between firms can take place. Johor does have chapters of the largest Malaysian business associations such as the Malaysian International Chamber of Commerce and Industry as well as the ethnic Chambers of Commerce. However, with the exception of the Chinese Chamber of Commerce and Industry, their membership does not tend to come from the manufacturing sector. There is a large branch of the Federation of Malaysian Manufacturers, which has been in Johor since 1968 and currently has 370 members. It regularly carries out surveys, produces business guides, and is also a training provider recognized by the federal government.[16] However, while these associations do liaise with the federal and state governments at a technical level, they are not involved in planning.

The most promising instance of dialogue and cooperation for firms in the manufacturing sector is carried out by the SME Association of

South Johor. Established in 2001, the association has 500 members in a range of sectors, including E&E. Its goals are to give SMEs a bigger voice, make linkages between the government and the private sector, and bolster capabilities through information sharing. The association holds training sessions in English and Chinese on government grants and loans, marketing techniques, and accounting techniques. In addition, its President is an adviser for the Johor State Government's training institute, the Johor Skills Development Centre.[17]

For much of the manufacturing sector, there is a feeling that means of communicating and articulating their issues need to be improved. A survey found that 89 per cent of respondents supported the establishment of an E&E Cluster Association to articulate issues facing the sector as a whole (RMA 2006).

Addressing Market Failures: Over the past three decades, the Johor state government planned and acted on a number of fronts to address perceived market failures. These include: making strategic investments in new sectors; promoting entrepreneurship; addressing skills shortages; and providing collective facilities and knowledge-generating failures.

Strategic Investments

Using JCorp as an institutional vehicle, the Johor state government made a range of investments in new sectors. From its initial focus on securing land in the 1970s, JCorp expanded significantly during the 1980s and early 1990s, coming to comprise some 15 divisions by 1997. This spanned primary, secondary and tertiary sectors, with a range of investments in Malaysia and also across the globe. While priorities changed over time and the Corporation underwent a large-scale restructuring post-1997, it had a number of sectors where it made a concerted attempt to attain market leadership.

> *Palm Oil* — JCorp placed a clear priority on this sector through acquiring plantations both for revenue generation and conversion into industrial parks. The Corporation was able to obtain majority control of one of Malaysia's largest palm oil interests, and purchased plantations in Indonesia, Solomon Islands, and Papua New Guinea. With majority stakes in six plantations and minority shares in another five, JCorp became the world's largest producer of palm oil seeds. The Corporation also invested in down-stream industries such as refining and storage.

Food and Quick Service Restaurants — JCorp owns a large network of franchises of fast food outlet, including more than 550 KFC restaurants in Malaysia, Singapore, Brunei, and Cambodia, with planned outlets in India, as well as more than 250 Pizza Hut outlets in Malaysia and Singapore. There are other down-stream activities such as corporate catering, poultry farming, a feed-mill business, and a range of food products (JCorp 2009; 2010).

Healthcare — JCorp has the largest network of private hospitals in Malaysia, with 22 in the country, a further two in Indonesia, and a head-count of almost 7,000 staff. Spin-offs from this include a College of Nursing and Health Science, with campuses in Kuala Lumpur and Johor Bahru (JCorp 2009).

Property — JCorp owns a number of well-placed commercial premises including four hotels, a resort, a number of shopping complexes, and the state's largest convention centre; as well as prime industrial land in all major urban centres in the state (JCorp 2010).

However, it is noteworthy that none of these initiatives targeted either the E&E sector or any supporting sector. This is despite repeated plans to foster excellence in specific areas of the electronics sector, as well as the automation and precision engineering industries (RMA 1996). JCorp did have a number of manufacturing subsidiaries such as steel, aluminium, and tube-making that were privatized in the wake of the financial crisis (Johor Corporation 1997).

Enabling Entrepreneurship

In addition to seeking to diversify Johor's economy and generate revenue, a central component of JCorp's organizational remit is to foster a *bumiputera* commercial and industrial community.

A key strategy is the "intrapreneuring" concept, where promising entrepreneurs have a minority stake in a firm and are responsible for day-to-day management. In return, JCorp has a majority stake and supplies the entrepreneur with technical support and capital. JCorp managers are eligible to be intrapreneurs, or those outside may apply for this status in return for selling a majority stake to the Corporation. This strategy is perceived to correct a key market failure as it selects the most able entrepreneurs and does not make the possession of capital a "pre-condition" for entrepreneurship (Ali 1993).

The largest and most successful intrapreneur operations then come under the Sindora holding company, which is listed on Kuala Lumpur's main board. At present, Sindora has 11 enterprises, which are active in shipping for the oil and gas sectors; medical device manufacture; timber; bio-fertilisers; edible oil; insurance; a business processing outsourcing firm; and a chain of parking lots. In addition, many of the Managing Directors of the Corporation's listed companies were selected through this method (JCorp 2009).

Skills Provision

The federal government has established two technology universities in the state and there are a number of technical and vocational colleges. Despite this, feedback from industry has persistently signalled the need for skilled workers in greater quantities and with more industry-relevant content.

To this end, the Johor state government funded the creation of 12 community colleges throughout the state, as well as an Industrial Technology Institute which provides diplomas in a range of engineering disciplines in conjunction with the Technology University; as well as a number of qualifications of a vocational nature (ITPYPJ 2011). The links with the local university are promising, however, there is little evidence of industry input into the curriculum or choice of courses.

JCorp has also been active in this area through its subsidiaries. Key institutions in its stable of firms include the Entrepreneurial Development Unit and the Institute of Management Development, both of which are geared to providing a range of technical and managerial courses to Bumiputeras.

Of more direct relevance to the E&E sector, JCorp established its own industry training centre, the Johor Skills Development Centre, in 1993. Inspired by the Penang Skills Development Centre, the Johor equivalent was set up with a grant and premises from JCorp in the Pasir Gudang industrial park. Over the past 18 years, the JSDC has trained some 35,000 workers in a range of industry-relevant courses. At present, the Centre has a turnover of RM7 million and obtains a considerable amount of funding and equipment from the federal government. Its programmes target current workers who need to upgrade skills; school leavers; and unemployed graduates. While the JSDC does cater to the electronics sector and has worked with firms such as Flextronics and Jabil Circuit, it is more oriented to the heavy industry sector — particularly oil and gas.[18]

Thus, both the Johor state government and JCorp have been proactive in trying to ameliorate the crucial skills shortages issue by establishing training centres and colleges for the E&E sector. However, as with the broker policies, these initiatives are state-led, as opposed to benefiting from input by the private sector.

Collective Facilities and Knowledge-generating Services

Policy-makers in Johor have been aware of the need to improve the local-level business environment through providing collective facilities for firms and improving their access to industry-relevant knowledge. As it stands, public facilities such as libraries, technology centres, or laboratories are under-provided, and survey responses indicate little interaction with the existing infrastructure for generating knowledge (RMA 2006).

Over the years, Johor's economic plans have proposed a number of initiatives to address these issues. In particular, the Johor Operational Master Plan proposed an ambitious plan to improve the environment for innovation. This included an information centre for the manufacturing sector; an industrial park for SMEs with an incubator, publicly-available equipment, training facilities and an information centre; and a high-technology park located next to the Technology University for the most technologically-advanced operations (RMA 1994*a*; RMA 1994*b*).

The first two initiatives were not implemented. However, JCorp did pursue the idea of a high-tech park aggressively, with the establishment of Johor Technology Park in 1996. This was meant to attract technology-intensive operations by providing R&D facilities, consultancy services, and a pool of skilled labour. Following in the footsteps of the federally-funded Technology Park Malaysia and Kulim High-tech Park, the Johor equivalent was the only such facility funded by a state government.

To this end, the Technology Park was located next to the Technology University as well as the airport. Original plans foresaw that the Park would have high-end telecommunications infrastructure routed through Singapore, as well as two business incubators with an array of supporting services (management, venture capital, patent protection); an R&D centre housing leading public sector bodies such as the Malaysian Institute of Microelectronic Systems (MIMOS) and the Standards and Industrial Research Institute of Malaysia (SIRIM) as well as high-end MNC operations; and a business centre. The targeted sectors would be electronics, information technology, and biotechnology (RMA 1994*a*).

However, the potential of the Park has been hamstrung by a number of issues. The first was the 1997 financial crisis, which scuttled plans to provide collective facilities for tenants. The second is that planned investments by the federal government have not materialized to the extent anticipated. The planned anchor tenants, particularly MIMOS, have not set up facilities in Johor, and while SIRIM has a facility, it is small and does not generate much demand for supplier services. The third is that efforts on the part of the Technology University to engage with the private sector are still in their infancy.[19]

As a result, the planned exclusive use of the Park for the targeted sectors was abandoned. At present, the Technology Park is 90 per cent full. It does have some anchor tenants in the E&E sector, such as Classic Advantage and Seagate. However, the remainder are comprised of firms in the oil & gas and steel fabrication sectors as well as some retailers. Efforts to strengthen ties between tenants in the Park have stopped.[20]

Following the establishment of the Iskandar region, some private sector developers are marketing industrial parks with an integrated approach, targeting specific sectors for investment and complementing this with sector-specific facilities for tenants. Thus, the Senai High Tech Park targets technology-intensive E&E manufacturing in sectors such as semiconductors, photonics, optoelectronics, and nano-technology, as well as green technology; research and development operations; as well as higher-end services such as operational headquarters and regional procurement centres. Upon completion, the Park promises a mix of vacant lots, ready-built factories, incubators for small businesses, as well as a laboratory and facilities for rental (SHTP 2011).

CONCLUSION

Using a Regional Innovation Systems approach, this chapter has sought to analyse Johor's electronics sector, and the extent to which meso — or state — level organizations and policies have sought to foster its development.

Johor houses a substantial number of electronics firms which account for some 20 per cent of employment in its manufacturing sector. However, over the past decade, the sector has contracted substantially. Analysing by sub-sector, it would appear that many of the operations in Johor are more labour-intensive than those in Penang. In addition, firms in Johor are less networked and have lower levels of technological capability. The relative abundance of large, self-reliant firms with an external orientation has meant that little has emerged in the way of a local culture or locally-rooted networks.

Insofar as the regional supportive infrastructure, the efforts of the Johor state government and its subsidiary, JCorp, loom large. Regarding providing an enabling environment for business, the most visible area of state government activity has been in the provision of land and industrial parks. In certain aspects, particularly securing strategically-located land, providing ports, and linking the two main parks together, JCorp has been very proactive. In the case of Pasir Gudang, assuming management functions of the local authority also put the Corporation in a key position to enable effective service.

However, in other areas of providing an enabling environment, Johor's record is rather more mixed. Its acceptance of federal policies promoting privatization entailed relinquishing its control over key utilities. This is particularly marked with regard to water provision which saw Johor's competitive advantage erode significantly.

Regarding brokering activities, in particular value-enhancing dialogue, the state government has invested significant resources in developing plans. However, planning has been exclusively state-led, and has not benefited from firm-level technical information regarding issues. In particular, established business associations have been by-passed. No technology or capability mapping exercises have been conducted, meaning that state government officials do not have the wherewithal to target priority areas for investment. In addition, land in state-owned industrial parks is sold uniquely with commercial criteria in mind, as opposed to seeking to create synergies and foster relationships.

The state government has been proactive with regard to fostering the emergence of a *bumiputera* entrepreneurial community, and established a series of initiatives to this effect. However, an exclusive focus on this community meant that the majority of the manufacturing sector was by-passed. As a result, the necessary base of social capital between the government and local firms has been absent, preventing any collective attempt to foster upgrading.

The state government has been a very important player in the local economy, seeking to address a range of market failures. Using the proceeds from its oil palm plantations, sales of industrial land, and profitable subsidiaries, JCorp was able to expand and make strategic investments in a range of sectors from restaurants to healthcare — attaining market-leading positions in many cases. This was been coupled with a concerted effort to groom the ablest JCorp managers to become entrepreneurs. However, none of these efforts connected with the E&E sector or any of its supporting industries.

Regarding the need to address market failures, both the state government and JCorp invested considerable resources in establishing specialized facilities for industrial training. As with other initiatives, they are exclusively state-led efforts, which may hamper their effectiveness. The Johor Technology Park represented the state government's most ambitious attempt to encourage the growth of more value-added investment in the state. However, the financial crisis curtailed its full implementation.

Thus, the Johor state government has been an active and proactive agent in the development of its economy. Despite ceding control over important services, it has seriously attempted to provide the necessary infrastructure and service delivery for the manufacturing sector. While it has, in some measure, attempted to provide more specialized infrastructure and human resources, these efforts have been surpassed by those oriented to pursuing other priorities, notably seeking to directly occupy strategic economic sectors and cultivate the emergence of a select group of entrepreneurs.

Notes

1 Under the 8th and 9th Malaysia Plans, the federal territories of Kuala Lumpur, Putrajaya, and Labuan received — in per capita terms — four times the average development expenditure allotted to the 13 states.

2 Data provided by the Department of Statistics, Malaysia. This refers to those firms classified under the Malaysian Standard Industrial Classification (2000) codes of 300, 321, 322, and 323.

3 Interview with President of a small electronics SME association in Singapore, Singapore, 5 May 2010.

4 However, there are examples of local firms in Johor that have begun to climb the value chain, notably that have moved from the plastics sub-sector into providing CEM services particularly for the consumer electronics sub-sectors. VS Industry is a Malaysian firm that supplies to some of the larger Johor-based flagships such as Sharp, Brother, and Panasonic, as it makes vacuum cleaners, remote controls, and fax machines, and also does PCB assembly, sub-assembly, and R&D for remote controls. Toyoplas is another local firm that is following the same path, moving from providing plastic injection moulding to firms such as Toshiba and Epson into CEM. ATA Industrial also began supplying plastic injection moulding services to Hewlett-Packard, Dyson, and Kenwood, before moving into the precision engineering sector. It maintains its headquarters in Johor Bahru, and only has a sales office in Singapore.

5 Interview with former State Secretary of the Johor State Government, Johor Bahru, 29 June 2010.

6 Interview with former CEO of JCorp, Kuala Lumpur, 5 September 2011.

7 Actually, this policy framework is limited to southern Johor only. The Johor State Government has commissioned an Economic Development Masterplan for the whole of the state for the period 2010–20. However, this document has not yet been approved by the State Executive Council. Interview with

private economic consultant for the Johor State Government, Kuala Lumpur, 7 September 2011.

8 Interviews with: JSEPU senior official, Nusajaya, 28 May 2010; JSIC senior official, Nusajaya, 18 May 2010.

9 Interview with a senior official of Johor Technopark, Johor Bahru, 23 June 2010.

10 Interview with a business association representative, Johor Bahru, 21 May 2010.

11 Interview with JSEPU official, Nusajaya, 28 May 2010.

12 Datasheet "Total Proposed Investments 1990–March, 2010", supplied by the Malaysian Industrial Development Authority.

13 Interview with a business association representative, Johor Bahru, 21 May 2010.

14 Interview with senior official of TPM Technopark, Johor, 9 September 2011.

15 Interview with senior Johor Enterprise Corporation manager, Larkin, 27 July 2010.

16 Interview with FMM SME Working Group, Johor, 22 April 2010.

17 Interview with SME Association of South Johor President, Pasir Gudang, 21 April 2010.

18 Interview with Johor Skills Development Centre official, Pasir Gudang, 28 April 2010.

19 Interview with a senior TPM Technopark official, Johor Bahru, 9 September 2011.

20 Interview with a Johor Technology Park tenant, Senai, 28 June 2010.

References

Secondary Sources:

Ali Hashim. "The Johor State Economic Development Corporation as a Successful Corporate Organisation and an Islamic Business Institution". Paper presented to the Islamic Development Bank, Nomination for the Islamic Development Bank Award in Islamic Economics, March 1993.

Andersson, M. and Charlie Karlsson. "Regional Innovation Systems in Small and Medium-sized Regions: A Critical Review and Assessment". CESIS Working Paper Series, No. 10. Stockholm: Centre of Excellence for Science and Innovation Studies, 2004.

Anuar, Abdul Rahim. "Fiscal Decentralization in Malaysia". *Hitotsubashi Journal of Economics* 41 (2000): 85–95.

Best, M.H. "Cluster Dynamics in Malaysian Electronics". In *Malaysian Industrial Policy*, edited by K.S. Jomo. Singapore: NUS Press, 2007.

Best, M.H. and Rajah Rasiah. "Malaysian Electronics: At the Crossroads". Small and Medium Enterprises Branch, Technical Working Paper No. 12. Vienna: UNIDO, 2003.

Celestica. <http://www.celestica.com/Worldwide/Worldwide.aspx?id=984> (accessed 14 September 2011).

Cooke, P. "Regional Innovation Systems, Clusters, and the Knowledge Economy". *Industrial and Corporate Change* 10, no. 4 (2001): 945–73.

First Engineering. <http://www.first-engr.com.sg/about.php> (accessed 15 September 2011).

Grunsven, L. van, M. van de Heijden, and P. Sluys. "Foreign Investment and Industrial Structure in Johor, Malaysia: Final Report of Research Findings". Utrecht: Faculty of Geographical Sciences, Utrecht University, 1995.

Guinness, P. *On the Margin of Capitalism: People and Development in Mukim Plentong, Johor, Malaysia.* Singapore: Oxford University Press, 1992.

Hew, D. "The Malaysian Economy: Developments and Challenges". In *Southeast Asian Affairs 2008*, edited by Daljit Singh and Tin Maung Maung Than. Singapore: Institute of Southeast Asian Studies, 2008.

Hutchinson, F.E. "'Developmental' States and Economic Growth at the Sub-national Level: The Case of Penang". In *Southeast Asian Affairs 2008*, edited by Daljit Singh and Tin Maung Maung Than. Singapore: Institute of Southeast Asian Studies, 2008.

ITPYPJ. <http://itpypj.edu.my/v3/bm/kursus> (accessed 29 September 2011).

Johari Mohamed. "The Role of the Johor State Economic Development Corporation (JSEDC) in the Industrial Development of Johor". In *Improving Urban Environment in South East Asia: Managing Industrialisation through Satellite Communities*, edited by Azman Awang, Mahbob Salim, and John F. Halldane. Johor: Institute Sultan Iskandar, 1997.

JSEPU. *Economic Report 2008/2009.* Johor Bahru: State Government of Johor Darul Ta'zim, 2009.

Johor State Investment Centre Industry Directory (JSICID), available at <http://www.jsic.com.my/database/industry/read_list.php> (accessed 15 September 2011).

Khazanah Nasional. *Comprehensive Development Plan for South Johor Economic Region 2006–25.* Kuala Lumpur: Khazanah Nasional, 2006.

Khor Y.L. "Iskandar Malaysia: Policy, Progress, and Bottlenecks". Malaysia Update: September 20011. Singapore: RSIS, Nanyang Technological University, 2011.

LTHCorp. <http://www.lcthcorp.com/> (accessed 15 September 2011).

Oughton, C, M. Landabaso, and K. Morgan. "The Regional Innovation Paradox: Innovation Policy and Industrial Policy". *Journal of Technology Transfer* 27 (2002): 97–110.

Rasiah, R. "Government-business coordination and the development of Eng Hardware". In *Industrial Technology Development in Malaysia: Industry and Firm Studies*, edited by K.S. Jomo, G. Felker, and R. Rasiah. London: Routledge, 1999.

———. "Industrial Clustering of Electronics firms in Indonesia and Malaysia". In *Production Networks and Industrial Clusters: Integrating Economies in Southeast Asia*, edited by Ikuo Kuroiwa and Toh Mun Heng. Singapore: IDE/ISEAS, 2008.

———. "Expansion and slowdown in Southeast Asian electronics manufacturing", *Journal of the Asia Pacific Economy* 14 (2 May 2009): 123–37.

———. "Systemic Coordination and Human Capital Development: Knowledge Flows in Malaysia's MNC Driven Clusters". INTECH Discussion Paper 2002–07. Maastricht: INTECH, 2002.

Santiago, C. "Public-Private Partnership: An Alternative Strategy in Water Management in Malaysia". In *Reclaiming Public Water,* Amsterdam: Transnational Institute, 2005.

SDaletech. <http://www.sdaletech.com/en/our_vision.php> (accessed 15 September 2011).

Seagate. <http://www.seagate.com/ww/v/index.jsp?locale=en-US&name=malaysia-culture&vgnextoid=c1d5856c23a1e010VgnVCM100000dd04090aRCRD> (accessed 15 September 2011).

SHTP. <http://www.senaipark.com/target_ind.html> (accessed 28 September 2011).

Singh, C. "The PDC as I Know It (1970–90)". In *Malaysia: Policies and Issues in Economic Development.* Kuala Lumpur: Institute of Strategic and International Studies, 2011.

Trocki, C.A. *Prince of Pirates: Temenggongs and the Development of Johor and Singapore, 1784–1885.* Singapore: NUS Press, 2007.

Venture, <http://www.venture.com.sg/contact/worldwide.htm> (accessed 15 September 2011).

Yoon C.L. "Penang's Technology Opportunities". In *Catching the Wind: Penang in a Rising Asia,* edited by Francis E. Hutchinson and Johan Saravanamuttu. Singapore: Institute of Southeast Asian Studies, 2012.

Yusof, Z.A. and Deepak Bhattasali. "Economic Growth and Development in Malaysia: Policy Making and Leadership". Working Paper 27. Commission on Growth and Development. Washington D.C.: World Bank, 2008.

Periodicals:

Business Times
New Straits Times
The Edge

Primary Sources:

The Constitution of Malaysia.

Iskandar Malaysia. *Flagship D: Important Facts and Details on Eastern Gate Development.* Johor Bahru: Iskandar Malaysia, 2008*a*.

Iskandar Malaysia. *Flagship E: Important Facts and Details on Senai-Skudai.* Johor Bahru: Iskandar Malaysia, 2008*b*.

Iskandar Regional Development Authority Act, 2007.

Johor Corporation/JCorp. *Annual Report*. Johor Bahru: Johor Corporation/JCorp. Various years.

Johor State Economic Development Corporation Enactment, no. 4, 1968.

Johor State Economic Planning Unit (JSEPU). *Pelan Ekonomi Negeri Johor, 1990–2005* [Economic Plan for Johor, 1990–2005]. Johor: State Government of Johor Darul Ta'zim, 1989.

Malaysian Industrial Development Authority (MIDA). "Cost of Doing Business in Malaysia". Kuala Lumpur: Malaysian Industrial Development Authority, 2009.

Malaysian Institute for Economic Research (MIER). *Johor Industrial Master Plan Study*. Kuala Lumpur: Malaysian Institute for Economic Research, 1997.

Ministry of Plantation Industries and Commodities (MPIC). *Statistics on Commodities 2009*. Putrajaya: Ministry of Plantation Industries and Commodities, 2010.

RMA Perunding Bersatu. "Background Industrial Surveys for SJER Development Master Plan Study", 2006.

RMA Perunding Bersatu. "Johor Operational Master Plan Study". Johor Bahru: State Economic Planning Unit, 1996.

RMA Perunding Bersatu. "Pre-feasibility Study Johor Technopolis". Johor Bahru: State Economic Planning Unit, 1994*a*.

RMA Perunding Bersatu. "Pre-feasibility Study for an SMI Industrial Park". Johor Bahru: State Economic Planning Unit, 1994*b*.

Interviews:

President, SME Association of South Johor, Pasir Gudang, 21 April 2010.

FMM SME Working Group, Johor, 22 April 2010.

Official, Johor Skills Development Centre, Pasir Gudang, 28 April 2010.

President, small electronics Singaporean SME association, Singapore, 5 May 2010.

Senior Official, JSIC, Nusajaya 18 May 2010.

Representative, Johorean business association, Johor Bahru, 21 May 2010.

Senior Official, JSEPU, Nusajaya, 28 May 2010.

Senior Official, Johor Technopark, Johor Bahru, 23 June 2010.

Former State Secretary, Johor State Government, Johor Bahru, 24 June 2010.

Tenant, Johor Technology Park, Senai, 28 June 2010.

Manager, Johor Enterprise Corporation, Larkin, 27 July 2010.

Former CEO of JCorp, Kuala Lumpur, 5 September 2011.

Consultant, Johor State Government, Kuala Lumpur, 7 September 2011.

Senior Official, Johor Technopark, Johor Bahru, 9 September 2011.

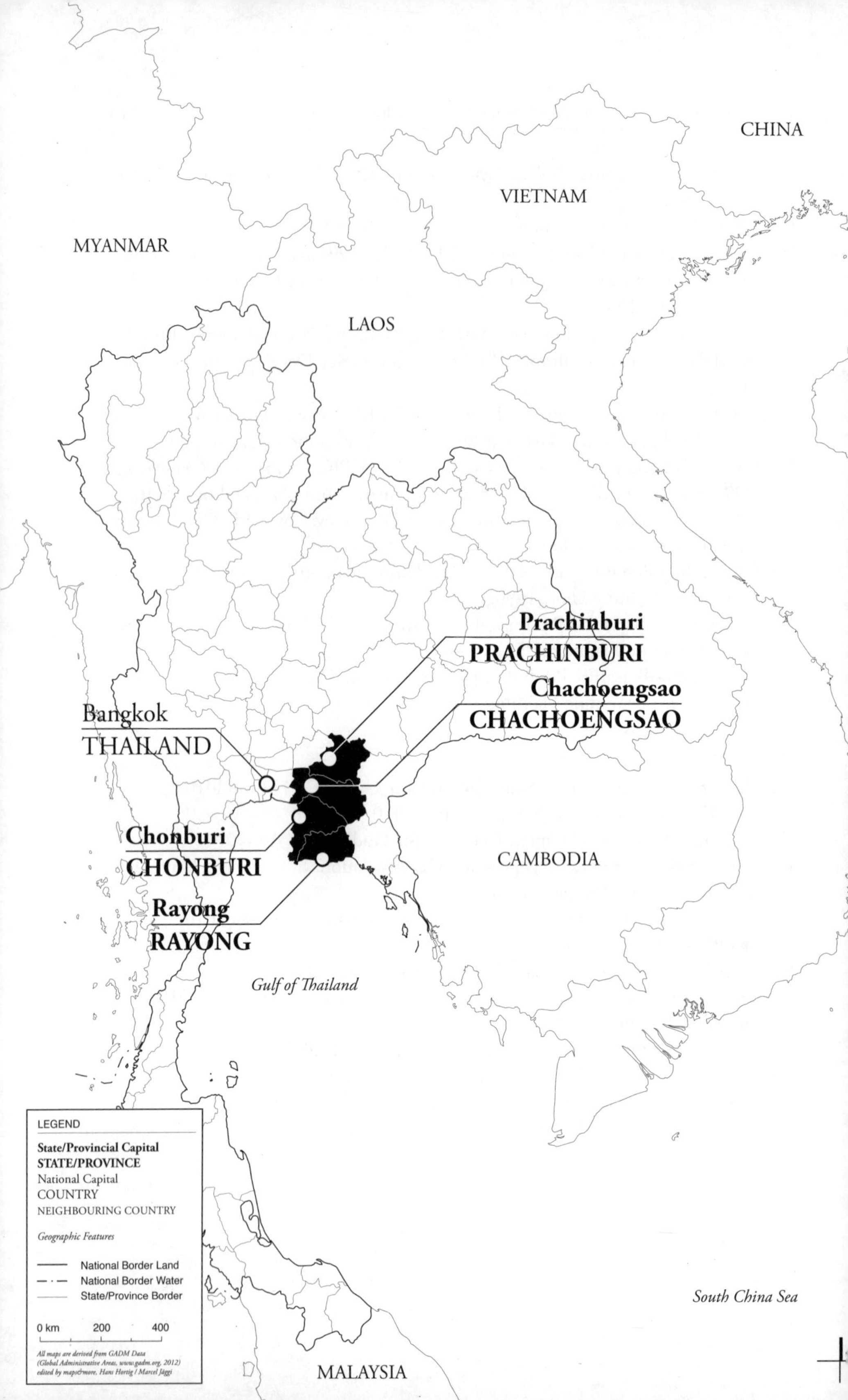
CHINA
VIETNAM
MYANMAR
LAOS
Prachinburi
PRACHINBURI
Chachoengsao
CHACHOENGSAO
Bangkok
THAILAND
Chonburi
CHONBURI
CAMBODIA
Rayong
RAYONG
Gulf of Thailand
LEGEND
State/Provincial Capital
STATE/PROVINCE
National Capital
COUNTRY
NEIGHBOURING COUNTRY
Geographic Features
National Border Land
National Border Water
State/Province Border
0 km
200
400
All maps are derived from GADM Data
(Global Administrative Areas, www.gadm.org, 2012)
edited by maps&more, Hans Hortig / Marcel Jäggi
South China Sea
MALAYSIA

5

WHY SUB-NATIONAL GOVERNMENTS IN THAILAND ARE NOT CREATING ELECTRONICS INDUSTRY CLUSTERS

Danny Unger and
Chandra-nuj Mahakanjana

INTRODUCTION

At an intensifying pace over the past generation, Thailand's structure of production and, in particular, its export profile has been shaped significantly by a relatively small group of firms in the electronics industry. When foreign multinational assemblers began to produce new electronic products in Thailand 20 years ago, it was conventional to think of the challenges of catch-up industrialization, at least in part, in terms of a particular national economy finding means to implicate itself to its advantage within a global division of labour. While that perspective remains meaningful, today, as a result of the increasing "fragmentation" of production, we might better think of the goals for a national economy in terms of finding advantageous positions in multinational firms' global (in the case of the electronics industry, this increasingly means East Asian) value chains.

The implications of the new global geography of production can be seen in the differential rates of economic growth within Thailand. In recent decades, traditional secondary urban centres such as Chiang Mai and Had Yai have grown relatively slowly. In contrast, its Eastern Seaboard provinces have come from nowhere in the 1980s to become Southeast Asia's most important concentration of mostly heavy industry manufacturing (Webster 2006; Unger 1998).

Thailand has been highly successful at embedding itself within the East Asian regional production strategies of major electronics producers. However, the potential positive dynamic effects, namely, the spillovers flowing from that success have been more modest than we might expect. We explore possible explanations for this mixed record below and point to indicators that some Thai state agencies are becoming more successful in supporting the electronics industry than in the past.

What are the means available to political jurisdictions to attract the attention and, ultimately, commitments from the major firms that define the geographic contours of the different segments of the global electronics industry?

These means include the conventional ones of maintaining stable macroeconomic conditions and providing a favourable business climate. Also important, of course, is the quality of physical infrastructure as well as the cost and skills of the labour force. In addition to these more straightforward factors that foreign firms weigh in deciding where to establish production facilities, there are variables that are specific to particular industries engaged in cross-border fragmented production. These include access to a variety of services that support production and supplies of highly specialized parts and components. The small size and weight of most of these components mean that rapid and assured access need not entail immediate geographical proximity. Many firms, however, anticipate benefits (positive externalities) flowing from propinquity and as a result there are self-reinforcing dynamics at work that suggest that once an industrial cluster is established in a particular location it may tend to perpetuate itself there. Thailand is now a major beneficiary of these externalities.

The wide variety of factors that go into making a particular site attractive to firms weighing production location decisions suggests a correspondingly diverse set of institutions that may be able to influence those decisions. Those institutions include, of course, the government of the host country. Given the global epidemic of government decentralizing initiatives, including many cases in Southeast Asia, "government" may refer to institutions at different levels. In the case of Thailand, these include elected provincial, municipal, and sub-district governments, provincial and sub-district arms of the central government, and the central government itself.

Not only public institutions, but established private firms also influence the decisions of firms pondering production location choices. Entrenched firms may help to attract subsequent private investments by providing specific services, elements of economic governance, or as critical buyers or suppliers of the firm considering its investment decision. With Seagate Technology in Thailand assembling hard disk drives (HDDs), for example, its many suppliers, some of them smaller firms, also were drawn to Thailand.

In this chapter we elaborate on these cluster dynamics in the Thai case. We begin below by offering a brief sketch of the key institutions of Thailand's political economy and their apparent commitments and capacities. In the second section, we trace the history of foreign direct investment (FDI) in Thailand in general and in the electronics industry in particular (focusing on the production of HDDs). The third and longest section describes a variety of initiatives by different levels of Thai government and the private sector (primarily foreign) to promote electronics industry clusters. The fourth section of the chapter tries to account for the patterns of promotional activism, or its absence, in Thailand among governments at provincial and central levels, as well as firms.

The chapter concludes by examining the implications for Thailand's political economy of the limited extent to which it is benefiting from FDI spillovers associated with a massive commitment of resources by foreign firms in the electronics industry. That said, over the past eight years, we have seen unaccustomed flexibility and innovation on the part of Thai state agencies in trying to anchor the HDD industry in Thailand. We also suggest that sub-national governments will play more important parts in promoting electronics and other business investments in the years ahead.

DISTINCTIVE FEATURES OF THAILAND'S POLITICAL ECONOMY

It is striking that a country that was still predominantly rural 15 years ago is the world's leading exporter of several relatively capital-intensive and high-technology goods. Thailand is the world's leading exporter of various agricultural goods, hard disk drives, and is second in motorcycles and pick-up trucks. It is among the top three per cent of all economies of any size in the world in the rate at which it has boosted average levels of affluence over the past half century (Commission on Growth and Development 2008).

Thailand sustained a run of appreciable economic growth over four decades before it crashed in 1997, and the Thai baht was, until the same year, one of the world's most stable currencies. In short, the country has a stunning and, in some ways, surprising record of economic growth, structural shifts, and economic and social stability.

Thailand's impressive economic record does not result in any direct fashion from the quality of its governance. Support for industry has generally not been well designed or effective.[1] Even in agriculture, the government contribution has not been great. Strikingly, Thailand's huge number of rice producers has one of Asia's lowest levels of productivity per unit of land. The pairing of a large and, in key areas, low-productivity agricultural sector with extensive exports of technology-intensive goods tells us a lot about the contemporary global political economy, as well as about Thailand's economic and social "dualism".

Due to its land abundance and the fact that it was not colonized, global capitalist forces impinged less dramatically on most Thai villagers than elsewhere in Southeast Asia. Not having to throw off the yoke of foreign rule, Thai elites had no incentives to organize and mobilize the masses for political action (far from it). Colonial rule in the region meant that the state no longer faced military threats other than those posed by France and the United Kingdom themselves, and for the most part developing the military capabilities necessary to fend off those powers was not actively contemplated. Instead, the military was used largely for purposes of internal colonization. The local business class was dominated by immigrant Chinese — as opposed to indigenous elites — so that for a time their political influence was far less than structural factors would suggest.

From the mid-twentieth century, economic growth was rapid enough and social mobility sufficiently widespread to accommodate the ambitions of what remained very small numbers of elites and middle classes until the 1980s. And the effectiveness of the campaign to build the charisma of the current king, Bhumipol Adulyadej, who has reigned throughout the era of Thailand's rapid social changes, helped to underpin, until recently, a durable but flexible social and elite consensus concerning the nature of a desired Thai polity and society.

Thus, Thailand's economic success over the past half century is a result of some favourable initial conditions (land, water, and forest abundance); macroeconomic, social, and policy stability; immigrant Chinese; a degree of political pluralism that prevented any narrow business coalition from entrenching its power too deeply, and the absence — until recently — of significant political mobilization that might have undermined fiscal stability. Thailand's natural resource wealth engendered a diverse agricultural base and encouraged the development of a large agro-processing manufacturing sector. Over time, the country has built up significant industries. The textile industry was for a time Thailand's most important, but it never became a world beater in the order of those in South Korea and Taiwan. The auto industry only very gradually developed into a major international player, which was no mean feat, but today still has limited local abilities to produce more sophisticated components and parts.

Thailand is becoming richer, but not through a disciplined, state-supported mobilization of scarce resources. Elites have not cohered to create a powerful and flexible state or political institutions. Thailand should not be mistaken for a Taiwan, Singapore, or even a Malaysia. In very loose outline, with its market energies and assimilation of risk-taking immigrant entrepreneurs, Thailand looks perhaps more like the United States.

FDI AND THE ELECTRONICS INDUSTRY IN THAILAND

In this section, we describe the prominent role that FDI played in the development of the Thai economy, focusing on recent decades, the prominence of the electronics industry, and looking particularly closely at HDD manufacture.

Until the mid-1980s, levels of FDI in Thailand were modest. The absence of a colonial past meant that foreign capital played a less prominent role in Thailand than it did in most other parts of Southeast Asia. Foreign investments were concentrated in sugar, rice mills, or logging, and in many instances these were abandoned during the 1930s or early 1940s (Suehiro 1989). Those industries that did exist in the mid-1950s were overwhelmingly in the hands of Chinese or the state.

Industrial growth expanded and the state's direct role in the economy receded under the import substituting regime established under Field Marshal Sarit Thanarat in the late 1950s. Japanese and U.S. firms established and expanded modern industries in Thailand. The textile industry grew with investments from Toray, Teijin and others and, by the late 1970s, the country was beginning to export garments in significant numbers (Unger 1998).

In 1970, manufactured exports accounted for only five per cent of total exports. That figure had reached 74 per cent by 2001 and electronics exports accounted for 30 per cent of the total (UNCTAD 2005). Today, manufacturing accounts for 34 per cent of gross domestic product, matching the level in China, though Thailand's share of manufacturing investment and production has fallen within East Asia as well as Southeast Asia (Webster 2006, p. 15). FDI as a share of gross fixed capital formation averaged about three per cent in the 1960s and was above five per cent in the 1980s (Kohpaiboon 2006, p. 107). Thereafter it grew rapidly, beginning with the radical restructuring of the East Asian economy in the latter 1980s as FDI flowed first from Japan, and subsequently from South Korea and Taiwan, to sites in Southeast Asia to establish production facilities for export (Unger and Bowie 1997).

Helped by rapid growth of FDI and exports, over the decade up to 1996 Thailand recorded the world's fastest rate of economic growth. By 1996, with a substantial FDI boost, Thailand's per capita income was six times its level in 1960. The share of manufacturing in Thai exports rose from about 30 per cent in the early 1980s to about 75 per cent in 1996 (Jongwanich and Kohpaiboon 2008).

The rate of capital inflows took off again in the late 1990s following the drastic devaluation of the baht that began in mid-1997. The ratio of FDI to gross fixed capital formation in Thailand has been high since the mid-1990s relative to other high growth East Asian economies. The figures are higher only in Hong Kong and Singapore (Kohpaiboon 2006, pp. 110–12).

By 2006, electrical appliances and electronics goods together accounted for close to half of all Thai manufactured exports (Jongwanich and Kohpaiboon 2008, p. 5) and were among the largest industrial employers in Thailand (Ministry of Industry 2011). The country is also a major producer and exporter of semiconductors and integrated circuits, involving firms such as LG Electronics and Philips Electronics. Its total electronics trade in 2008 was over $50 billion (Board of Investment 2011).

HDD production is particularly important in Thailand, as firms in this sub-sector alone employ well over 300,000 workers (Afzulpurkar and Brimble 2004). In 2006, the Thai HDD industry surpassed Singapore to become the world's leading exporter. Among the key firms in Thailand producing HDDs are Fujitsu, Hitachi, Seagate Technology, and Western Digital, all among the world's largest assemblers. Key producers of parts include Alps Electric, Magnecomp Precision, Minebea, NHK, Nidec Electronics, San-Ei, and Nitto Denko (Board of Investment 2011). HDD production now accounts for ten per cent of value added in the entire industrial sector. Total exports in 2008 were about $12 billion, giving Thailand about 17 per cent of global exports (Kohpaiboon 2011).

The electronics sector took a hit in 2008 with the global economic slowdown and then contracted by about ten per cent in 2009 before booming the following year.[2] It was on course to stabilize in 2011 when disastrous flooding submerged much of the industry. Nonetheless, electronics exports in 2012 were likely to account for about 15 per cent of all Thai exports with the largest markets being China and then ASEAN. Total exports reached about $30 billion in 2007, $33 billion in 2010, and a projected $36 billion in 2012 (Electrical and Electronics Institute 2011).

Despite the sector's impressive growth, there are worries about the limited acquisition of technological capabilities by local firms. Over half of all FDI in Thailand is in manufacturing and almost half of all manufacturing plants are foreign-owned. Foreign plants account for close to 60 per cent of manufactured exports. Foreign firms' "export spillovers" make Thai firms more apt to export as well, presumably as a result of market information externalities and demonstration effects (Jongwanich and Kohpaiboon 2008). In the case of electrical machinery, foreign-owned facilities account for 85 per cent of production and a slightly larger share of exports (Kohpaiboon 2006, pp. 119–20).

In general, the higher the level of technology required for production of the goods, the greater is the dominance of foreign firms in the sector. And, compared to the industry in either Malaysia or Singapore, fewer Thai firms have developed into significant suppliers. In "high technology" export industries, however, Thailand specializes in processes in which the capital-labour ratios are on a par with those for export industries such as toys, garments, and processed foods (Kohpaiboon 2006, p. 125).

For example, by the mid-1990s, some firms in the electrical appliance industry were sourcing parts and components in Thailand extensively. However, in microelectronics, local firms are generally not significant players, as the industry relies on imported semiconductor devices, diodes, transistors, and integrated circuits. Local value-added in HDD production in 2003 was not much above 30 per cent, but nonetheless accounted for about three per cent of gross domestic product (Afzulpurkar and Brimble 2004; Danish Trade Council 2006).

Most analyses of the industry conclude that FDI spillovers in the electronics sector in Thailand have been modest and that low local skill levels and limited and ineffective government support inhibit industrial upgrading.

A 2005 UNCTAD report, however, is more upbeat. It suggests that Thai firms not only use foreign technologies but also "upgrade and improve assembly processes" and are starting "to acquire manufacturing capabilities". The report does, however, note that few firms have been able to upgrade or do reverse engineering, fewer are engaged in design or developing new products, and that technology diffusion is limited largely to multinational firms' networks.[3]

Thus, while FDI has expanded sharply in recent decades, it always has been accompanied by concerns for Thailand's longer term competitiveness. That said, the country has been able to provide a generally welcoming context for business. Thailand's broad trade policies generally have supported foreign investors. The country adopted important tariff restructurings in the 1980s and again in the mid-1990s (Kohpaiboon 2011). Thailand is a member of the World Trade Organization, the Association of Southeast Asian Nations (ASEAN) Free Trade Area (FTA) and a wide array of multilateral, regional, and, bilateral (including Australia and Japan) trade agreements. Particularly important in this regard was the Information Technology Agreement that went into effect in 2000 (interview, Official B, Ministry of Industry, 13 July 2011).

Thailand's infrastructure is generally regarded as quite good, although congestion at Laem Chabang port is growing worse and is a concern among auto firms (interview, Kohpaiboon, 7 July 2011). Thailand relies heavily on road transport (88 per cent of the total, by weight), lacks a national expressway system, and incurs high logistics costs (Webster 2006, pp. 30–33). The World Economic Forum puts Thailand 39th among 142 countries in economic competitiveness. Its ranking on infrastructure is only slightly lower, at 42 (World Economic Forum 2012).

However, according to the World Economic Forum's ranking, areas where Thailand does more poorly are in the areas of innovation (54), higher education and training (62) and technological readiness (84). Many quantitative measures and more qualitative assessments concur in a fairly negative evaluation of Thai public education generally, including in technical and scientific fields (Jongwanich and Kohpaiboon 2008; Lall 2000, pp. 41–49; Dodgson 2000, pp. 236).[4] Thailand's "national technological capability"[5] (Lall 2000) does not seem to be high. In 2003, private firms accounted for 44 per cent of research and development spending in Thailand (the government contributed 22 per cent and universities 31 per cent). There were until recently few technology linkages between universities and private firms (Intarakumnerd 2011, p. 41). Thailand's total research and development spending of around 0.24 per cent of GDP ranked it 40th on a list of countries that spend more than $100 million a year. The National Science, Technology, and Innovation Policy Office set goals of raising total research and development spending to two per cent by 2021 and the private sector's share of that spending to 70 per cent (Wongsamuth 2012).

PROMOTING INDUSTRIAL CLUSTERS IN THAILAND

A variety of both public and private institutions are involved in the promotion of the electronics industry through a wide array of different policies. Below we consider the roles of the Thai government at national and provincial levels as well as those of private, primarily foreign, firms. First, however, we briefly sketch the division of responsibilities across Thailand's different levels of government.

Administrative and Political Decentralization in Thailand

Minister of Interior Prince Damrong Rajanupab created Thailand's modern administrative structures late in the nineteenth century. The impetus behind these reforms was to create a modern state that controlled a territory and a population and protected the key commercial interests of foreigners. Such a state held out the hope of not offering imperialists a pretext for seizing control of the kingdom and administering it themselves. These motives, and the less than complete administrative integration of the kingdom at the time, dictated that the new structures would be highly centralized as the court tried to impose its authority over even distant regions that had previously operated under loosely federal arrangements.

No significant decentralization of Thailand's administrative apparatus followed until the early 1990s, culminating in the major shifts mandated by the 1997 constitution. Subsequent implementing legislation called for the creation of new elected bodies at provincial and municipal or sub-district levels, and the transfer from the central government of first 20, and then 35 per cent (by 2006) of its revenues to these new bodies. Newly created Provincial Administrative Organizations (PAOs) featured an assembly with directly elected representatives serving four-year terms. While resource transfers to local governments, starting from around five per cent in 1996, initially grew quite rapidly, they stalled with the figure at about 25 per cent. The new 2007 constitution embraced the prior decentralization targets for revenue distribution, but did not resuscitate the earlier 35 per cent target.

Elected bodies at the local (municipal and sub-district) level have assumed extensive new responsibilities over the past dozen years and more. And while they also now disburse substantial revenues, they typically complain that the allocations of the bulk of these funds are mandated by various central government-imposed regulations. This tends to leave them with limited scope for investments or launching new initiatives dependent on substantial local financing. These local governments can raise revenue, but few have done so in significant quantities. Furthermore, the central government sets tax rates at the local level and also plays oversight roles in personnel policies. The new elected governments receive more funds from the central government than they raise locally. The PAOs have significant resources, but directly administer very little. They have assumed many of the central government's previous oversight functions and are responsible for passing on funds they receive to municipal and sub-district governments. At both the local and provincial levels there remain substantial areas of overlapping authority among the newly created elected governments and the centralized administrative structure under the control of the Ministry of Interior (Mahakanjana 2009).

The Ministry of Interior fought decentralization vigorously. And while it ultimately failed to stop it, the ministry took steps to try to retain its former authority at the provincial and local levels. One result has been extensive overlap in responsibilities among the new local governments and the central government's delegates (governors, *nai amphor*, and *kamnan*) at the local levels. This overlap has been clearest at the provincial level where appointed governors retain key coordinating roles. At the provincial level, then, elected PAOs operate in parallel with the apparatus of the central government working under the appointed governors. The new entities are squeezed from above and below, and command only modest resources.

We expect to find local governments at the meso-level with the resources and authority to contribute to the creation of industrial clusters rather than at the local level. In Thailand, this means that we look to the provincial rather than the municipal or sub-district level. However, Thailand's decentralization largely shifted resources down to the municipal level and left comparatively limited resources to PAOs. Furthermore, the new PAOs faced the most direct competition from the centralized structures of the central government. The Ministry of Interior nudged the design of decentralization in ways that guaranteed that no new powerful entities would result (interview, Jitpiromsri, Deputy Dean, Prince of Songkhla University, 9 June 2010).

How the Central Government Promotes Industrial Clusters

Since the early 1960s, the central government has promoted FDI through a variety of policy instruments and government organizations. Broadly speaking, these efforts focused on providing fiscal incentives, removing impediments in Thailand's import tariff structures, and improving energy, communications and transport infrastructure. Excepting the auto sector, they did not tend to include more narrowly-targeted efforts to promote FDI in particular economic sectors or to specifically promote industrial clusters.

For decades, Thai economists voiced concerns that FDI was not creating enough jobs (until the 1980s) or that the local value-added in Thai manufactured exports was not rising rapidly. In response to the latter concern, in 1992, the Board of Investment established its Unit for Industrial Linkage Development aimed at fostering assembly-supplier links largely through the collection of information and matching services. It organized supplier visits of several dozen firms to some of the major HDD assemblers such as Fujitsu and IBM Storage. These programs aimed to help assemblers identify sources of parts and components, working with an ASEAN database of over 20,000 suppliers. They also aimed to help suppliers to boost their productivity and to link them to foreign assemblers (Board of Investment 2011; UNCTAD 2005, p. 21). It is not clear that this service has been particularly effective (Afzulpurkar and Brimble 2004).

The government created the Ministry of Science, Technology and Energy in 1979. The government later added environmental issues to its portfolio before breaking it up into three separate ministries in 2002, including the Ministry of Science and Technology. The ministry established the National Science and Technology Development Agency (NSTDA) in 1991 which in turn oversees and gives one-third of its budget to the National Electronic and Computer Technology Center (NECTEC), established in 1986 (Kondo 2008).

NECTEC employs 30 scientists and engineers and supports research and development and links between firms and research institutes. It has worked with universities to establish closer links between universities and industries. Working with the King Mongkut Institute of Technology at Thonburi, it set up the Institute of Field Robotics that aims to boost firms' capabilities in the HDD industry. The Engineering Faculty at Khon Kaen University worked with NECTEC to boost training in producing HDD components. And NECTEC worked with KMIT at Ladkrabang to develop firms' capacities in data storage technologies (Board of Investment 2011). NECTEC has been less active in working with private firms (Afzulpurkar and Brimble 2004).

In 1992 the NSTDA established, as part of its Industrial Technology Assistance Program, Industrial Consultancy Services in a bid to get local firms to make more use of consultants and research institutes in dealing with technology challenges, going so far as to help pay for consulting costs (Intarakumnerd 2011, p. 48). Also in 1992, the government set up the National Information Technology Committee, chaired by the prime minister and including several cabinet ministers and private firms. In 1997, the NSTDA set up Software Park Thailand with a technology transfer service, working with Oracle, IBM, Intel, and Microsoft (UNCTAD 2005, p. 20). In the 1990s, NECTEC established the Thai Microelectronics Centre Project to be the first facility fabricating integrated circuit wafers in Thailand and to provide training in technology-related skills (Afzulpurkar and Brimble 2004; Board of Investment 2011). Plans for the integrated circuit production line were disrupted by the economic collapse beginning in 1997 (Afzulpurkar and Brimble 2004; interview Punyasai, Senior Researcher, NECTEC, 21 February 2012).

In 1999, the Electrical, Electronic and Allied Industries Club's[6] work with the Ministries of Commerce, Industry, Science and Technology, and Environment resulted in an Information Technology Industry Development Master Plan. The Cabinet approved the plan for the electrical, electronics, and information industries, laying the foundation for the creation of the Electrical and Electronics Industry Institute (EEI). The Institute operates an industrial standards centre, an industrial clustering service centre, and an information and technology service centre (Board of Investment 2011; UNCTAD 2005, p. 21). The Master Plan called for the development of firms providing services that would support the HDD industry, expanding its exports, and increasing local value-added. It was not clear, however, how the government envisioned pursuing those goals, what was to be done concretely, or indeed just what was the status of the plan (Afzulpurkar and Brimble 2004). As for the EEI, its activities were not generally important to the HDD industry (Kondo 2008).

There are few ongoing contacts between private firms and the Thai government (McHendrick et al. 2000, p. 197). The firms themselves tried on several occasions to push cooperative ventures, but these failed. Seagate Technology, for example, tried to push testing and failure analysis at the Asian Institute of Technology, as well as various industrial training programmes, but got nowhere initially. KR Precision similarly fell short in its bid to establish a tool and die institute and Micropolis' bid to set up a magnetic testing facility through the King Mongkut Institute of Technology went nowhere (McHendrick et al. 2000, p. 198). Eventually, however, these and other initiatives began to bear fruit. It is not clear to us whether the gestation period for these projects was simply long or if their eventual emergence came in response to the concentrating of minds that started over a decade ago when the HDD industry was in serious trouble.

McHendrick et al.'s research in Thailand from the late 1990s suggested that foreign firms in the HDD industry did not believe that the Thai government had done a great deal to draw in and support firms in the industry.[7] Thailand was attractive largely due to low labour costs, attractive tax and tariff policies, macroeconomic and political stability, and its proximity to Singapore (McHendrick et al. 2000, pp. 185, 188). In contrast to the Malaysian government's work in support of its electronics industry, for example, the Thai government established few new institutions to support the industry. The most important local institution supporting the industry apparently was the International Drive Equipment and Manufacturers Association (IDEMA), a private international group that established a branch in Thailand in 1999 (McHendrick et al. 2000, p. 132). The Thai government was slow to develop expertise in the industry or to consider crafting policies to support it (McHendrick et al. 2000, p. 197).

When Prime Minister Thaksin Shinawatra was in office, starting in 2001, the government moved toward adopting more sector-specific promotional policies. It set forth a ten-year Science and Technology Action Plan for 2003–13 (Intarakumnerd 2005, p. 29) and identified five key strategic sectors (not including electronics). The Board of Investment gave more attention to investments that would foster local technological capacities. Particularly important in the latter regard was the decision to extend investment promotion privileges under the Skills, Technology and Innovation scheme to suppliers as well as assemblers in the HDD industry (more details below). Intarakumnerd writes that "The cluster concept was used as the main industrial policy of the Thaksin government at national, regional and local levels" (2011, pp. 45–48). The Thaksin government adopted elements of Results Based Management reforms of the state sector and one apparent effect was to give stronger incentives to universities to work with the private sector (Intarakumnerd 2011, p. 51; Kondo 2008).

Since 2007, both the Board of Investment and the Ministry of Industry have at least nominally intensified their commitment to promoting industrial clusters through a variety of means (Kohpaiboon 2011). Some of this more recent effort has been in response to concerns that absent the right array of supporting firms and services, some important HDD assemblers would shift their production to sites where such firms and services were in greater supply or, failing that, wages were lower (interview, Official A, Ministry of Industry, 2 July 2011). The largest HDD assemblers were indeed reconsidering the benefits of large production facilities in Thailand (McHendrick et al. 2000).

It was in this context of very salient threats to the important HDD industry over a decade ago, that a small number of key actors pressed for critical changes in government policies. The resulting "uniquely good case of triple helix collaboration" initially involved, most centrally, IDEMA, Seagate Technology, and NSTDA (Kondo 2008). For years, Seagate Technology had been launching a series of research and development and technology training collaborations with Thai universities. IDEMA's first chair was a Seagate Technology Senior Vice-President and it was effective in making the HDD industry's pitch to the Thai government (Afzulpurkar and Brimble 2004). NSTDA was strongly supportive and commissioned the Asian Institute of Technology and a private consultant to carry out a study of the industry and its needs. This led to the establishment of The Hard Disk Drive Institute (HDDI) in 2004. Various government agencies such as NECTEC, NSTDA, and the Board of Investment were involved, as well as research and industry institutes such as the Asian Institute of Technology, EEI, as well as private firms such as Fujitsu, Seagate Technology, and Western Digital. The HDDI set out to promote a HDD cluster in Thailand by building human resources in specific technologies such as contamination reduction, automation, and high precision moulds and dies.

The HDDI served as a research technology intermediary between the needs of the assemblers and government research and technology institutes (Intarakumnerd 2005, pp. 23–24), though local supplier firms were not actively engaged. From the outset the new organization benefited from the expertise of its manager. He was ideally positioned for this role, having worked with both Seagate Technology and Western Digital, having had his own high technology firm (Siliconcraft) and taught occasionally at the Asian Institute of Technology (interview, Poyai, Principal Researcher, NECTEC, 21 February 2012).

IDEMA also worked closely with the Board of Investment. In May 2004, the Board announced its new policy on Investment Promotion of the Hard Disk Drive Production Industry. The following month Seagate Technology, Hitachi Global Storage, and Western Digital announced major new investments totalling over $600 million (Afzulpurkar and Brimble 2004).

Private Sector

As is evident in the discussion above, many firms (all the examples we report involve foreign firms) have initiated or cooperated in a variety of initiatives aimed at promoting clusters in the electronics industry. Seagate worked closely with NECTEC and, together with Western Digital, seems to have been at the forefront of efforts to fundamentally shift the Thai government's approach to regulating the electronics sector. Both firms were leaders within IDEMA and, subsequently, HDDI. Seagate Technology has long been active seeking means of strengthening the capabilities of its Thai operations and, apparently, often has been frustrated with the limited and ineffective contributions of the state (McHendrick et al. 2000). Seagate also worked closely with several different Thai universities, including Khon Kaen and Suranaree. Western Digital worked with NECTEC to set up a Technology Training Institute (Board of Investment 2011). Japanese firms were less active initially in IDEMA and HDDI and maintained relatively low profiles (interview with Poyai, Principal Researcher, NECTEC, 21 February 2012; interview, Intarakumnerd, Advisor, Science and Technology Agency, 1 March 2012).[8] It seems clear that the major assemblers were the most critical actors in creating the HDD industry in Thailand and in attempting to foster a cluster to serve their production needs. In doing so, they apparently were hoping to reproduce the more positive production experiences they had in Penang and Singapore (Afzulpurkar and Brimble 2004).

Union Technology, owned by Saha Union, a large Thai conglomerate that got its start in the textile industry, is the only large Thai firm to enter into the assembly of HDDs. It did so at the behest of Hitachi Global Storage, running the latter's assembly operations. Seagate supported the creation of KR Precision in the early 1990s. Initially started by Thai and Japanese partners (Afzulpurkar and Brimble 2004), it then merged in 2005 with Magnecomp Precision Technology. KR Precision was active in several initiatives that aimed to boost the Thai HDD industry's technological base (Afzulpurkar and Brimble 2004). In general, however, few Thai firms were important actors in the industry. And while foreign assemblers and the central government hoped to boost participation in the industry by local suppliers, few of those showed a strong interest. They saw the challenges of supplying the industry as including high competition, low profits, shortages locally of engineering skills, and high training costs, as well as the more general difficulties of slow and costly customs operations, corruption, and the lack of any tariff protection under the Board of Investment's duty drawback arrangements (Afzulpurkar and Brimble 2004).

Provincial Governments

The Thai electronics industry is concentrated in an arc running from the north to the northeast and east of Bangkok, within about 250 miles of the capital (much of it is considerably closer). There is significant electronics production in Lamphun Province farther north, concentrated in an Industrial Estate Authority of Thailand (IEAT) facility, the Northern Region Industrial Estate. The northern region's electronics exports in 2006 came to over $750 million, primarily semiconductors and parts and components. The industry employed over 30,000 workers in the north.

There are also important concentrations of electronics production in the east, in the provinces of Rayong, Chonburi, Chachoengsao, and Prachinburi. Some of these are located in or near the Eastern Seaboard development zone that includes a couple of industrial estates, deep sea ports, and much of Thailand's heavy industry, including its petrochemical, steel, and auto firms. Seagate Technology has plants in Samut Prakarn, just south of Bangkok and Nakorn Ratchasima, in the Northeast but not far from Bangkok.

The industry, however, is concentrated primarily in the provinces immediately north of Bangkok, Pathum Thani and Ayutthaya. Much of the industry is housed within a half dozen industrial estates in the two provinces. The largest of these, Rojana Industrial Estate in Ayutthaya, has an enormous client list. The vast majority of these are Japanese firms. One estimate suggested that between 80 and 90 per cent of all firms located in the industrial estates in these two provinces are Japanese (Svasti 2012). Western Digital operates in a Pathum Thani industrial estate, having got its start there by buying up Toshiba's operations as the latter abandoned the production of 3.5 inch HDDs (Afzulpurkar and Brimble 2004).

In over a dozen discussions with local elected officials, local agents of the central government, people working in the electronics industry, and central government officials, no one suggested that provincial governments play important roles in attracting investment in the electronics industry generally. There was no specific instance mentioned in which local officials or politicians had worked to bring investment to their jurisdiction. However, several did refer to provincial cooperation in unsuccessful efforts to protect industrial estates from flooding and subsequent efforts to bar its recurrence.

As the governor of Pathum Thani Province put it, provincial authorities lack authority to induce firms to invest. Such powers rest with central agencies such as the Ministry of Industry or the Board of Investment. Governors simply implement central government policies. It is central government agencies that do road shows abroad seeking to attract foreign investment to Thailand, not provincial governments. He went on, however, to acknowledge that the elected PAOs have larger budgets and, therefore, more potential to try to lure foreign investors (interview, Governor, Pathum Thani Province, 15 March 2012).

Industrial Estates

A state enterprise, the Industrial Estates Authority of Thailand, operates industrial estates, as do private firms. These are scattered around Thailand, but heavily concentrated in the belt stretching from the north of Bangkok to the east, where electronics firms are clustered. They are important in attracting some foreign firms and help determine the specific locations of those firms' investments. Those estates operated by the Authority began in what are now heavily industrialized regions to the east of Bangkok, but now are often found in regions that are less successful in attracting investment such as the North and the Far South. Many foreign manufacturers elect to produce in industrial estates, as the estates offer established and reliable supplies of power and water and transportation links. They also are likely to have some urban amenities, including housing, that make it easier for firms to attract workers (interview, Punyasai, Senior Researcher, NECTEC, 21 February 2012). And many of them have ties with powerful military figures (interview, Intarakumnerd, Advisor, Science and Technology Agency, 1 March 2012).

As we noted above, at least some of the industrial estates north of Bangkok in Pathum Thani and Ayutthaya house primarily Japanese firms. NMB-Minebea Thai offers an illustrative example. Minebea was a high tech pioneer in Thailand, starting the production of miniature bearings in a plant in Ayutthaya in 1980. A decade later, it had expanded and diversified and had production facilities in the Bang Pa-In, Navanakorn, and Rojana industrial estates north of Bangkok, as well as farther north in Lopburi (www.minebea.co.th, accessed 16 March 2012).

ACCOUNTING FOR OBSERVED PATTERNS OF INDUSTRIAL CLUSTER PROMOTION

What factors account for variations in levels of motivation among politicians and state officials to promote economic expansion generally or, more specifically, to draw it to their jurisdictions? Why do some politicians work hard to attract investment and create jobs while others invest their energies in other pursuits? Our understanding of these issues is limited. It seems clear, however, that the structural power of capital is felt more by politicians in some contexts than in others.

One popular argument sees strong political and state institutions as products of intensive elite commitments to building those institutions. Why do some elites, but not others, have such commitments? Perhaps due to different intensities in the political and security pressures that confront policy makers (Tilly 1992). Creating strong institutions, or inducing firms to invest, can be hard work and elites may not undertake it if easier options are available. Elites may instead devote themselves to the pursuit of rents. Whether due to severe external security threats (Johnson 1982) or as a result of pressing threats of extensive and disruptive domestic political conflicts, politicians and state officials may, as a result, be motivated to work harder to gain the wherewithal they need (including revenue for the state and foreign exchange for the economy) to survive in office (Doner, Ritchie, and Slater 2005; Slater 2010). While these explanations are useful, we find them only partially satisfying because they locate all variation in structural variables. They tend to go too far in imagining that the behaviours of different groups of elites are likely to respond in like fashion to similar arrays of incentives.

Politicians may not be motivated to provide collective goods. Particularly in contexts of clientelistic political competition, there may be few incentives for them to do so. Clientelism has been pervasive in Thailand at the national level, though it may now be diminishing. It seems likely that it remains dominant at the provincial level. Politicians are more apt to be motivated to offer broad collective goods if they are embedded within encompassing institutions, such as strong national political parties. In such cases, they not only face stronger pressures to be accountable, but they also have hopes of benefiting electorally from the collective goods that they help to supply. Thailand does not have strong political parties. Those in Malaysia and Singapore, for example, are stronger. The importance of politically integrating institutions such as political parties is particularly great at the provincial level in Thailand because PAOs have few means of effecting policies except through the mobilization of cooperation among the many municipal and sub-district governments they supervise. Another obstacle to collective goods provision in clientelistic contexts is that citizens are less likely to engage in socio-tropic retrospective voting (Lyne 2007). Rather than vote for or against parties based on their previous policy performance, voters are more likely to support candidates who offer them relatively tangible benefits.

Provincial politicians in Thailand cannot expect to garner greatly increased revenue by inducing investments in their province, although they do receive a share of value-added and other revenues that would rise with economic activity. They do not tax land and property. Those taxes are levied by municipalities and sub-districts, although only within guidelines established by the central government.

Thai politics long were noted for their Bangkok-centrism and Bangkok still is an extreme version of a primate city, attracting disproportionate shares of investment as well as concentrating cultural and higher level educational activities. As we noted above, even the central government has been slow to provide active support to the electronics industry. It therefore may not be entirely surprising that in our research we encountered no case of provincial authorities playing important roles in attracting or supporting investment in the HDD industry or the electronics industry in general.

When we asked one central government official why provincial governments do not, for example, attempt to boost local education and training in a bid to attract investments by providing a larger pool of skilled workers, he replied, "They don't think along these lines." The prevailing pattern in Thailand, he suggested, was for initiative and tutelage to flow from the centre to the periphery. "Thailand is not like China," he added (interview, Official A, Ministry of Industry, 2 June 2011).

An official of the Ministry of Commerce working at the provincial level acknowledged that the province had access to funds to promote investments. In describing what form such hypothetical promotion might take, however, he referred only to the maintenance of infrastructure and utilities and possibly the building of modest local transport infrastructure (interview, Official, Ministry of Commerce Official, 15 March 2012). An official of the Ministry of Industry, also at the provincial level, suggested that provinces could help to attract business by repairing roads and providing U-turn bridges. In discussing ways in which provinces could boost their local labour skills and thereby attract investment, this official suggested that provinces were more interested in offering vocational education and placement programs for local job seekers (interview, Ministry of Industry official [Pathum Thani], 15 March 2012). When asked about steps taken to attract electronics industry investment, another provincial-level official referred only to the formation of a committee to consider how to attract investment (interview, Ministry of Industry official [Saraburi], 15 March 2012).

Governors are not expected to be policy entrepreneurs. Their principal roles are as coordinators of the many central government agencies operating at the provincial, municipal, and sub-district levels. Even if governors were inclined to launch new initiatives, they might be stymied by their generally short tenures and the limited discretionary funds at their disposal, on the order of $3 million in the case of Pathum Thani (interview, Governor, Pathum Thani, 15 March 2012). And both agents of the central government and elected provincial politicians referred to the central government's development guidelines within which provincial actors are supposed to focus their development strategies (interviews, Ministry of Industry official [Pathum Thani] 15 March 2012, and Acting President, Ayutthaya PAO, 23 March 2012). PAOs are responsible for generating development plans but those are supposed to conform to these central government guidelines.

Various pieces of legislation governing decentralization have specified the responsibilities of PAOs. These acts refer to various activities that could enable provincial activism in support of electronics industry investment. PAOs, for example, are expected to support education, the development of suitable technology, promote tourism as well as commerce and industry more generally, build and maintain transportation within the province, promote family industries, maintain and promote people's livelihoods, and city planning and construction management. The list is extensive enough that ambitious provincial politicians intent on attracting investment probably would not be overly constrained in their efforts to do so.

However, these acts seem to reflect prevailing expectations in putting particular emphasis on activities such as law and order, maintaining facilities and markets, public health and disaster management, promoting cultural activities, vocational and other education, and sports (Mektrairat 2007). Only one person with whom we spoke emphasized specific examples of provincial government activism. Instead, respondents explained to us that such activism and leadership rested with the central government. The only exception cited was in reference to the governor of Chachoengsao Province. He apparently has been active seeking technological assistance with relatively small scale agro-processing and water management initiatives (interview, Poyai, Principal Researcher, NECTEC, 21 February 2012).

Elected PAOs are still relatively new and, as noted above, they command only modest resources. If we ask ourselves why a provincial politician motivated by the desire to retain office would devote efforts to luring (foreign) electronics firms to invest in the province, several possibilities suggest themselves. Investments would create jobs and would boost revenue to some degree, both of which might redound to the politician's electoral fortunes. New business investments could contribute to the province's prosperity. Such a diffuse and long term achievement might not have great impact on elections, but politicians might be motivated to serve their vision of the general good. We do not see indications at this time that such factors are swaying the behaviour of provincial level politicians.

Provincial politicians do not face expectations that they will promote investment, and if they attracted a substantial investment, it is not clear that they would benefit much. Furthermore, opportunities to gain rents might be fewer in working with large, internationally competitive firms than with some alternative lines of business.

As for still powerful provincial governors and other agents of central government ministries, they are, in essence, officials of line agencies. They implement policies rather than crafting visions or broad strategies. They are rotated frequently and therefore are not likely to develop deep roots in any particular locality. They have no career incentives for taking on developmentally activist roles.

The geographical location of electronics producers seems to be shaped primarily by the following factors. First, perhaps, is the influence of the location decisions of other firms in the industry. Many of these tend to be located either in industrial estates that are relatively close to the major assemblers. Some of the pioneers in the electronics industry such as Minebea, in Ayutthaya, probably shaped the spatial distribution of firms in the industry. Seagate's choices could have been influenced by the Board of Investment's zoning system. For decades, the Board of Investment has been using a system of graded fiscal incentives to encourage firms to locate in outlying and less developed areas of the country. Firms also of course will be influenced by other factors such as the relative cost and efficiency of, as well as access to, transportation links and supplies of critical inputs such as water or skilled labour. A particular location may serve well as a distribution centre to serve markets within Thailand. All of these factors, for example, make the provinces immediately north of Bangkok attractive (or at least they did until the floods of 2011). Access to Thailand's neighbours' markets is important for some investors and hence tends to favour some locations in Thailand over others. However, such concerns are not likely to be important to firms in the HDD industry.

ARE THE GOOD TIMES JUST ABOUT OVER?

Given some of the weaknesses in labour skills, state and political institutions, and local firms' technological capacities discussed above, is Thailand likely to be vulnerable to the middle income trap? Richard Doner argues that the barriers to participation in global production networks are rising. Together with Thailand's weaknesses — particularly high dependence on foreign firms' technologies — these rising barriers suggest that Thailand may find an exit from its current middle income status elusive (Doner 2011). Thailand meanwhile confronts for the first time high levels of sustained political mobilization with only weak political parties to manage it. And sharp changes in its demographic profile in coming decades will add further strains as will heightened competition from neighbours with the implementation of the ASEAN Economic Community agreements in 2015.

Alternatively, might Thailand continue to enjoy the benefits of living in a good neighbourhood? It was, in substantial part, Thailand's proximity to Singapore that led to its becoming a major producer within the HDD global value chain. Athukorala and Kohpaiboon suggest that ASEAN as a whole now enjoys major agglomeration advantages. ASEAN now accounts for about one-fifth of world trade in electrical machinery and semiconductors (Athukorala and Kohpaiboon 2010). Cambodia, Laos and Vietnam, and now perhaps Myanmar too, will continue to grow rapidly and Thailand is well positioned to benefit from their economic expansion. This suggests that geography may continue to work in Thailand's favour.

The sub-national focus reflected in this volume is compelling, but we conclude that it is not yet important in understanding the electronics industry in Thailand. We have described the modest roles of the Thai government in general, particularly until more recently, in promoting the electronics industry. We noted the absence of any significant roles at the provincial level in promoting the industry. We sketched factors that help to explain the limited roles of local Thai governments in boosting the industry. We conclude our discussion, however, by suggesting that regional factors will be increasingly important in Thailand's political economy in the relatively near future.

While essential in most countries, a sub-national focus on Thailand's economic institutions remains premature. We believe, however, that such a focus will be more relevant in coming decades. Business people and government officials, and sometimes politicians, talk increasingly frequently of Thailand's economic geography. They refer, for example, to the Greater Mekong region and the new major infrastructure investments being made, mostly by international financial institutions and the Chinese, to integrate it. Such discussions point to new opportunities being created by enhanced access to neighbouring markets. Similar points are made about business opportunities in Yunnan Province and, most recently, to the west in Myanmar.

Another factor that may induce more attention to the geographic distribution of investment within Thailand is the rising force of regionalism in Thai politics. The current governing party, for example, has its strongest electoral bases in the north and northeast. The leading party of the prior government fared best in Bangkok and the south. No doubt in response to these political realities, the Cabinet in 2012 was making plans to establish the Industrial Estate Authority of Thailand's first estate in the northeast. Political discussion in Thailand also has focused more on income and wealth gaps than in the past. Given the regional concentration of poverty in the country, that too could encourage more attention to issues of economic geography.

It remains likely, however, that new sensitivity to these issues will be exploited by national level rather than provincial level politicians. More activist and developmental PAOs probably will require changes in relations between central and provincial and local governments, as well as changes in Thailand's political parties and the bases of political competition. In the meantime, the maintenance and expansion of clusters in the electronics industry will be shaped largely by foreign multinational firms and the central government.

Notes

1 A former Minister of Industry referred to one of his then officials as "a freak" (as in an unusual exception) because of his extensive technological training and commitment to supporting the textile industry's development (Unger 1998).

[2] These trends refer to both the electrical and electronics industries.

[3] It also notes that a McKinsey 2002 study, using workers in the U.S. electronics industry as a benchmark of 100, found worker productivity in Singapore of 72 but only eight in Thailand (UNCTAD 2005, p. 20).

[4] Thailand's modest endowments of skilled technical and scientific workers have on at least one occasion worked to its advantage. One television producer decided to locate in Thailand rather than China due to the belief that problems of counterfeit production would be less troublesome in Thailand (Kohpaiboon, 2011).

[5] The "nonmarket-system of interfirm networking and linkages, ways of doing business, and the web of supporting institutions".

[6] The Club is part of Thailand's peak industrial organization, the Federation of Thai Industries. The Club is dominated by appliance producers and does not represent electronic industry concerns energetically (interview with official A at the Ministry of Industry, 2 June 2011).

[7] A UNCTAD report offers a more complimentary picture of the Board of Investment (UNCTAD, 2005/6). The difference in appraisals of the BoI's performance may reflect UNCTAD's need to be diplomatic, but probably also reflects changes in Thai government promotion activities over the intervening years.

[8] Kondo (2008), McHendrick et al. (2000), and Intarakumnerd (2011) concur in the view that U.S. firms were central to these developments. It is possible, however, that this finding reflects the access these researchers had, and did not have.

References

Afzulpurkar, Nitin and Peter Brimble. "Final Report. Building a World-Class Industry: Strengthening the Hard Disk Drive Cluster in Thailand. A Blueprint from Industry/Government/Academia". Report prepared for the National Science and Technology Development Agency, September 2004.

Athukorala, Prema-chandra and Archanun Kohpaiboon. "Intra-Regional Trade in East Asia: The Decoupling Fallacy, Crisis, and Policy Challenges". Working Papers in Trade and Development, No. 2009/09. Canberra: Australian National University, August 2009.

———. "East Asia in World Trade: The Decoupling Fallacy, Crisis, and Policy Challenges". Arndt-Corden Division of Economics Working Paper, No. 52. Canberra: Australian National University, 2010.

———. "Production Networks and Trade Patterns in East Asia: Regionalization or Globalization?" Asian Development Bank Working Paper Series on Regional Economic Integration, No. 56, August 2010.

Board of Investment. "Thailand, The World's Electrical and Electronics Investment Destination". Bangkok: Board of Investment, 2010.

———. "Why Thailand?" Bangkok: Board of Investment, 2011.

Bowie Alasdair and Unger, Danny. *The Politics of Open Economies*. New York: Cambridge University Press, 1997.

Commission on Growth and Development. *The Growth Report. Strategies for Sustained Growth and Inclusive Development*. Washington, D.C.: The World Bank, 2008.

Danish Trade Council. "Sector Overview, The Electronic Industry in Thailand". Bangkok: Royal Danish Embassy, 2006.

Dodgson, Mark. "Policies for Science, Technology, and Innovation in Asian Newly Industrializing Economies". In *Technology, Learning, and Innovation. Experiences of Newly Industrializing Economies*, edited by Linsu Kim and Richard R. Nelson. New York: Cambridge University Press, 2000.

Doner, Richard. "Politics, Institutions and Performance: Explaining Growth Variation in East Asia". Unpublished manuscript. 2011.

Doner, Richard F., Bryan Ritchie, and Dan Slater. "Systemic Vulnerability and the Origins of Developmental States: Northeast and Southeast Asia in Comparative Perspective". *International Organization* 59 (Spring 2005): 327–61.

Electrical and Electronics Institute. "Status of Electrical and Electronics Industries". Bangkok: Electrical and Electronics Institute, 2011. (In Thai)

Hobday, Michael. "East versus Southeast Asian Innovation Systems: Comparing OME- and TNC-led Growth in Electronics". In *Technology, Learning, and Innovation. Experiences of Newly Industrializing Economies*, edited by Linsu Kim and Richard R. Nelson. New York: Cambridge University Press, 2000.

Intarakumnerd, Patarapong. "The Roles of Intermediaries in Clusters: The Thai Experience in High-tech and Community-based Clusters". *Asian Journal of Technology Innovation* 13, no. 2 (2005): 23–43.

———. "Thaksin's Legacy: Thaksinomics and Its Impact on Thailand's National Innovation System and Industrial Upgrading". *Institutions and Economics* 3, no. 1 (2011): 31–60.

Jitsuchon, Somchai and Chalongphob Sussangkarn. "Thailand's Growth Rebalancing". ADBI Working Paper Series, No. 154. Tokyo: Asian Development Bank Institute, October 2009.

Johnson, Chalmers. *MITI and the Japanese Economic Miracle*. Stanford: Stanford University Press, 1982.

Jongwanich, Juthathip and Archanun Kohpaiboon. "Export Performance, Foreign Ownership, and Trade Policy Regime: Evidence from Thai Manufacturing". ADB Economics Working Paper Series No. 140. Manila: Asian Development Bank, 2008.

Kohpaiboon, Archanun. *Multinational Enterprises and Industrial Transformation. Evidence from Thailand*. Cheltenham, UK: Edward Elgar, 2006.

———. *The Production Fragmentation Phenomenon*. Mimeographed. Bangkok, 2011. (In Thai)

Kondo, Masayuki. "Triple Helix Collaboration in High-Tech Industries of Developing Countries: Case of Thai Hard Drive Industry". The Seventeenth International Conference on Management of Technology. Dubai, 6–10 April 2008 Proceedings.

Lall, Sanjaya. "Technological Change and Industrialization in the Asian Newly Industrializing Economies: Achievements and Challenges". In *Technology, Learning, and Innovation. Experiences of Newly Industrializing Economies*, edited by Linsu Kim and Richard R. Nelson. New York: Cambridge University Press, 2000.

Lyne, Mona M. "Rethinking Economics and Institutions: the Voter's Dilemma and Democratic Accountability". In *Patrons, Clients, and Policies. Patterns of Democratic Accountability and Political Competition*, edited by Herbert Kitschelt and Steven I. Wilkinson. New York: Cambridge University Press, 2007.

Mahakanjana, Chandra. "Municipal Government and the Role of Cooperative Community Groups in Thailand". In *Local Organizations and Urban Governance in East and Southeast Asia: Straddling State and Society*, edited by Benjamin Read and Robert Pekkanen. London: Routledge, 2009.

McHendrick, David G., Richard F. Doner, and Stephan Haggard. *From Silicon Valley to Singapore. Location and Competitive Advantage in the Hard Disk Drive Industry*. Stanford, California: Stanford University Press, 2000.

Mektrairat, Nakharin. "Report on Development of Decentralization in Thailand". Bangkok: United Nations Development Program. 2007. (In Thai)

Ministry of Industry. "The Electronics Industry: Today and in the Future". PowerPoint presentation, given on 16 February 2011. (In Thai)

Office of Industrial Economics, and Electrical and Electronics Institute. "Status Report on Industrial Business, Electrical and Electronics Sectors". Bangkok: Ministry of Industry of Thailand, May 2011. (In Thai)

Otsuka Keijiro. "Cluster-Based Industrial Development: A View from East Asia". *The Japanese Economic Review* 57, no. 3 (2006): 361–76.

Slater, Dan. *Ordering Power. Contentious Politics and Authoritarian Leviathans in Southeast Asia*. New York: Cambridge University Press, 2010.

Suehiro Akira. *Capital Accumulation in Thailand 1855–1985*. Tokyo: Center for East Asian Cultural Studies, 1989.

Svasti, Pongsvas. Public talk delivered at the Bangkok Post Forum 2012 "Roadmap to Recovery", 23 February 2012.

Tilly, Charles. *Coercion, Capital and European States: A.D. 990 to 1992*. Oxford: Blackwell Publishers, 1992.

Trakulhoon, Kan. Public talk delivered at the Bangkok Post Forum 2012 "Roadmap to Recovery", 23 February 2012.

UNCTAD. "A Case Study of the Electronics Industry in Thailand". Transfer of Technology for Successful Integration in the Global Economy Series. UNCTAD/ITE/IPC/2005/6. New York and Geneva: United Nations, 2005.

Unger, Danny. *Building Social Capital in Thailand.* New York: Cambridge University Press, 1998.

Webster, Douglas. "Supporting Sustainable Development in Thailand: A Geographic Clusters Approach". Bangkok: National Economic and Social Development Board/World Bank, 2006.

Wongsamuth, Nanchanok. "Businesses Press for Flood Plans". *Bangkok Post,* 27 February 2012, p. 3.

World Economic Forum. *Global Competitiveness Report, 2011–2012.* Geneva: World Economic Forum, 2012.

Interviews

Acting President, Provincial Administrative Organization, Ayutthaya, 23 March 2012.

Governor, Pathum Thani Province, 15 March 2012.

Official A in the Ministry of Industry, 2 June 2011.

Official B in the Ministry of Industry, 13 July 2011.

Official in the Ministry of Industry, Saraburi Province, 15 March 2012.

———, Pathum Thani Province, 15 March 2012.

Official in the Ministry of Commerce, Saraburi Province, 15 March 2012.

Intarakumnerd Patarapong, Advisor to the Science and Technology Agency, 1 March 2012.

Srisomphob Jitpiromsri, Deputy Dean, Faculty of Political Science, Prince of Songkhla University, 9 June 2010.

Archanun Kohpaiboon, Faculty of Economics, Thammasat University, 8 July 2011.

Amporn Poyai, Principal Researcher, National Electronic and Computer Technology Centre, 21 February 2012.

Chumnarn Punyasai, Senior Researcher, National Electronic and Computer Technology Centre, 21 February 2012.

CHINA
Hanoi
VIETNAM
LAOS
Da Nang
DA NANG
THAILAND
CAMBODIA
Gulf of Thailand
South China Sea
LEGEND
Provincial/Municipal Capital
PROVINCE/MUNICIPALITY
National Capital
COUNTRY
NEIGHBOURING COUNTRY
Geographic Features
National Border Land
National Border Water
State/Province Border
0 km
150
300
All maps are derived from GADM Data
(Global Administrative Areas, www.gadm.org, 2012)
edited by maps&more, Hans Hortig / Marcel Jäggi

6

REGIONAL ECONOMIC DEVELOPMENT AND PERSPECTIVES FOR THE ELECTRONICS SECTOR IN VIETNAM: THE CASE OF DA NANG

Tran Ngoc Ca

INTRODUCTION

Industrial development in Vietnam is concentrated in the two main economic centres of the country, namely Hanoi, the nation's capital in the north, and Ho Chi Minh City, in the south. However, over the last 15 years, Da Nang, situated roughly in the middle of the country, has begun to emerge as an economic hub. A major port city and long a cultural centre, Da Nang has received significant amounts of investment in infrastructure, real estate, and tourism, and its industrial sector has begun to develop rapidly.

This chapter will examine the case of Da Nang and the challenges it has faced in its quest to become an important industrial centre. Following this introduction, the second section will examine how national-level policies regarding foreign direct investment have affected the distribution of industrial activity across the country. The third section will look at the recent development of Da Nang's industrial sector with specific reference to electronics. From there, the chapter will examine the case of Da Nang municipality, seeking to understand how local government authorities have sought to promote its industrialization. The fourth section will conclude.

BACKGROUND: REGIONAL INDUSTRIAL DEVELOPMENT AND FDI IN VIETNAM

In 1986, the Vietnamese government implemented a series of sweeping reforms to move from a planned economy to one more open to market forces. In 1987, the Foreign Direct Investment (FDI) Law was approved by the National Assembly. This policy change opened up Vietnam to investment and spurred its industrialization process. This was taken further in 2005, with the introduction of the Investment Law that further facilitated foreign investment. And, with Vietnam's accession to the World Trade Organization in 2007, registered FDI increased even more rapidly. Thus, over the period 1988–2008, Vietnam attracted around 11,000 FDI projects with a total registered capital of US$164 billion (GSO 2008).

Presently, around 100 multinational companies have invested in Vietnam. Nearly 70 per cent of FDI is from Asia and over 60 per cent has been in the industrial sector. Overall, FDI in Vietnam's economy plays an important role in terms of contribution to GDP growth, total investment, and job creation. In 2008, the foreign-invested (FIS) sector contributed approximately one-fifth of GDP (Figure 6.1). FIS has contributed to the country's

Figure 6.1
Contribution of FIS to GDP, 1995–2008
(at 1994 constant prices)

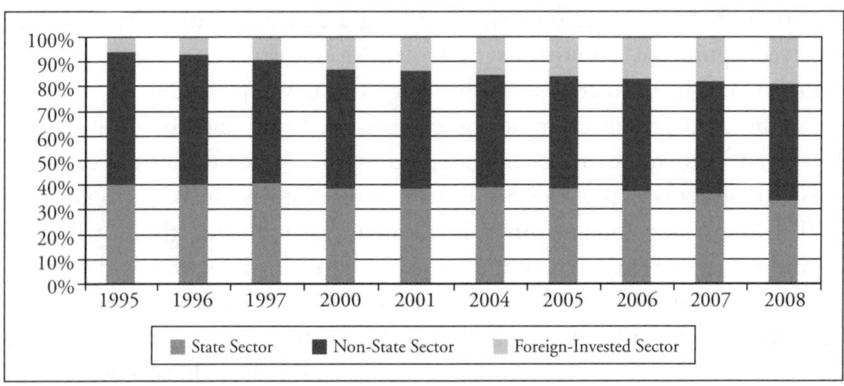

Source: GSO 2008.

structural change toward industrialization, as it has been dominant in high value-added industries such as computers, electronic products and spare parts, and steel. In provinces such as Vinh Phuc, Dong Nai, and Binh Duong, this sector made up 65 per cent to 70 per cent of the total industrial output.

In 2008, FDI projects in the industrial sector accounted for about 55 per cent of total invested and registered capital. Concerning the structure of registered capital by sector, the figures show the largest share going to manufacturing (73.8 per cent for 2003 and 45.2 per cent for 2008), real estate and renting business (9.7 per cent and 37 per cent), and mining industries (2 per cent and 10.7 per cent respectively).

Although the heavy and light industries constitute the largest share in total registered projects and capital, implemented projects in these industries remained low, representing only 30 per cent in 1988–2007 (Figure 6.2). This is a large contrast to the higher implementation rates of projects in other less technically-intense sectors such as food processing and construction. Thus, as regards industry, it would appear that despite an increasingly conducive policy environment for investment, it is difficult to implement projects in this sector due to poor infrastructure and a lack of skilled labour. Overall, the flow of FDI is still focused on heavy and capital-intensive sectors as a result of Vietnam's industrial policies, which prioritize the development of heavy industry. More specific allocation of FDI in economic sectors is provided in Figure 6.3.

Figure 6.2
Share of FDI by Industrial Sectors 2005 and 2008

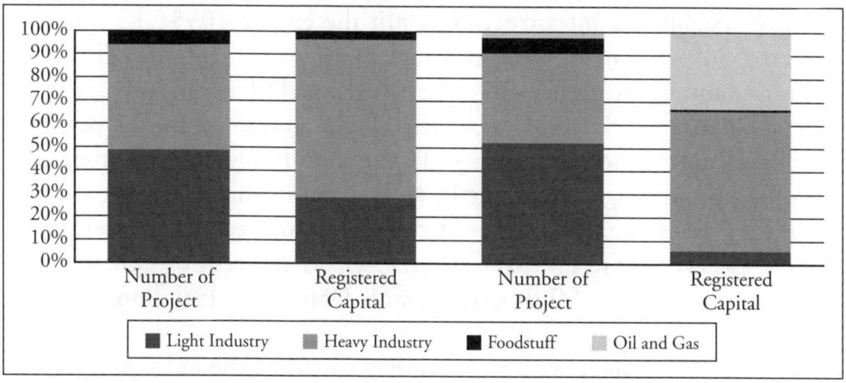

Source: Foreign Investment Agency 2009.

Figure 6.3
FDI Disbursement Ratio by Sectors, 1988–2007

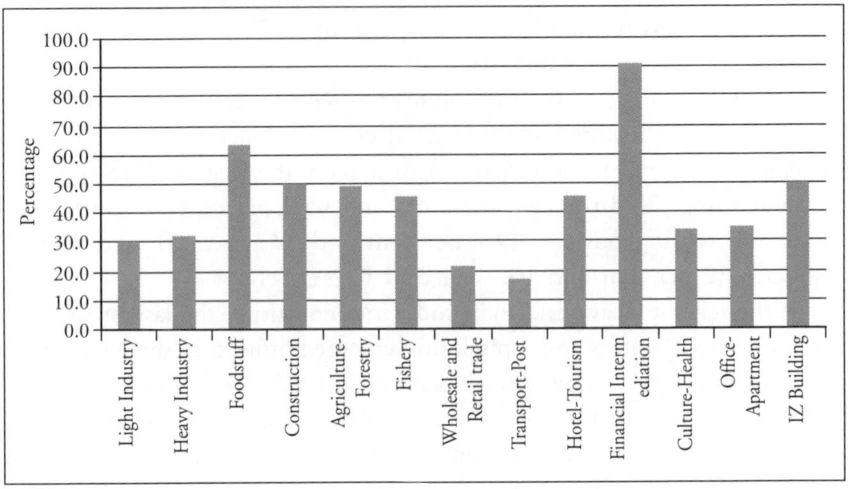

Source: Foreign Investment Agency 2008.

In the past 5 years, FDI has flowed more into export-oriented sectors, which are largely labour-intensive sectors such as garment, footwear, and electronic parts due to the Bilateral Trade Agreement with the U.S., and Vietnam's WTO commitments on trade and investment. Overall, the foreign-invested sector has continued to increase its share in the total turnover of exported goods.

Geographically, FDI is concentrated mostly in the southeast, near Ho Chi Minh City, and the Red River Delta, near Hanoi, which have the most established infrastructure. Until the end of 1995, the southeast attracted 56 per cent of total registered FDI, with Ho Chi Minh City alone accounting for 35 per cent. The Red River Delta comprised 28.5 per cent of total registered FDI, of which Hanoi accounted for 21 per cent. This trend did not seem to change despite the introduction of investment incentives promoting investment in regions with difficult socio-economic conditions. From the perspectives of cities and provinces, the concentration of FDI was even more unbalanced. In 1995, 79 per cent of FDI was only in five cities: Ho Chi Minh City, Hanoi, Dong Nai, Haiphong, and Ba Ria-Vung Tau.

Prior to 1995, there were no incentives regarding the location of investments. However, amendments to the Law on Foreign Investment in 1996 introduced many incentives that encouraged FDI inflows into areas with difficult socio-economic conditions and reduced the concentration of FDI in others. Some specific incentives included the reduction of land rent or exemption from corporate income tax. Since the Investment Law of 2005 took effect, incentive policies by geographic areas have been applicable equally to both foreign and domestic investors. Under this law, there are investment incentives that are being applied in 54 areas with difficult socio-economic conditions and severely difficult socio-economic conditions in Vietnam. But this incentive system is not specifically designed for any single province or city, but is applicable to all. In some more specific cases, different cities or provinces may provide some extra incentives applicable to activities in their location. The establishment of economic zones, industrial zones, export processing zones, and high-tech zones in many provinces also has influenced the allocation of FDI. The Investment Law stipulates that all projects investing in economic zones and high-tech zones shall benefit from the incentives equal to that of areas with severely difficult socio-economic conditions. In addition, investors investing in these zones will be provided with physical infrastructure and business services of good quality.

In terms of sectors, prior to 1996, policies for attracting FDI mainly focused on import substitution or export-oriented industries. This led to a growing imbalance of FDI allocation, which largely centred on heavy and capital-intensive industries. Since 1996, incentive policies have been in line with the national development strategy, paying more attention to the positive spillover effects of FDI. In addition, policy changes have had a bigger impact on the technology level of foreign-invested sectors. This is reflected in investment pledges of more than 100 MNCs, including high-tech giants such as Intel, Panasonic, Canon, Samsung, Nokia in high-tech fields such as construction materials production, ICT, oil refining technology and so on. Their investment pledges are expected to accelerate the transfer of advanced technology into Vietnam.

From 1997 to 2004, new attractive destinations for FDI such as Vinh Phuc, Bac Ninh, Ha Tay, Long An, and Da Nang have emerged. However, the South-East has continued its lead. In 2004, Ho Chi Minh City, Hanoi, Binh Duong, and Dong Nai accounted for 65.5 per cent of total projects and 61.7 per cent of registered capital. In 2008, the top 11 cities and provinces that have more than US$1 billion of registered FDI made up 85.8 per cent of newly registered and additional capital, of which 56.8 per cent belongs to five southeastern cities and provinces: Ho Chi Minh City, Ninh Thuan, Ba Ria-Vung Tau, Dong Nai, and Binh Duong (GSO 2008).

THE CASE OF DA NANG: ITS INDUSTRIAL DEVELOPMENT

With a population of around 820,000 inhabitants, Da Nang is the largest city in the central region of Vietnam and has historically been regarded as an important centre for economic, social, science and technology development.

Da Nang is among the top six richest provinces or municipalities in Vietnam with a GDP per capita (PPP) of US$3,700, compared to the national average of US$2,840. While high, this is significantly lower than either Ho Chi Minh City (US$4,834) or Hanoi (US$4,343). In addition to a high income level, the city also scores well in terms of education and health indicators. According to UNDP's Human Development Report, in 2008 Da Nang was the fourth most highly-ranked province or municipality in Vietnam, behind Ba Ria-Vung Tau, Ho Chi Minh City, and Hanoi. While Da Nang's GDP per capita is somewhat below the other three cities, it scores particularly well on education and literacy (UNDP 2011).

While Da Nang has sought to compete with Hanoi and Ho Chi Minh City for investment, its lower population base means that it has a smaller domestic market, a limited array of supporting services, and a relatively under-developed firm base. However, it does have lower labour costs, cheaper land — particularly in relation to Ho Chi Minh City — and relatively well-developed infrastructure in the way of an international airport, seaport, and extensive road connections (Dapice 2003).

In recent years, the city's rapid economic growth has consolidated its importance. With an annual growth rate of 11.6 per cent since 1997, Da Nang has grown faster than the national average — even outpacing Ho Chi Minh City and Hanoi. While growth rates were high in the immediate post-1997 period, they climbed even higher following the enaction of the Enterprise Law in 2005 (Table 6.1).

Da Nang's economy has evolved in tandem, with its industrial sector expanding significantly. From constituting 35 per cent of GDP in 1997, the industrial sector expanded to 50 per cent by 2006. In contrast, its agricultural sector contracted from 10 per cent to 4 per cent over the same period, with the services sector similarly shrinking from 55 per cent to 46 per cent. While Vietnam's economy has witnessed similar aggregate trends over the same period, Da Nang has substantially larger industrial and services sectors relative to the national average (Table 6.2).

To date, FDI into Da Nang has largely been concentrated in the real estate sector, which accounts for 72 per cent of the total. However, the

Table 6.1
GDP Growth Rate of Cities, On Average
(in %)

	1997–2000	2001–06	1997–2006
National	6.4	9.3	7.1
Da Nang	***10.2***	***12.5***	***11.6***
Ho Chi Minh	9.0	13.6	10.3
Hanoi	10.2	13.8	10.8
Hai Phong	9.2	13.8	10.5
Binh Duong	14.2	18.5	14.8

Source: Statistical Yearbook 1996–2006.

Table 6.2
Structure of Vietnam and Da Nang's Economy
(%)

Years	Total	Agro-Forestry-Seafood	Industry-Construction	Services
		Vietnam		
1997	100	28.1	31.8	40.1
2000	100	24.5	36.7	38.7
2006	100	20.4	41.5	39.0
		Da Nang		
1997	100	9.7	35.2	55.1
2000	100	7.9	40.6	51.6
2006	100	4.2	50	45.9

Source: Statistical Yearbook 1997–2006; Dinh 2010.

processing and manufacturing sector is the second largest recipient of investment, receiving US$382 million in investment, accounting for 12 per cent of the total (Danang IPC 2011).

Foreign investment in the manufacturing sector is concentrated mainly in the garment, textiles, agricultural processing, and electronics sectors.

Table 6.3
Exports of Manufactured Commodities from Da Nang
(USD million)

	2002	2005	2010
Garments and Textiles	52	80	155
Leather Goods	28	55	106
Car parts, mechanical, electrical, electronic, and high-tech	0.4	15	200
Home appliances	3.8	11	45
Plastic Items	10.2	25	40

Source: Le n.d.

Da Nang has six industrial parks with more than 200 firms that employ 50,000 workers. Fifty-seven firms are funded by overseas investors and 157 operate with domestic investment (DOST 2010).

The city's manufacturing profile is evolving rapidly. While all sub-sectors are expanding quickly, the manufactured parts subsector — comprising car parts, mechanical, electrical, electronic, and high-tech components — has expanded the fastest. From generating a mere US$400,000 in exports in 2002, in 2010 it accounted for US$200 million, surpassing more established activities such as textiles and garments and leather goods. The manufactured parts sub-sector now accounts for the same quantity of exports as the largest agricultural processing activity, dried and frozen aquatic products.

In 2011, South Korea, Japan, and Singaporean firms were the largest investors in the city (Danang IPC 2011). Japan has historically been an important participant in the electronics sector in the city, particularly for the production of motors and transformers through investments from firms such as: TOKO (Electronics Viet Hoa); TKR Corporation; and Seto Electronics. In particular, Mabuchi Motors, employs 6,000 people and invested some US$40 million in its facility there (Mabuchi 2005).

The bulk of manufacturing is carried out by foreign concerns. The local private sector in Da Nang is small, but growing. The 2005 Enterprises Law and progress in improving the business environment have led to greater enterprise creation. In 1999, there were 630 private enterprises, and in 2003, this was up to 2,756 firms (Tran 2011).

Despite these achievements, the private sector is still not very competitive, due to outdated technology, high production costs, uneven production quality and a poor assortment of products. Most firms are small, with around one-third of them having capital of less than US$5,000. The main sectors of activity are consumer products, seafood, textile/garments, trading, and tourism. There are relatively few working in technology-intensive sectors such as ICT or electronics (Le 2005).

In addition, local firms have difficulties accessing capital due to a large state sector that competes with the private sector. This is particularly an issue in Da Nang, as according to the Vietnam Provincial Competitiveness Index of 2005, in that year it had the highest proportion of capital being lent to state-owned enterprises (84 per cent) (Malesky 2005).

Da Nang has a branch of the Vietnam Chamber of Commerce and Industry (VCCI), the country's premier business association, in the city since 1989. Catering to members from Da Nang and ten surrounding provinces, the branch has some 1,000 members. VCCI seeks to act as an intermediary between the government and the business community. Its remit also encompasses: representing employers in tri-partite negotiations with the government and trade unions; providing training activities, particularly for SMEs; and carrying out trade promotion activities (VCCI 2012).

DA NANG AND ITS MUNICIPAL GOVERNMENT

Vietnam has a multi-levelled government structure. At the apex is the central government with the Prime Minister at its head. In addition to a collection of ministries with functional tasks such as health and education, there are 58 provincial and 5 city governments (or municipalities) that are under the direct control of the Prime Minister.[1] Each province or municipality is governed by a People's Council (*Hoi Dong Nhan Dan*), whose members are elected every five years by residents. The Head of the People's Council is the Chairman. There is also a People's Committee (*Uy Ban Nhan Dan*), which is selected by the People's Council in coordination with the central government and headed by a Chairman, who is at the same time a member of the People's Council. This Chairman (which is equivalent to a Mayor) is in charge of the day-to-day management of each province or municipality. Depending on the importance of the province, one or both of the Chairmen of the People's Council and Committee are members of the Party Central Committee.

Over time, the number of provinces and municipalities has risen as a result of Vietnam's growing population as well as periodic reorganizations. The number of provinces and municipalities has increased from 38 in 1978 to 63 in 2011.

Along with provincial governments, municipalities are tasked with revenue collection. People's Councils are responsible for deciding taxation rates, after consulting centrally-drafted plans. Depending on their level of income, they are allowed to keep a different proportion of their taxes according to a centrally-defined formula.

A number of changes in recent years have meant that now, more than ever before, provincial and municipal governments are important players in economic development, particularly as it relates to foreign direct investment. Up until the early 1990s, central control meant that provinces were precluded from acting on their own to attract large amounts of FDI. Ultimate approval of any FDI project was taken by the central State Committee for Cooperation and Investment, and provinces had to lobby investors to request the Committee to grant approval for particular projects to be sited in particular locations. During this time, Ho Chi Minh City, Hanoi, and Dong Nai were the prime beneficiaries of investment approvals (Malesky 2004).

However, during the 1990s, key legislation changed. Of particular importance was a decree in 1996, which stipulated that FDI decisions were to be taken by the provincial level People's Committees. In addition, provincial and cities were allowed to sign FDI agreements themselves, rather than through the central government. At present, provincial and municipal governments are allowed to approve projects in particular non-strategic categories and below a certain threshold (Malesky 2004).

Figure 6.4
FDI Allocation by Region, 2008

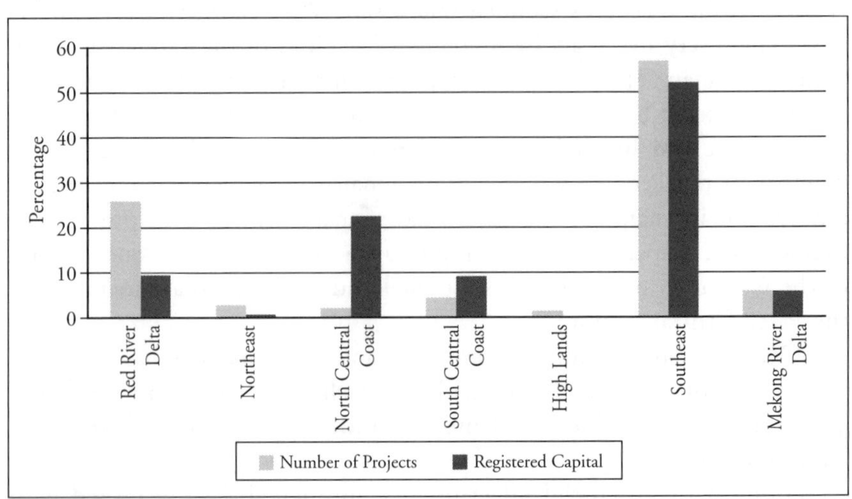

Source: Foreign Investment Agency 2008.

This has resulted in competition between provinces for investment, although within a context decided by the national government. While more open and proactive than in the past, the overall national approach has the following characteristics: national policies to attract FDI with technology content and technology transfer have focussed mainly on financial incentives; and policies to promote innovation activities such as the establishment of science and technology development funds have only recently been implemented. As a result, supporting industries are still underdeveloped. Thus, both domestic and foreign enterprises have to depend on imported components for production. Furthermore, the low qualification of the labour force constrains processes of technology absorption and innovation.

Despite the above, provincial and municipal governments can play an important role in creating an environment conducive to business. This includes providing cheap land and incentives such as reduced taxes, but also can encompass: streamlining and expediting investment approval processes; investing in skills training; establishing dialogues with the private sector; and also seeking to matchmake between foreign investors and local entrepreneurs.

Despite its economic importance, Da Nang has historically not enjoyed a significant level of autonomy or financial independence. Unlike Hanoi and Ho Chi Minh City, that were already municipalities prior to 1997, Da Nang was a city under the authority of the Quang Nam-Da Nang provincial government. As a result, its finances were subsumed under the provincial government. Thus, Quang Nam-Da Nang's contribution to the government budget was VND1.2 billion (US$60 million) per year. Of this, only 45 per cent was paid back to the province, which was then divided among all 16 administrative units (Nguyen 2003).

In 1997, the central government decided to elevate Da Nang city to the status of a municipality under the direct supervision of the prime minister. In addition to Da Nang, four other cities — Hanoi, Ho Chi Minh City, Can Tho, and Hai Phong — have this status, which is equivalent to a province.

This decision substantially changed the amount of funds available for investment by the municipal government. Figure 6.5 shows investment by the central and municipal governments in Da Nang over the 1995–2002 period. As can be seen, investment by the municipal government almost doubled from 1996 to 1997, and by 2000 represents almost four times the amount invested by the central government. Following a contraction in 2001, in 2002, spending by the municipal government is again substantially higher than central government spending.

Figure 6.5
Investment in Da Nang by Source, 1995–2002

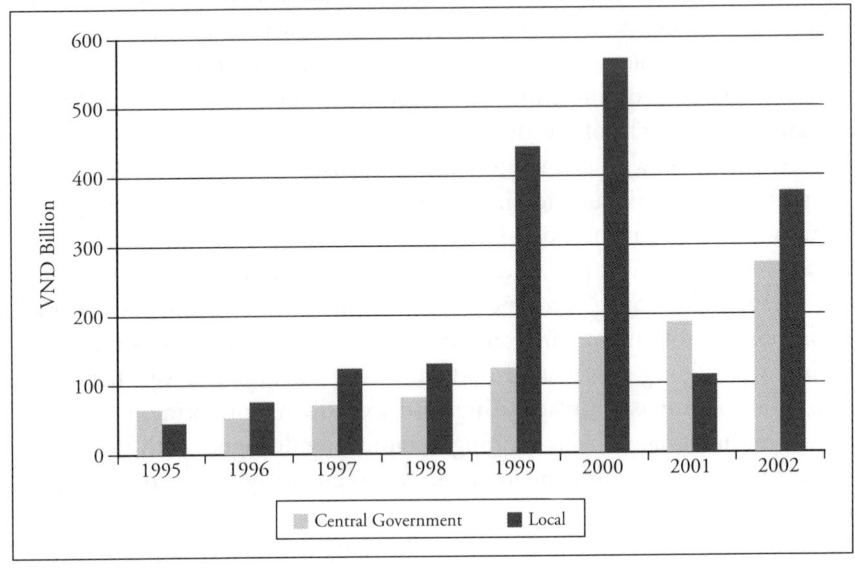

Source: Nguyen 2003.

In many municipalities and provinces, while Mayors and Governors occupy the highest administrative position, they are usually subordinate to the Party Chief in the local branch of the Communist Party. However, in Da Nang's case, the first Mayor, Nguyen Ba Thanh, was both Mayor and Party Chief.

Mr Thanh was first Mayor (Chairman of the People's Committee) until 2004, when he became Party Chief and left the Mayorship to Mr Tran Van Minh. But in the admistrative structure of all cities and provinces, the position of Chairman of the People's Council are usually held by the Party Chief. In Da Nang's case, after leaving the Mayorship, Mr Nguyen Ba Thanh became Party Chief and Chairman of the People Council. As such, he still retains the highest leadership position in the city.

This unusual juxtaposition of administrative and political authority enabled Nguyen to move quickly and implement reforms swiftly. This was bolstered by close ties with the prime minister and deputy prime minister.

During Nguyen's tenure, the key axes of municipal government action were as follows: providing hard and soft infrastructure; ensuring an enabling environment for business; and implementing market-complenting measures.

Hard and Soft Infrastructure

During the first few years of the Nguyen Ba Thanh administration, investment on infrastructure was increased six-fold. Explicitly targeting Hanoi and Ho Chi Minh as comparators, the municipal government undertook to build a new bridge and port, upgrade the airport and existing port, and establish three industrial parks. In addition, some 60 roads were built and a significant number of urban upgrading projects were undertaken.

This drive was financed through a variety of means. First, its new status entailed significantly more revenue that could be retained by the municipal government. Second, additional central government funding was made available for infrastructure projects as, over the 1998–2000 period, the central government transferred VND120 billion to Da Nang. Third, changes in municipal legislation also entailed higher revenue flows. Through adjustments in taxation on land more than VND7,600 billion was earned, which accounted for one third of the municipality's budget, enabling its development budget to more than double (Tran 2007).

This additional funding also enabled a variety of tax incentives to be offered to investors. This included an exemption from corporate income tax for the first two years of operation, and 50 per cent for the following two years. Additional incentives were rolled out for more industrial activities, and import taxes on fixed assets or inputs for production were eliminated.

During 1997–2003, 30,000 households were re-located to make way for many of these infrastructure improvements. This was made possible, in large part, through the introduction of a local voting mechanism to expedite the clearance of land for infrastructure projects. If more than 80 per cent of affected households supported the clearance, the project was then implemented (Nguyen 2003).

Over time, Da Nang has built up a network of industrial parks. At present, the municipality has five parks compared to Hanoi's 11 and Ho Chi Minh City's 13 (Dinh 2010). In 2010, it also established a High-Tech Park for priority sectors such as: microelectronics; nano-technology; bio-technology; and environmental technology (VNS 2012). In October 2011, the Da Nang Municipality also concluded an agreement with a U.S.-based firm to develop an ICT park to cater to the city's growing BPO sector.

Rankings for industrial parks in Vietnam have found Da Nang's parks to be of higher quality than the national average (Dinh 2010). In addition, the municipal government has moved to ensure that the land is available for private investors, in some cases, as in the Hoa Khanh Industrial Park, re-titling land that had been given to state-owned enterprises to ensure that land would be available (Nguyen 2003). However, despite having a relatively small population base, Da Nang's land prices are quite high, even more than Hanoi's (although lower than Ho Chi Minh City). This may well be due to the municipality's smaller agricultural hinterland.

The Provincial Competitiveness Index is an annual ranking exercise carried out every year by a number of agencies, including the Vietnamese Chamber of Commerce and Industry. Calculated from the responses of foreign and local firms as well as a range of quantitative indicators, the Index looks at the local-level environment for business including aspects such as entry costs to setting up a business; access to land; access to information; and the proactivity of provincial leaders.

In 2011, according to the Index, Da Nang obtained the second-highest ranking of all provinces and municipalities for the quality of its physical infrastructure, including its industrial zones; roads; energy and telecommunications supply; and internet service (Malesky 2011).

Insofar as soft infrastructure is concerned, Da Nang has some ten universities and colleges, which collectively produce some 30,000 graduates a year. This includes the University of Da Nang; a specialized ICT university established by FPT, the largest ICT group in Vietnam; and an IT college. Da Nang municipality also has approximately 80 science and technology development programs, with various focuses such as ICT, new materials technology, biotechnology, and automation (DOST 2010).

To meet increasing needs for skilled labour, the municipal government has begun to cooperate with some foreign organizations such as APTECH (India) to train local residents. APTECH has some 1,600 centres in 33 countries in various areas of software development, and the Da Nang municipality is working with the firm to train some 1,000 programmers per year (DOST 2010).

In addition, there are several new international universities in the planning such as the Vietnam UK University to be located in the Da Nang High-Tech Park, and the American Pacific University, which will be founded with a US$100 million investment (VoV 2010).

Enabling Environment for Business

Despite the municipal government being created in 1997, by 2005 when the first Provincial Ranking exercise was carried out, Da Nang was among the highest performers. This has continued up until the present, with the municipality consistently ranking at or near the top. In addition to high scores for its physical infrastructure, the municipality consistently ranks highly insofar as its regulatory environment for businesses.

For three consecutive years (2008–10), the Da Nang municipal government obtained the highest ranking in the country. Key aspects of its competitive advantage include a good record of reaching out to the private sector, including holding workshops to brief businesses on new legislation; low time costs for regulatory compliance; good access to land; and a good track record of proactivity on the part of provincial leaders.

In 2011, Da Nang slipped to fifth place overall. However, this is still considered an "excellent" ranking, and one that is far higher than either Ho Chi Minh City (20th place) or Ha Noi (35th place). Of particular note, the Da Nang municipal government is noticeably more efficient at processing business registrations than either of the two big cities (Malesky 2011).

This is not to say that Da Nang does not face challenges in providing an enabling environment for business. It traditionally has had a large state-owned enterprise sector, which has had the effect of crowding out smaller firms vying for real estate and capital. Indeed, in 2002, a full 45 per cent of the municipality's GDP came from the state sector (Nguyen 2003). In addition, the municipality's business environment is characterized by quite high informal costs and long wait times to clear customs (Malesky 2011; PCI 2012).

The Da Nang municipal government has undertaken some recent initiatives that may have an impact on these rankings in the coming years. It has established an ICT Transaction Centre to facilitate its e-commerce and e-government activities. By moving more business processes onto e-platforms, the amount of informal charges levied on businesses should decline.

The municipality is also introducing a more focused investment policy, by specifying its interest in: (1) investment projects which have higher levels of value-added manufacturing; and (2) relatively advanced technology. The city is also not encouraging firms to lease land for use as warehouses, and has begun to deny licences to large-scale polluting industries (DOST 2010). These measures should limit the number of applications that need to be processed, free up land for more productive uses, and also limit negative externalities.

In the 2012 World Cities Summit in Singapore, out of the top ten Asia-Pacific "Cities of the Future", Da Nang ranked third, based on its scores across a range of categories, including business friendliness, economic potential, human resources, prices, quality of living, and strategy to attract FDI (*Da Nang Newspaper* 2012).

Market-Complementing Measures

In addition to providing the necessary infrastructure and an enabling environment for firms, the Da Nang municipality has implemented a number of market-complementing measures.

First, given the importance of Japanese investment to Da Nang, the municipal government opened its own representative office in Tokyo in 2004. The office carries out trade promotion activities for the municipality across Japan. In addition, in collaboration with the Japan External Trade Organization, the municipality is working with potential investors in Japan to secure needed human resources in Da Nang prior to project start-up. Through the representative office, investors from Japan have been able to cut the time to procure investment licenses from ten to three days (*Young People Newspaper* 2012).

Second, a number of support programs for local firms has been established. Over the past five years, the Da Nang Department of Planning and Investment has been carrying out a training programme with a small number of local small-scale firms to enable them to export. In addition, the Municipality is working with local firms to help them use IT in their every-day operations, in addition to working with Microsoft to support wi-fi provision (Thanh 2011).

That said, Da Nang does not do as well in the category of business support services in the Provincial Competitiveness Index as it does in the other areas. In 2012, the Municipality scored 3.72 out a maximum of 10, a score far below that of competitor cities. In comparison, Hanoi and Ho Chi Minh City scored 7.26 and 6.79 respectively (PCI 2012).

Thus, while Da Nang has been a leading performer in the provision of infrastructure and a good regulatory environment for business, it would appear that in relative terms, it has been less able to establish a range of business support services that help local and international firms. This, in addition to a large state-owned enterprise sector that competes with the private sector for resources and attention may mean that the local context is not conducive for the birth and development of a diverse firm base. As the competition between provinces deepens, and they begin to experiment with additional policy innovations that go beyond the provision of hard infrastructure and efficient business processes, this could emerge as an issue.

CONCLUSION

Da Nang's economy has grown rapidly and has industrialized quickly. For the past decade and a half, it has expanded at a rate higher than the national average, outpacing Vietnam's key economic centres of Ha Noi and Ho Chi Minh City. The industrial sector is now the municipality's most important, and the production of manufactured commodities now accounts for the bulk of its exports. In particular, manufactured parts and components, including electronics, now rival traditional staples such as dried and processed aquatic products for export income.

Key to this success has been Da Nang's drive to invite foreign direct investment. However, this strategy was only pursued following the central government's decision to upgrade the city's status to that of a municipality. This new status and autonomy was accompanied by additional financial wherewithal, by allowing the city to retain a higher proportion of the revenues it collected, as well as supplementary central government support. This was further bolstered by innovations in tax collecting, than enabled more revenues to be collected.

This, in turn, enabled a highly targeted drive to overhaul and upgrade the city's infrastructure — with the explicit goal of matching Ha Noi and Ho Chi Minh City — as well as a range of financial incentives to be offered to investors. Insofar as infrastructure is concerned, this target was met, with Da Nang rivalling the country's two commercial centres for the quality of its infrastructure.

This was accompanied by the provision of a high-quality regulatory environment for investors. Thus, low entry costs, high levels of transparency, and rapid approval of business registration and licenses made the municipality one of the easiest places in which to do business in Vietnam. Relative to Ha Noi and Ho Chi Minh City, Da Nang had a far more conducive environment for business.

However, looking forward, it would appear that the municipality needs to do more to groom its local private sector. Its small population base, tradition of state-owned enterprises, and relatively under-developed business support services mean that while foreign investors may choose to locate in Da Nang, they will find it hard to establish partnerships with local firms and/or procure supplies locally. Given the deepening competition between provinces in Vietnam, it is likely that more sub-national governments will be experimenting with ways to help their local firms increase their capabilities.

Note

1 Below this are districts, communes, and villages. For a more extensive discussion on this issue, Kerkvliet (2004).

References

Da Nang Newspaper. "Da Nang Ranked Third of Top Ten Cost-effective Cities", 4 July 2012, available at <http://www.ubnd.danang.gov.vn/TabID/76/CID/696/ItemID/28566/default.aspx> (accessed 8 July 2012).

Dapice, D. O. "Vietnam's Economy: Success Story or Weird Dualism? A SWOT Analysis". Research Report. Cambridge MA: Vietnam Program, Center for Business and Government, Harvard University, 2003.

DIPC. *Danang Investment Bulletin*. Da Nang Investment Promotion Center, February 2011.

Dinh Hien Minh. "Survey on Business Costs in Major CLMV Cities: Case of Vietnam's Hanoi, Ho Chi Minh and Da Nang Cities". In *Investment Climate of Major Cities in CLMV Countries*, edited by Masami Ishida, BRC Research Report No. 4. Bangkok: Bangkok Research Centre, IDE-JETRO, 2010.

Dinh Van An and Nguyen Thi Tue Anh. "Implemented Foreign Direct Investment after Vietnam's WTO Accession: Evidence from a Survey of 140 FIEs". Central Institute for Economic Management. Working Paper. Hanoi, 2008.

Doan Gia Zdung and Le Van Chon. "Vietnam's Private Economic Sector". *Journal of Science and Technology* 36, no. 2 (2005): 46–49.

DOST. "Report on the Scientific, Technological and Socio-economic Basis for the Creation of a High-tech Park in Da Nang City". Da Nang: Department of Science and Technology, 2010.

FIA. *Report on 20 Years of Foreign Investment in Vietnam*. Ha Noi: Foreign Investment Agency, 2008.

General Statistics Office. *Statistical Yearbook*. Ha Noi: Statistics Publishing House, 2008.

Kerkvliet, B.J.T. "Surveying Local Government and Authority in Contemporary Vietnam". In *Beyond Hanoi: Local Government in Vietnam*, edited by B.J.T. Kerkvliet and D.G. Marr. Singapore: Institute of Southeast Asian Studies and Nordic Institute of Asian Studies, 2004.

Le Hanh. "Da Nang Targets to Become an Export-Oriented Economy". Vietnam Business Forum website, n.d., available at <http://vccinews.com/news_detail.asp?news_id=2610> (accessed 11 July 2012).

Le The Gioi. "Improving Investment Climate to Develop Private Economic Sector in Da Nang city". *Journal of Science and Technology* 36, no. 1 (2005): 82–87.

Mabuchi Motor Co. "Mabuchi Held a Ground-breaking Ceremony of its Da Nang Plant". News Release, 27 October 2005, available at <http://www.mabuchi-motor.co.jp/en_US/news/n2005_1027.html> (accessed 8 July 2011).

Malesky, E. *The Vietnam Provincial Competitiveness Index 2005: Measuring Economic Governance for Private Sector Development.* 2005 Final Report. USAID/VCNI Policy Paper #4. Ha Noi: Vietnam Chamber of Commerce and Industry and United States Agency for International Development's Vietnam Competitiveness Initiative, 2005.

———. *The Vietnam Provincial Competitiveness Index: Measuring Economic Governance for Business Development.* 2011 Final Report. USAID/VCNI Policy Paper #16. Ha Noi: Vietnam Chamber of Commerce and Industry and United States Agency for International Development's Vietnam Competitiveness Initiative, 2011.

———. "Push, Pull, and Reinforcing: the Channels of FDI Influence on Provincial Governance in Vietnam". In *Beyond Hanoi: Local Government in Vietnam*, edited by B.J.T. Kerkvliet and D.G. Marr. Singapore: Institute of Southeast Asian Studies and Nordic Institute of Asian Studies, 2004.

Nguyen Hong Cu. "Da Nang's Economic Growth is Stable or Not?". *Journal of Science and Technology* 5, no. 28 (2008): 125–34.

Nguyen Xuan Thanh. "Da Nang: Policy Choice for Investment and Economic Development". Policy Working Paper. Ho Chi Minh City: CIEM-Fulbright Economic Program, July 2003.

PCI. Da Nang Provincial Competitiveness Profile, available at <http://www.pcivietnam.org/province_profile_detail.php?page=1&province=28&cboYear=2011> (accessed 8 July 2012).

Pham Thi Ngoc Anh, Le Thi My Linh, Do Thi Thuy Trinh, and Nguyen Doan Thanh Van. "Economic Restructuring in Da Nang City in the Trend of International Economic Integration". *Proceeding of Student Scientific Research Conference.* Da Nang: Da Nang University, 2010.

Thanh Tan. "Developing Key Industrial Products". Vietnamese Chamber of Commerce and Industry website, posted 9 December 2011, available at <http://vccinews.com/news_detail.asp?news_id=24924> (accessed 8 July 2012).

Tran Thi Thuy Trang. "Private Trade Development in Da Nang City". *Journal of Science and Technology* 1, no. 42 (2011): 193–201.

Tran Van Minh. "Da Nang's Dream of Reaching the Sea Realised". Vietnam Business Forum, posted 21 February 2007, available at <http://vccinews.com/news_detail.asp?news_id=8995> (accessed 10 July 2012).

UNDP. *Social Services for Human Development: Vietnam Human Development Report 2011.* Ha Noi: United Nations Development Programme, 2011.

Vietnamese Chamber of Commerce and Industry (VCCI), available at <http://www.vccidanang.com.vn/default.aspx> (accessed 7 July 2012).

VNS. "Da Nang Hi-tech Park Seeks Capital", 15 March 2012, available at <http://vietnamnews.vnagency.com.vn/Economy/Business/222122/da-nang-hi-tech-park-seeks-capital.html> (accessed 10 July 2012).

VoV. "Foreign Invested University Being Built in Da Nang", 10 September 2010, available at <http://english.vov.vn/Home/Foreigninvested-university-being-built-in-Da-Nang/20109/119451.vov> (accessed 10 July 2012).

Young People Newspaper. "Doing Business with Japan", 10 November 2006, available at <http://www.ubnd.danang.gov.vn/TabID/72/CID/731/ItemID/13252/default.aspx> (accessed 8 July 2012).

SECTION III

Cases from China and India

RUSSIA
KAZAKHSTAN
MONGOLIA
KYRGYSTAN
Bejing
CHINA
NORTH NOREA
TAJIKISTAN
Chengdu
SICHUAN
SOUTH KOREA
AFGHANISTAN
PAKISTAN
JAPAN
NEPAL
BHUTAN
INDIA
MYANMAR
VIETNAM
LAOS
TAIWAN
East China Sea
N
LEGEND
State/Provincial Capital
STATE/PROVINCE
National Capital
COUNTRY
NEIGHBOURING COUNTRY
Geographic Features
National Border Land
National Border Water
State/Province Border
Border Dispute
0 km 250 500
All maps are derived from GADM Data
(Global Administrative Areas, www.gadm.org, 2012)
edited by maps&more, Hans Hortig / Marcel Jäggi

7

THE EVOLUTION OF CHENGDU AS AN INLAND ELECTRONICS "BASE" IN CHINA AND ITS LOCAL STATE

Leo van Grunsven and
Cassandra C. Wang

INTRODUCTION

With economic reform and the opening up of the country to the global economy, China has witnessed the entry and vigorous growth of an array of globalizing industries. A recent account of China's electronics industry compiled by the Hong Kong Trade Development Council observes that "Massive relocation of the global electronics production over the last decades has turned China into a world electronics factory, producing more than a quarter of the world's total output in value terms. China is the world's major producer of end-consumer products including AV equipment, computers and mobile phones" (HKTDC 2011*a*, p. 1).

Concurrent with industrial transformation in structural terms, a geographical bifurcation has taken place. While domestically-oriented industries continued to be quite widely distributed across the country, new export-oriented industrial agglomerations have developed in the coastal areas. A substantial volume of research has provided insights into the patterns, drivers and effects of industrial geographies, as seen in the first half of the previous decade (He Canfei 2009; He Canfei and Wang 2012;

He Canfei et al. 2008; Fan and Scott 2003; Thun and Segal 2001; Thun 2004; Thun 2006, Wei et al. 2007; Wang and Lin 2008; and Wang and Mei 2009; Wang 2010).

The distribution of domestically-oriented industries in China has often been linked to entrepreneurial local governments coming to the fore in the 1980s and early 1990s coupled with deepening decentralization processes. In particular, the coastal agglomeration of export-oriented industries has been linked to several phenomena. The first is the top-down fostering of territorial complexes in the form of special economic zones (SEZs) and industrial parks (Zeng 2010). This reflects national industrial policy and planning as a determinant of location and regional industrial development. This has allowed initial growth following the establishment of firms from outside. The second is bottom-up industrial development as specialized industrial districts (often labelled as "clusters") developed through locally-born firms, institutionally supported by the local state (Wang 2010, see also Gereffi 2009). Third, again at the local level, the engineering of agglomeration economies undergirded the effective operation of SEZs. Thus, there is agreement that, since decentralization measures were enacted, the sub-national level has substantially gained significance vis-à-vis the national level.

In terms of industrial geography, China continues to be dynamic. Since the middle of the last decade, the growth of export industries in the coastal areas has slowed substantially as new foreign investment is gravitating to new inland locations. The emergence and growth of the latter is further fed by increasing domestic relocation from the coastal areas. A discussion of the drivers of this process as such is outside the scope of this contribution. However, it is important to note that inland centres have already gained prominence vis-à-vis coastal ones in several industries. This is contributing to a new dynamism of inland 2nd and 3rd tier cities that are rapidly emerging as new engines of growth in the central, northeast and inner west parts of China (e.g., Jones Lang Lasalle 2008).

Evidently, China's electronics industry has joined the recent sizeable geographical shift inland (HKTDC 2011*a*,*b*,*c*). While not implying the demise of coastal (specialized) agglomerations, inland electronics centres have emerged, by capturing — or being able to capture — a large part of the geographical shift, rather than by "endogenous" development. Recent press reports and consultancy research (e.g., Digitimes 2011*a*,*b*) refer to "new bases" of electronics manufacturing inland. Such references suggest a spatially selective and concentrated redistribution. The industry still appears to agglomerate in selected places that offer more favourable conditions or comparative advantages than others. Selection thus indicates the varying success of aspiring and established locations in capturing investment.

Chengdu, China's "gateway" to the west of the country is upheld as a "winner inland city" (Walcott 2007) in the context of the processes indicated above. It is considered as one of the "new bases" of China's electronics industry. This chapter analyses this particular case. The discussion is directed to two aspects. First, it seeks to outline the characteristics of the recent development of electronics in this city in the context of nation-wide recent dynamics and shifts of the industry. Second, it attempts to develop an understanding of the city's recent surge in the industry. Chengdu clearly has pursued a strategy of "coupling" with investing firms from outside. The main question here is "What explains the apparent success of this strategy?" Comparative advantages of the city have, on one hand, provided incentives and, on the other, enabled it to successfully grasp opportunities when they presented themselves. What is the nature and the source of these advantages? We deem two linked questions relevant here. First, what has been the rationale behind the city targeting electronics as a suitable sector? Second, what can explain the "choice" of the strategy? As to these questions, we are specifically interested in the agency of the sub-national state, its "choices" and expressions in policies and actions. We define sub-national as local and provincial. Another question then is, to what extent and in what way have the local and provincial states prevailed over their central counterpart?

According to perspectives offered by consultancy firms, one could suggest that the industry is (in part) reacting to uneven new local and regional initiatives that have differentially produced, inter alia, dedicated infrastructure in the form of various types of economic zones and parks, transport/logistics facilities, human capital, labor-migration arrangements and so on. While this demonstrates the operation of sub-national states (part and parcel of government and governance in China for several decades now, since the reforms) through policy-making and implementation, varying success may well reflect differences in the suitability of local industry policies and instruments. As places try to out-compete others, success may require more than imitation and duplication of programmes, projects, and so on (especially relative to coastal agglomerations). This is suggested by the fact that electronics growth in the new "bases" is occurring in varying configurations. Success may be enhanced through local specialization that may also serve the purpose of avoiding too much competition. However, in these aspects it is noteworthy that, in a number of success cases, the local electronics industry predates the recent wave of development. This might suggest that a successful strategy also builds on pre-existing industrial development and associated institutions in general, and more specifically the presence in some measure of an established electronics industry.

Our starting point in resolving the question is a brief consideration of the consensus in the literature of the significant role of factors at the sub-national level, in the form of agglomeration economies and institutions, in shaping dynamics of industrial geographies in China since reform. Next, we outline the characteristics of the electronics industry in Chengdu. The fourth section will provide our current interpretation of Chengdu now assuming a position as national electronics base, with a focus on the role of sub-national factors. We end the chapter with some concluding observations on the case considered.

SUCCESSFUL POST-REFORM LOCAL INDUSTRIAL DEVELOPMENT IN CHINA: THE ROLES AND WORKINGS OF SUB-NATIONAL FORCES

In the pre-reform period in China, the geography of the manufacturing sector was largely determined by central policy. Industries were centrally directed to specific locations. Since 1978, the central government has continued to play a significant role in propelling sub-national development, which is not surprising for China, given its history of central planning. The centre has exerted direct influence on location by designating areas for FDI and export manufacturing in the form of SEZs, export processing zones or industrial parks with the necessary infrastructure and incentives, including regulations influencing the availability of labour. Fan and Scott (2003) note that through specific policies, the central government has fostered rural-urban migration; engendered a migrant labour regime; and increased the supply of cheap rural labour to globalized export production spaces. The central government has continued to exert influence through policies and programmes devised by the National Development and Reform Commission, which also has the prerogative to attribute specific status and/or roles to a given territory. In addition, from the 1980s, a triple process of economic liberalization (coupled with privatization), globalization, and decentralization has taken place, triggering significant changes.

Marketization and globalization have, in part, engendered the notion of agglomeration by market forces, related to comparative advantages of places and agglomeration economies that have been relevant especially to globalized export-oriented industries (like electronics) and firms. Fan and Scott (2003, p. 315) state: "Sectors and places undergoing economic liberalization have been those also most prone to the formation of agglomeration economies." These economies are categorized as localization economies (i.e., efficiency-enhancing phenomena that come from the physical proximity of firms in a given sector) and as urbanization economies (i.e., efficiencies that result from the agglomeration of many different kinds of activities in a given region). Fan and Scott (2003, p. 297) further note that "regions that have some or all these attributes, combined with dynamic learning capacities, stand some chance of becoming significant articulations of value-adding activity and entrepreneurial energy in the new global economy". Producing and fostering these advantages and economies have become targets of government intervention — both central and local — in China.

Urbanization economies have developed as local governments have built on initial interventions and initiatives devised from the centre in a myriad of ways. Decentralization, in particular, greatly enhanced the role of sub-national governments. In a nutshell, most literature (He et al. 2008; He Canfei and Wang 2012; Fan and Scott 2003; Thun 2004; Thun 2006) perceives decentralization — particularly the fiscal variety — as instrumental to engendering entrepreneurial government at provincial and local levels, which have been directed to fostering urbanization and localization economies, among other things. Clearly, jurisdictions in the coastal areas have successfully devised and implemented local development programmes that "produced" such urbanization and localization economies which were suited to the attraction of specific sorts of firms. This is because the industrial organization in SEZs and industrial parks was initially marked by a dominance of relatively large, foreign, externally implanted firms (only later followed by localization as the co-development of local firms increased). In this case, bureaucratic structure and competence was congruent with the management of complexes in such a way as to enhance urbanization and localization economies relevant to dominant firms (and vice versa). These economies are suggested to explain local industrial success in large part.

As places engage in inter-city competition and rivalry, a tendency has been apparent to duplicate or imitate other places that are successful in an attempt to achieve similar growth. Duplication concerns not only policies but also concrete initiatives and instruments. This is evident in the so-called "development zone fever" (Thun 2004; J. Zhang 2011), that had to be redressed in the 1990s by the central government in view of increasing inefficiencies. Related to these practices are the ideas of the possibility of emulation of bureaucratic structure and competencies that have brought success to others.

We move here to the second aspect to consider, as to successful local industrial development, namely local industrial strategy choice, in terms of industry branch and mode of development. Thun and Segal (2001), Thun (2004, 2006) believe that in decentralization, imitation actually has been difficult to achieve; likewise the duplication of industrial development and industrialization profiles has in fact been quite difficult. In addition, policy imitation has seldom produced the same degree of success. They argue that one can frequently observe significant place variation. This entails highly differential sets of urbanization and localization economies; varying performance of the same industry in different places despite identical policies and instruments; distinct attraction of firms operating in the same industry; differing performance of industries in the same place; dissimilar modes of development of the same industry (e.g., endogenous or exogenous), and so on. They attribute this place variation to localized, place-specific institutional configurations and the presence or absence of fit between local institutions and firms in a particular industry. They argue that needs of (firms in) industries differ, and therefore require specific institutional structures to meet industry- and firm-level needs. At the same time, actual institutional structures are mostly not only local and place-specific, but also path-dependent. This produces on the one hand spatial variation in development outcomes and structures as a "fit" is sought between local institutions and firms in different sectors. On the other hand, it also produces constraints to change. Thus, a place is hardly ever successful in developing a wide range of industries, given highly different structural characteristics, as industrial policies adopted with respect to one industry are highly unsuitable for other industries. In addition, it is hardly possible for places to "run" a range of policies at the same time. Conversely, for the same reasons, industries are likely to blossom in one place but not another.

Differences in institutional configurations make it hard to successfully imitate policies and strategies. Ideally, places invoke differential strategies with respect to the same sector, congruent with their institutional configuration. Outcomes are almost necessarily differential (i.e., the same industry will have different characteristics in different places) as not every strategy is able to serve specific needs of specific sets of firms in sectors. We add that outcomes nevertheless can be positive for a range of places. To nuance Thun's argument, we suggest that this is also because sectors are seldom completely homogeneous with respect to most characteristics, especially in terms of composition of products, processes and firms, and therefore needs. Rather, heterogeneity/variety is the norm, allowing differential modes of development in the same industry.

In electronics, the latter is indeed revealed in success generated by such differential strategies. One is the initial attraction and presence of relatively large externally implanted, often foreign, firms in the industrial organization and their rise to dominance in a "choice" labelled by Yeung (2009) as "strategic coupling" between institutional and firm actors. This "choice" or "type" of development has been associated with bureaucratic structures and competences in which policies directed to agglomeration economies thrive. While in a later phase localization through co-presence and co-development of local firms grows (Lu and Wei 2007; Meyer et al. 2012), this institutional configuration tends to develop in a path-dependent way.

The other is "bottom-up" development of local small firm ensembles in industrial districts, with local or regional orchestration, driven even more by sub-national factors. Again, much of the literature here has emphasized the agency and place-specificity of local institutional factors — both formal and informal. Local institutions, this time grounded in social capital (local social relationships), foster ties of proximity and network association, collective efficiency, and trust. They are thus held responsible for fostering local entrepreneurial effort; specialization of enterprises in particular tasks in a production chain (division of labour); and networked organization that promote learning processes, diffusion of information and knowledge, high levels of efficiency and minimization of risks (Fan and Scott 2003; Wang and Mei 2009). A different institutional configuration here privileges local SMEs and their internal, localized, networks. Conversely, such networks need different institutional configurations. Wang (2009, 2010) also refers to the role of local policy, however of a different nature compared to the first mode, in privileging localization economies above urbanization economies, through assisting association and collective efforts, collaboration, addressing bottlenecks in networks and so on.

This thinking not only reinforces a focus on the sub-national level, but also broadens the scope of consideration of "the local" by including characteristics and abilities of local institutions (institutional endowments). Success is linked to strategies congruent with the existing institutional fabric and arrangements which allow a full exploitation of the latter, rather than simply imitation. The adoption of such strategies often requires — often deep — changes in the institutional fabric (Segal and Thun 2001; Thun 2004). Like Thun; Fan and Scott (2003); He et al. (2008) and He (2009), among others, point to formal and informal institutional factors amongst sub-national stimuli, and, on the other side, the place-specificity of these factors. They underlie inter alia the specific shape of policies, programmes, actions and instruments, and patterns of specialization visible in export production structures. Generally, institutional factors are seen to consist of some fundamental elements of the business system, especially industrial organization as indicated by the dominance of specific types of firms (by size or ownership), as well as bureaucratic structure (unified or fragmented), disposition and competences (Thun 2004, 2006). An interesting question that is not addressed in conceptualization along institutional lines (at least in the literature cited above) is how these institutional configurations have come into being in the first place? We believe that the ensuing analysis of the Chengdu case can shed light on this question as well, besides demonstrating the operation of constructs discussed above.

CHENGDU: THE RISE OF A NEW SUB-NATIONAL ECONOMY AND A PROFILE OF ITS ELECTRONICS INDUSTRY

Chengdu is the provincial capital of Sichuan and is less than three hours by plane from Guangzhou, Shanghai, Beijing and Hong Kong. It is one of fifteen sub-provincial cities in China which means that, while it is part of a province, it is managed independently with regard to economic and legal issues. The population of the city is some 13 million, 3 million of whom are non-permanent residents (http://www.chengdu.gov.cn). Its GRP is still about one-third that of Shanghai and its per capita GRP amounts to 45 per cent of that of Shanghai. Reflecting its growing importance as an economic centre in Western China, average annual GRP growth has outpaced the national average since 2005, 13.9 per cent versus 11.4 per cent (KPMG 2010).

The economy's major pillars are manufacturing, services, communications and transport, and real estate. Its manufacturing base is diverse, comprising machinery, chemicals, pharmaceuticals, metallurgic products, and electronics. These industries are dominated by large-scale firms, domestic and foreign, with many of the domestic firms having a history of state-ownership. Another important industry is defence. The Chengdu Aircraft Industry Corporation, producing several types of fighter jets, is a major player in China's military aviation technology, also through research and development (China Knowledge 2007). In services, Chengdu is a growing destination for Business Process Outsourcing. It is the second largest banking market in China, and enjoys significant growth of (international) business services (Wang and Meng 2007). On the back of manufacturing and services growth, Chengdu has become a logistics centre and aviation hub.

Over the past five to ten years, Chengdu has developed a strong position in a number of industries as it has experienced an accelerated incorporation in the relocation, both domestic and global, of industrial activity. The city has become a magnet for FDI in China, not just in manufacturing industries but also in services, logistics and so on. Many Western and Asian multinational companies in a range of sectors and branches have made investments in the city. About one third of Global Fortune 500 companies have operations in the city (KPMG 2010).

Next to Wuhan in Hubei province and Changsha, Hengyang, and Zhuzhou in Hunan province (Central China), the regional triangle that comprises Chengdu, Chongqing and Xi'an has risen to prominence in the electronics industry (HKTDC 2011*b*,*c*; Digitimes 2011*a*). As Table 7.1 shows, Sichuan has the largest electronics industry output in western China. Sales have almost doubled to nearly US$16 billion during 2007–09 (HKTDC 2011*b*). Much of this is concentrated in Chengdu as a target of the shift inland.

Regarding the definition used for the electronics industry and ICTI, literature and statistical sources refer to the former as comprising electronics components, industrial equipment (incl. optical electronics), consumer electronics, and electronics-related service activities such as design. The latter encompasses some aspects of electronics manufacturing, along with a range of service activities, particularly those that are telecommunications- and computer-related, as well as IT services and software. Following the demarcation often used locally, namely electronics manufacture and software, we concentrate on electronics manufacture and ignore software, although the latter is sizeable in Chengdu (Quan 2011; Invest Chengdu).

Chengdu's electronics manufacture took off after 2004. At the end of 1999, Chengdu produced an output value of less than RMB9 billion with a loss of RMB0.5 billion. But profits started from 2000, despite output value remaining rather low (RMB11.4 billion in 2000, RMB16.6 billion in

Table 7.1

The Performance of the Electronics Industry in China's Inland Provinces, 2009

2009 figures	Sichuan (incl. Chengdu)	Chongqing	Shaanxi (incl. Xi'an)	Hubei (incl. Wuhan)	Hunan (incl. Changsha)
Sales (09/07% growth)	US$15.8 bn. (+77%)	US$2.4 bn. (+197%)	US$3.6 bn. (+17%)	US$9.8 bn. (+52%)	US$3.5 bn. (+46%)
Export sales/total sales	18%	3%	11%	23%	9%

Source: HKTDC 2011*b*.

2004). Profit earned in 2004 was almost three times higher than in 2000. In 2005, the manufacturing sector generated an output value of RMB22.13 billion, nearly twice as much as in 2000. Rapid growth is revealed by the fact that by 2008 output value reached RMB70 billion and total profit had increased to RMB4.4 billion (Chengdu Statistical Yearbook 2000–09). It should be noted that these figures include only so-called "above-scale" enterprises, which refer to firms with an annual revenue equal to or larger than RMB five million (http://www.stats.gov.cn).

Compared to output value and profits, growth in year-end employment in Chengdu's electronics manufacture appeared not very notable during the 2000s as revealed in Figure 7.1. Official figures indicate that, from 1999 until after the mid-2000s, Chengdu's electronics manufacture was marked by substantially lower employment, hovering around 60,000 persons. However, employment has been steadily increasing since 2005 with growth of the industry.

According to the latest official statistics, the business income of Chengdu's electronics manufacture over the last few years has increased by 45 per cent on a year-on-year basis, reaching RMB98.2 billion in 2010 (S. Zhang 2011). In early 2011, 489 "above-scale" electronic manufacturing firms were located in Chengdu. Figure 7.2 shows the current composition of the manufacturing

Figure 7.1
Year-end Employment in Chengdu's Electronics Manufacturing Enterprises (above-scale), 1999–2008

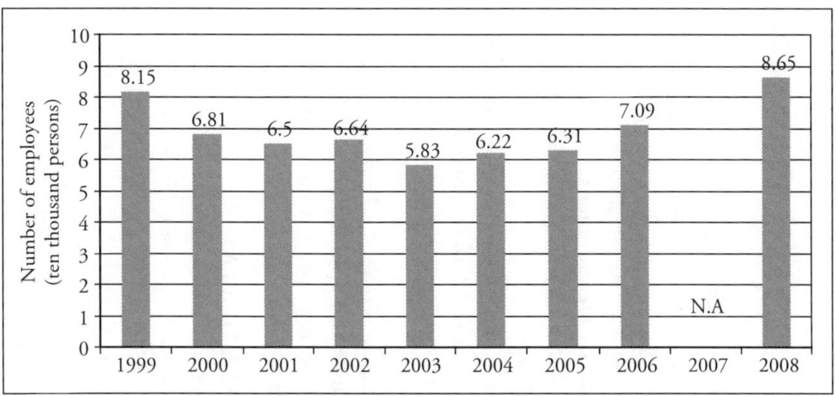

Note: The number of year-end average employees in 2007 is 8,132, according to Chengdu Statistical Yearbook 2008, but we suspect that it may be misspelled.
Source: Chengdu Statistical Yearbook 2000–09.

Figure 7.2

The Composition of Chengdu's Electronics Manufacturing Sector

INTEGRATED CIRCUITS

* **Core:** *Intel, AMD, Samsung, Toshiba, TI, ST, Infineon, Renesas, NXP, NEC, Shenzhen Segem, Norcent, ShanghaiHuahong, Shijiazhuang Jinglong, Beijing Daheng New Epoch, Zhenhua Electronics, China Resources Microelectronics, Jiangsu Xinchao, Jiangxi Electronics Group, NantongFujitsu*

* **Raw Material Development**

Xinguang Silicon, Chaolei Industrial Corporation, Ulvac, Ultra-pak Electronics, Linde Gas, Messer, Sumiko Electronics

* **R&D, Design**

Conexant, Fujitsu, Freescale, Overtop, O2, Sonix, Hua Wei, NSBIC, West Star, Panovasic, IPGoal, Guoteng, Rockship, Skytek, Dexie

* **IC manufacture (Wafer), Assembly, Packaging, Testing**

Cension, Intel, SMIC, Unisem, Xinyuan, Kewei Electronics, Phoenix, ZETEX

* **Service**

ULVAC, Sumiko, Linde, Air Products and Chemicals, Air Liquide, Messer, Emei Semiconductors, CREDEN

* **IC products and application**

Agilent, Nokia, Ericsson, Motorola, Alcatel, Flextronics, Changhong, TCL, Maipu, Jiuzhou, Guoteng, Da Tang, BOE, Huawei, Lenovo, China Wireless, Dell

FLAT PANEL DISPLAY

* **Raw Materials and Accessories, Cell and Module**

TG Chengdu Co.Ltd, Comm Unitex Glass (Chengdu) Co. Ltd, CSG Chengdu Industrial Base, ChengDu Guang-Ming Optoelectronic Information Material Co. Ltd., American Corning, Asahi Glass, NEG, NSG, Toppan Printing, Dai Nippon Printing, Fujitsu, Taiwan Sintek, Taiwan Novatek, HK Solomon Systech, 3M, LG Chemical, Canon, Nicon, Yongsheng Huatsing, Rainbow, Hisense, BOE Chatani; TFT-LCD of BOE, TFT-LCD of Shenzheng Tianma, OLED of Changhong, LCD module of TSMT, TFT-LCD module of Chengdu Space Photo-electronics Sharp, Samsung, Lejin-Philips, Taiwan AUO, Chi Mei, Chunghwa Picture Tubes, BOE Sci&Tech Group, SVA-NEC, Longteng Optic-electronics (Kunshan), Shanghai Tianma, Truly, Shenzheng Laibao

* **Production Terminal Products as a whole**
TCL, Changhong, Chengdu General Touch, Sichuan Direction Photo-electronics, Philips, LG, Sharp, Panasonic, Samsung, Sony, Hitachi, Fujitsu, AOC, Hair, Hisense, TCL, Skyworth, Konka, XOCECO

COMMUNICATION EQUIPMENT
* **Optical Communication (transmission materials, optical transmission equipment, network communication devices, data storage&processing)**
Futong Group, Chengdu Corning, Chengdu SEI Optical Fiber Co. Ltd, Chengdu Aoge Optical Glass Co. Ltd, Chengdu Crystal Technology Co. Ltd., Chengdu Datang Cable Co.Ltd, Chengdu Putian Telecommunication Cable Co. Ltd, Sichuan Guangheng Communication Technology Co. Ltd., Sichuan Huiyuan Sci & Tech Development Co.Ltd, Fiberxon Chengdu Technology Co. Ltd, Chengdu Aisi Wideband Technologies Co. Ltd., Siemens, Lucent, Rohde & Schwarz, Acterna, Shanghai Alcatel-lucent, ZTE, Huawei, Datang, UT Star, Fiberhome Telecommunication, Sichuan Huiyuan Communication
* **Communication research and production (Chengdu R&D Center)**
Alcatel, Ericsson, Motorola, Nokia, ZTE, Huawei, NEC, Flextronics, Freescale, O2Micro, NTS Technology, Bravo Tech, China Wireless Technology

COMPUTER EQUIPMENT **(sub-assembly and final assembly)**
Foxconn, Lenovo, HP, Dell

OTHER ELECTRONIC EQUIPMENT **(components, ancillary equipment, other final products)**
Agilent Qianfeng Electronic Technology (Chengdu) Co.Ltd, Chengdu Yaguang Investment Management Co. Ltd, Coship Electronics, Chengdu Zhongling Radio Communication Co. Ltd, Maipu, Chengdu Bell Communication Co. Ltd., Chengdu Guoteng (Group) Co. Ltd., Chengdu 30 Kaitian Communication Industry Co. Ltd., Sichuan King Von Electronic Technology Co. Ltd, Sichuan Video Electronic Co. Ltd, Chengdu Galaxy Magnetics Co. Ltd, Chengdu Tiger Microwave Tech Co. Ltd, Sichuan Tongda Electronics Company, Chengdu Jiuzhou Electronics System Co. Ltd, Chengdu Sifang Information Technology Co. Ltd, Chengdu Taiwei Wireless Tech Co. Ltd, Chengdu Sobey Digital Tech Co. Ltd, Chengdu 30 SAN Information System Co. Ltd, Chengdu Westone Information Industry Co. Ltd, Chengdu Guoxinan Base Co. Ltd, Dongfang Hitachi, Wuniu Science and Technology, Emicon Refrigeration

Source: EBN 2008; China Knowledge 2007; Digitimes.

industry, including the main firms (foreign and local) in each segment. This figure does not include military electronics, for which no information is available. The fields in which electronics firms are involved are diverse, covering: integrated circuits; photo-electronic display; communication equipment; computer manufacture; electronics components; digital media; information security; and so on.

What is noteworthy about Chengdu's electronics sector is that it includes entire production chains, rather than a few specialized product-groups. In the field of integrated circuits (IC), companies like TI, Intel, SMIC, Unisem, Diodes and Advanced Semiconductor enabled Chengdu to create its own chips. It is contended that one in every two computers in the world uses chips "made in Chengdu" (Digitimes 2011*b*). Chengdu has also attracted system manufacturers such as Agilent, Nokia, Ericsson, Fujitsu, Freescale, Huawei, and ZTE. The total investment in Chengdu's IC industry amounts to over US$3 billion.

The role of large foreign companies in the growth of the electronics industry should be highlighted. This is evident in IC production. In August 2005, Chengdu witnessed a significant moment in the development of its electronics industry when Intel opened its China chip-packaging factory here. The Intel operations, initially providing 600 skilled jobs, proved a catalyst to the further development of the local IC branch. It also propelled electronics manufacture onto the path of growth by attracting firms in other branches. Up to 2009, Intel had altogether invested US$600 million in Chengdu. In 2009, Intel made its third-round investment with jobs going up to over 3,000. Dr Tang Jiqiang, the director of Chengdu Hi-tech Zone Development and Planning Commission highlighted the contribution made by Intel to Chengdu's electronics industry: "Ten years ago, there were only a few state-owned electronics enterprises in Chengdu. They were unable to fabricate chips, conduct software design and introduce competitive electronic products to the market. However, the arrival of Intel attracted many other related enterprises to locate nearby, leading to the agglomeration of capital and high-qualified talents and eventually making Chengdu a new growth pole in western China" (B. Wang 2011).

Chengdu has developed into the largest R&D centre of mobile communication in China as the world's leading communication equipment manufacturers such as Motorola, Ericsson, Alcatel, and Nokia established R&D centres here. Leading firms have also located substantial manufacturing capacity in the city. Sizeable capacity has developed in flat panel display as leading enterprises have established here (Figure 7.2). Being a recent branch, the impetus to the development of computer manufacture chains has come from the investment of major firms. After MNCs like Lenovo, Wistron, Compal, TI, etc., invested in Chengdu in the second half of the 2000s, a much publicized milestone for the growth of this branch was the decision by Dell taken in early 2010 to establish its new China operations (expansion and relocation from Xiamen) in Chengdu. Dell's facilities in the city, opened in 2012, generates over 3,000 jobs directly (Digitimes 2011*b*). Another eye-catching case is much-plagued Foxconn (Hon Hai), which has facilities already in operation in Chengdu to produce computers, TVs and electronic game gears. Current expansion will significantly raise capacity; reportedly 120,000 workers will be employed in its Chengdu facilities in the years to come (HKTDC 2011*b*). Much of the expansion concerns relocation from its Shenzhen facilities that will be significantly downsized.

Support firms are following the lead firms. As to Taiwanese enterprises like Wistron and Compal, the Deputy Director of the Bureau of Investment Promotion of Shuangliu County conveyed: "Wistron and Compal may need up to one hundred supporting enterprises to locate in Chengdu. They bring us more than two individual enterprises: a complete value chain" (Luo 2011). He also pointed out that Wistron and Compal, among the world's biggest foundries, also will create thousands of jobs in Chengdu, exerting a positive influence on the local economy and labour market.

THE SUCCESSFUL DEVELOPMENT OF THE ELECTRONICS INDUSTRY IN CHENGDU: WHAT INFLUENCE OF THE LOCAL STATE?

From the discussion in the previous section, it is apparent that the recent development of the electronics industry in Chengdu has followed a top-down mode of "strategic coupling" marked by at least two elements: first, a dominance of the external implantation of firms; second, this implantation concerned relatively large operations of both foreign and domestic firms (for the moment engendering co-presence), particularly in manufacture. As yet, local-born firms and SMEs appear to play a minor role. Thus, the strategy has privileged the capture of firms in the industry shift inland. We endeavour to interpret the "choice" of this strategy and why it has proved successful. Is the background to this strategy principally local and, if so, what is fundamental at local level? Are other levels at play also? Should this be the case, to what extent and in what way?

Targeting Electronics: What Drivers?

A first noteworthy observation we made when delving into the case is that Chengdu has, in fact, a long history in electronics industry and implantation from elsewhere is not a recent phenomenon. During China's first Five Year Plan (1955–60), Chengdu, Beijing and Nanjing were designated by the central government to be the three national electronics industrial bases (Wang 2011). Around the same time, China established its first industrial research institute for telecommunications in Chengdu (the famous "No. 10 Research Institute") and built up the Chengdu Telecommunication Engineering College, now renamed as the University of Electronics Science and Technology of China (Wang and Meng 2007). The state constructed its own-designed factories in Chengdu during the second Five-Year Plan (Ding and Li 2001). During Mao's "Third Front" industrialization project, it also benefited from the strategic geographical shift of many electronics research institutes and manufacturers from the coastal regions to central and western cities (Naughton 1988). Sichuan province then hosted two thirds of the leading national electronics enterprises that were under supervision of the Ministry of Information Industry and over 20 per cent of research units of the whole of China (Digitimes 2011*a*).

Notwithstanding the dismantling of the project from the mid-1970s onwards, Chengdu maintained an important position in China's military industrial complex with e.g., the retention and expansion of military aircraft production. In the defence industry, research activities and technology development gradually moved into the core of activities (Ding and Li 2001). This is currently one of the sources of a substantial S&T base and capability as well as of high-tech talent for electronics players.

From the 1980s, serving the national defence machinery as well as the civil-oriented branches, rendered Chengdu's electronics industry the status of "strategic industry". In the process of reform, some state-owned enterprises and resources linked to the industry re-directed resources and talent to civilian applications in order to stay afloat. As they branched out, a civilian capacity developed constituting a supplementary source of autonomous growth of the electronics industry. The historical accumulation of activities in electronics implied the accumulation of human capital as research, science and technology, and educational institutions agglomerated here. Already in the 1940s, 27 universities were relocated from north China (Kang 2010). It also implied the accumulation of technological competences and capabilities.

Yet much of this remained tied to the core defence industry. Endogenous civil-oriented growth was constrained by location, domestic market limitations, and institutional factors as we shall see below. However, in the 2000s, since the 10th Five Year Plan, this inheritance or legacy prompted the idea of further developing electronics as a target industry on the basis of this stock of assets. By this time, conditions had changed, *inter alia* the introduction of new regional policy (see later). In view of technological relatedness and the possibility for new firms to avail themselves of spill-overs, electronics provided the opportunity to capitalize on existing industrial variety. It appears that electronics as a targeted industry in recent times cannot be understood without considering inheritance and capitalization on technologically related variety.

Local Forces in Strategy and Success

External "Coupling" as a Strategy and Local Institutions

Capitalizing on the volume and characteristics of the stock of human capital and competences generated by institutions and embedded in large state-owned enterprises and institutes referred to above, required an effective strategy. While in principle, both endogenous and exogenous strategies were "open" to local policy makers, in practice the "choice" quickly narrowed down to the mode of development indicated already. As suggested earlier, did the institutional setting prompt such a "choice"? We believe that industrial organization, bureaucratic disposition and competences, ingrained in institutions inherited from industrial development predating the recent growth of electronics, and institutional path-dependence in fact produced historical continuity and congruence of the "choice" of mode with institutional configuration. This resulted in favouring exploitation of the local base and capability by activities performed by exogenous firms with an equally large scale as already existing in the local base, drawn in by agglomeration economies.

As to industrial organization, industrial development and the characteristics of electronics industry from the 1960s in Chengdu meant a predominance of large firms, state-owned or not. An equally important aspect is asset-origin/ownership and management in core industries, in the past and now. From the 1960s, main firms in these core-industries were not only SOEs and externally implanted; firms and institutes in the heart of the military industrial complex (a fair part of the local non-services economy) were directly owned and managed by the central State. This ownership structure is still prevalent today. For example, the Chengdu Aircraft Industry Corporation, a manufacturing complex in the city employing about 18,000 workers combined with the 2,000 employees strong Research, Design and Engineering Institute, is part of the Beijing-based China Aviation Industry Corporation.

This circumstance forces the interpretation of local institutions against the background of local-centre relations. External implantation and substantial control of the central State over core industries produced locally what Thun (2006) refers to as fragmented external-firm-dominated bureaucracy, entailing little means and power to coordinate generation and allocation of resources and to embark on endogenous development. While openness was engendered by external control, it also narrowed competences in the local bureaucracy. Oriented to serving needs from the centre, competences were shaped around management rather than developing local firms in relevant industries. It is suggested that bureaucratic routines that developed over time revolved around rapid familiarization with, perception of, and responding to, needs of the core industries. The combination of routines, bureaucratic competence, the institutional set-up proving "sticky", thus marked by path-dependence, and institutional coordination deficit preventing a endogenous development course, led to congruency between continued reliance on firms from outside and institutional configuration.

Such bureaucratic disposition is corroborated by the emphasis on facility development, setting up of management agencies, and more recently the municipal government encouraging enterprise through the provision of direct and indirect capital support, technology and exporting programmes. For example, electronics enterprises could enjoy a 15 per cent reduction in corporate income tax rate within a certain period; the municipal government allocated more than RMB30 million per year to the development of software and IC design industry during the period of the 10th Five Year Plan in the form of tax breaks and other financial incentives; newly established IC design firms were exempted from income tax in the first and second years and allowed a 50 per cent reduction in the third to fifth years; software and IC design firms may list the entire amount of their actual personnel remuneration and training expenses as a before-tax expense for enterprise income tax purposes, and so on. The municipal government set up a "Guidance Team" for the information industry but only to solve operational problems encountered in the process of industry development (Digitimes 2011).

URBANIZATION AND LOCALIZATION ECONOMIES

There is a powerful notion in investment documentation that Chengdu's electronics industry has been growing on the back of an extraordinary amount and set of local assets "produced" by local government policies, initiatives, and programmes. These assets combined with other factors are labelled as the "investment environment". Two assets in particular are frequently advanced: infrastructure; and human capital suited to electronics.

Chengdu indeed offers an excellent infrastructure for business, superior to that of many other cities. Since the 1980s, a range of industrial parks have been completed. Prime is the Chengdu Hi-Tech Industrial Development Zone that was developed from the early 1990s and in 1991 received national-level status (implying eligibility for central funding). The local government has enlarged the zone to almost 90 km^2, comprising two main parks (West and South, see Figure 7.3). These have sub-

Figure 7.3
Chengdu's Hi-Tech Industrial Development Zone

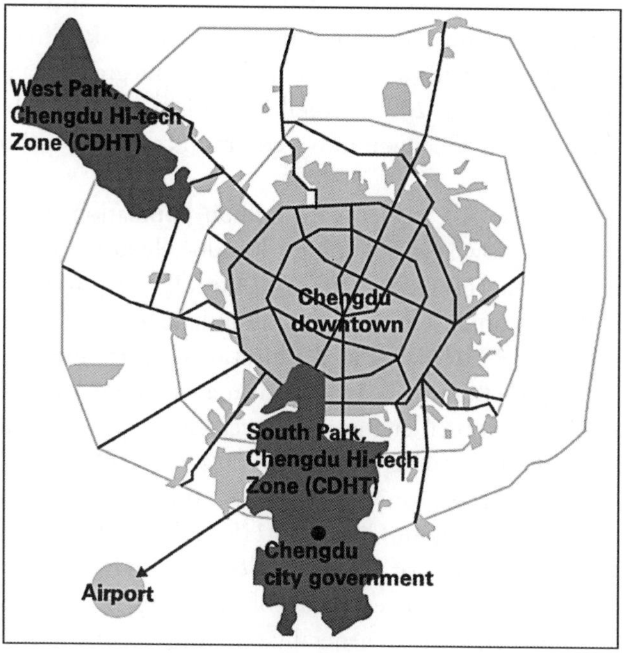

Source: KPMG Advisory 2010.

zones dedicated to specific activities; through continuous upgrading they now provide up-to-date facilities at a price that is competitive. Several sub-zones in the main parks and other new developments (e.g., the Chengdu Hi-tech Integrated Bonded Area that came into operation in early 2011) are geared to the electronics sector. Chengdu employs a "one-district, one-owner" philosophy to avoid duplication of infrastructure across counties and districts and to promote co-location, whereby similar firms can enjoy benefits from physical proximity. The "Plan on the Development of Chengdu's Electronics Industrial Clusters (2008–17)", depicts the spatial arrangement of Chengdu's electronics industry on this basis, in the form of sub-zone specialized agglomerations (Chengdu Municipal Government 2008).

Other infrastructure concerns aviation facilities, railways, expressways and communication. Upgrading, expansion and diversification have made them into strong attraction points by producing substantial efficiency benefits for business (Wang and Meng 2007). Chengdu is developing into China's fourth largest aviation hub as a second airport is constructed (Zhang 2011; Xiao 2009; CDMG 2009). It is also the largest railway hub in Southwestern China and has one of the highest densities of expressways (Wang and Meng 2007). The municipal government has made great effort to restore transportation and communication networks after the 2008 earthquake, safeguarding its position as the largest integrated transportation hub of Western China. In 2009, several local plans were launched for the further development of Chengdu into a communications hub (http://www.chengdu.gov.cn).

As to human capital, Sichuan's 80-million population forms a source of workers who are now keen to look for jobs locally rather than migrating to the coastal area (HKTDC 2011*b*). Equally important from the perspective of electronics, Chengdu has abundant science and technology resources as well as electronics-related educational institutions. It hosts 168 high-level educational and research institutes, state-owned independent research institutes, national engineering research centres and national key laboratories. With an enrolment of over 85,000 annually, these "produce" tens of thousands of electronics and related (under-)graduates per year who usually choose to stay in Chengdu (HKTDC 2011*b*). The city ranks number one in southwestern China in S&T and educational resources (Miu 2010; Zhang 2011). The quality of graduates is as high as that in coastal cities; however, salary is generally one-third to one-fifth lower than in coastal cities (Jiang 2010).

Local government in its human capital policy has striven to further develop this capital and invest in expansion and qualitative upgrading thus providing for the local availability of electronics engineering talent, hardly to be found elsewhere in the country. A boon in this respect is that Sichuan/Chengdu continues to be one of China's military science and technology bases with a wealth of talent engaging not only in IC design, but also display technologies, IT and communications applications and software (HKTDC 2011*b*). All this translates into an advantageous position of Chengdu compared to many other inland cities as to the quantity and quality of human capital (Jones Lang Lasalle 2008).

In general, generic policies and programmes in the areas of infrastructure and logistics, development zones, investment incentives, labour supply and costs, labour migrant policies (including Hukou), public and business services, environment, and so on have augmented urbanization economies.[1] The nurturing and development of electronics human capital signifies substantial localization economies. However, especially as to high-skilled human capital it was noted earlier — in the context of capitalization — that this asset actually long predates the recent surge of the electronics industry. The stock available in the early/middle 2000s constituted an "inheritance" from earlier industrial development in Chengdu.

STRATEGIC COUPLING FROM A FIRM PERSPECTIVE: CONGRUENCY, SUPPLY, AND FIRM NEEDS

Supply elements emanating from actions with respect to local conditions for investment (associated with strategy "choice"), proved to be congruent with the needs of firms in the electronics industry that were located externally. This engendered "strategic coupling" as it met firm needs. At the particular juncture of the second part of the 2000s and into this decade, there has been a growing interest by external firms in capturing these assets by investing in the place, as they were increasingly searching for suitable inland locations.

Thus far, mainly "anecdotal" evidence suggests this congruity or "fit" between the needs of external firms and asset endowments. Firm representatives have frequently stated Chengdu's attraction factors: strategic location; dedicated infrastructure in the form of various types of economic zones and parks; transport facilities; labour migration arrangements; quantity of labour; quality of the educational system; well-trained workforce; the pool of design engineering talent; local talent being conversant in English; and the social and cultural environment (Kang 2010; Miu 2010; Wang et al. 2007). As to the latter, Chengdu is well-known for its cultural vibrancy, leisure facilities and amenities. Under a good cultural environment, labour turnover is quite low: at least 10 per cent lower than that in coastal cities (Miu 2010). The availability of technology talent has facilitated upstream electronics production.

For many electronics investors, the opportunities to tap relational assets pertaining to supply (sourcing, contracting out) and demand (markets) are abundant. This relates to products (semi-finished and components), technology, information and knowledge beyond that embodied by human capital. We have noted already the presence in the city of operations that span entire production chains, engendering significant localization economies. Thus, "strategic coupling" should also be related to the presence of related variety through earlier rounds of investment and firm establishment. The recent entry of the computer industry, constituting a new variety, appears a case in point. The accumulation spanning a wide range of areas covered by firms in a range of branches (not exclusively under the umbrella of electronics) signifies spill-overs that — once a sufficient mass and diversity has been reached — become lucrative to exploit by firms in related areas.

It should be borne in mind however that the "strategic coupling" should be seen in the perspective of the particular time frame. As noted, firms became interested at the particular juncture of the second half of the 2000s. This is associated with circumstances external to Chengdu (or inland China in general).Without referring to accident, luck, or chance, it may be asserted that while the city had the "necessary" assets the strategy actually came to fruition when a range of circumstances began to coincide.

EXTRA-LOCAL: ROLES OF CENTRAL AND PROVINCIAL STATES

The above has already provided a glance at the role of levels of government/ governance beyond the local state. Thun (2006) *inter alia* has argued that no matter how much decentralization has empowered the local it does not imply detachment from other levels. Thus, influences from other levels remain. In the Chengdu case, interactions between local and higher (provincial and national) levels indeed appear relevant to some extent and in some aspects as to the subject matter under discussion. This seems more relevant for the central-local relationship than for the local-provincial relationship.

As to the former, from the beginnings of the 2000s, the major changes in regional development policy have affected Chengdu's development path as Beijing started to prioritize other parts of the country beyond the coastal area. The Western Development Policy was followed by Developing the Central Region Programme and Reviving the Northeast Programme. From Chengdu's perspective, the first two are relevant. Their implications were probably threefold. First, the central government programmes eased a financial constraint as substantial funds were made available to western and central localities for development purposes. Earlier, although fiscal decentralization had meant improved opportunities to fund local projects from retained tax revenues, fiscal contracts nevertheless imposed a limitation on the amount of retained revenue. Second, a core element of Western and Central Development has been the tremendous expansion and improvement of infrastructure throughout the Central and inner West regions (rail, expressways, airports) through central government investments. Cities like Chengdu have enjoyed spill-overs from this as connectivity with central and east coast parts of the country has tremendously improved. To a large extent, it removed the issue of distance (in terms of time and costs) for many inland locations. Third, from the central government perspective, the rationale behind huge investment in regional connective as well as urban business infrastructure was to "anchor" West China's economic advancement in the urban centres by improving conditions for the mobility of firms towards labour. As such, the Western Development policy gave an additional impulse to the "choice" of strategy of Chengdu argued earlier.

In 2008, the National Development and Reform Commission again identified Chengdu as one of the national hi-tech bases for the electronics sector. However, the relevance in the context of the argument here can be doubted as it likely was a response to the rise that was occurring already. Another — more relevant — aspect of the central-local relationship has emerged more recently. In the previous and current (12th) Five-Year Plan, a concerted effort appears in the making to "change lanes" in overall development by de-emphasizing exports as a pillar of growth in favour of domestic consumption. A large number of measures have already been taken, aimed at increasing domestic consumption. The more discernable shift towards the domestic market in central economic policy is part of the circumstances that made investment decisions of firms no longer as bound to east coast locations as before. It has started to become a new driver of inland location of especially foreign companies.

As to the local-provincial relations, a significant part of central government funds (made available to Sichuan) channelled through provincial bureaus and agencies, as well as provincial funds, have landed in the Chengdu metropolitan region. This is only in part explained by Chengdu being the dominant city in the province. As discussed extensively by Hong (1999), while the central government's treatment of Chongqing and Chengdu was biased in favour of the former throughout much of the reform period, Chengdu has enjoyed favoured treatment by the province since the 1990s and even more so since Chongqing became a Municipality directly administered by Beijing in 1997. Initially motivated by the wish to provide a counterweight to influence of the centre in Chongqing, central administration of the latter escalated into outright rivalry whereby Chengdu city leaders were served in a favoured way by provincial authorities.

Specifically with respect to the electronics industry, the province has acted as a catalyst to local policy formulation through setting provincial priorities. It was in the spirit of a policy made in 2000 by the Sichuan government that Chengdu City formulated a string of policies and measures to encourage the development of its electronic information industry. This included: the "Implementation Opinions of CPC Chengdu Municipal Commission and Chengdu Municipal People's Government on Accelerating Informatization and Information Industry Development"; "Policies and Opinions on Encouraging Software Industry Development"; "Implementation Details on Encouraging IC Design Industry Development"; and "Opinions on further Accelerating Software Industry Development" among others.

CONCLUSION

Chengdu has already assumed the status of one of the new "inland bases" of China's electronics industry. Besides describing quantitative growth and the profile of the local industry, an attempt has been made to provide a tentative interpretation of success. This has been done by addressing a number of questions: What is the nature and the source of comparative advantages? What has been the rationale behind the city targeting electronics as a suitable sector? What can explain the "choice" of the strategy? Finally, to what extent and in what way have local and provincial states prevailed over the central state? As to the first question, the specific interest is the "site" of the comparative advantages (apart from their origin) that enabled it to successfully grasp opportunities when they presented themselves. Is this "site" predominantly local indeed? Assuming an agency of state, has the local state been more important than central and provincial state?

Our scrutiny thus far produces the following observations. Chengdu is one case where the recent growth of electronics has not taken place "from scratch" or in a still rudimentary institutional environment and governance structure. The city has a longer history of manufacturing development, and the electronics industry has a longer presence in the city than the 2000s. Long before the recent surge of electronics growth, it also had established institutional arrangements and a governance structure in which networks with other levels of government were prominent. Consideration of these is necessary to correctly grasp the extent and nature of recent electronics development in the city and policies associated with it. The combination of inheritance, path-dependent institutions, existing variety on the one hand and local-provincial and local-central links under a favourable macro policy-environment on the other hand engendered policies and programmes that extended earlier developed agglomeration economies and industry organization. Congruency is apparent between these factors and the "choice" of development strategy of the industry in local context, aiming to capitalize on assets earlier developed and reproduced through the presence of technologically-related industrial variety. Congruency is also apparent between local strategy/policies and the needs of external firms that the strategy could reach producing "strategic coupling" from firms; however this "occurred" at a specific juncture, namely the second half of the 2000s and into the current decade when firms external to the locality and with rather large scale operations started to search for suitable alternative locations away from the coast. In this search, push factors played a large role. Finally, agglomeration economies accumulated as firms in different segments moved, subsequently further growing the industry as related variety became attracted to the place because of the potential of relational assets in accumulated localization economies.

Economic and fiscal decentralization, liberalization (marketization) and globalization meant significant empowerment of "the local" in China. Main elements of our understanding of the case indeed privilege "the local", however without excluding extra-local factors, now and in the past. As an industry such as electronics is heterogeneous, in principle policy "choices" present themselves at the local level. Our discussion points towards the operation of the local state at a particular time being shaped and circumscribed by a myriad of factors, and is contingent on time- and place-specific relations to other levels of government and governance. Local should not be framed just in policy terms. Thun's argument about institutional configuration appears a valid additional frame in view of the insights obtained in the functioning of institutions and the nature/scope of bureaucratic competences. For our case we are able to pinpoint the origin of the current local institutional configuration. This is related to the history of central-local relations, producing in several episodes development configurations that are inherited and constitute a legacy-shaping industrial organization and competences in the present. Thus, as to the understanding of institutional configuration and disposition, history and evolution matter.

Can the current success of the electronics industry development in Chengdu be maintained in the long run? Some argue that, due to the development path outlined above, the industry lacks a locally rooted structure and will become fragile and unable to get through when it faces more volatility in global markets and the retreat of foreign investment (Wang 2011). A different prospect is suggested by recent research into the evolution of the electronics industry in China's earliest developed SEZs (Meyer et al. 2012) that has demonstrated the possibility of localization following some period after the external implantation of (foreign) firms. It may be postulated then that what we see now in Chengdu is only the start of a virtuous development of the industry in — indeed — a new base in China. However, if a shift to a more endogenous development path can better guarantee long-term viability, it follows from the argument developed here that institutional change will be required.

Acknowledgement

The work described in this chapter was sponsored by grants obtained from the National Natural Science Foundation of China (No. 41101112); and from the Basic Research Fund of the Central Universities of China (Program No. 2011QNA3042 and Program No. 2011QNA3006).

Note

1 Hukou refers to the Mao-reinstituted household registration system that identifies the residency status of households and individuals according to urban/rural and place of birth. The practical significance of an individual's status is the entitlements to housing, healthcare, schooling, and other (government) services that are attached to it. Hukou has been blamed for the fact that many of the millions of migrants that, over the past decades, have moved from the rural areas to cities live on the margins of urban society. With several rounds of reform, individual provinces and cities are able to have their own scheme and criteria for granting local hukou to migrants. Chengdu has substantially relaxed its criteria.

References

Chengdu Municipal Government (CDMG)."Planning on the Development of Chengdu's Electronic Industrial Clusters". Internal Report, 2008.

———. "Chengdu is Making Significant Progress in the Construction of 'Two Hubs, Three Centers and Four Bases'", 16 December 2009, available at <http://www.sc.gov.cn/zwgk/zwdt/szdt/200912/t20091216_871207.shtml> (accessed 13 September 2011).

China Knowledge. *Chengdu Business Guide*. Singapore: China Knowledge Press Pte. Ltd., 2007.

Dell. "Dell to Build Flagship Manufacturing and Customer Support Center in Chengdu to Support Western China Growth". Retrieved from <http://content.dell.com/us/en/corp/d/secure/2010-09-16-chengdu.aspx> (accessed 1 September 2010).

Digitimes. "Sichuan's Five Advantages: Attracting Electronics Enterprises towards Western Regions in China". *Digitimes*, 23 May 2011, available at <http://www.digitimes.com.tw/tw/dt/n/shwnws.asp?CnlID=13&Cat=&Cat1=&id=234101> (accessed 7 September 2011*a*).

———. "Being the Fourth Biggest National Production Base in Portable Computer, Chengdu is Ranked Number One in the Production of Portable Computer in Interior and Western China". *Digitimes*, 22 July 2011, available at <http://www.digitimes.com.tw/tw/dt/n/shwnws.asp?Cat=&id=243332> (accessed 2 September 2011*b*).

Ding, R.Z. and Li, S.M. "Vitalize the Old Electronics Industrial Base of Eastern Suburban Chengdu and Construct the Strategic Pivot of Chengdu's Electronics Industry". *Journal of University of Electronic Science and Technology of China* (Social Sciences edition) 3, no. 3 (2001): 97–101. In Chinese.

European Business and Innovation Centre Network. *Guide to Invest in Electronic Information Industry of Chengdu*. 2008. Full text available at <http://www.ebn.be/assets/assets/pdf/events/>.

Fan, C. and A. Scott. "Industrial Agglomeration and Development: A Survey of Spatial Economic Issues in East Asia and a Statistical Analysis of Chinese Regions". *Economic Geography* 79, no. 3 (2003): 295–319.

Financial Times. "Foxconn to Move China Jobs Inland". 3 March 2011.

Gereffi, G. "Development Models and Industrial Upgrading in China and Mexico". *European Sociological Review* 25, no. 1 (2009): 37–51.

He Canfei. "Industrial Agglomeration and Economic Performance in Transitional China". In *Reshaping Economic Geography in East Asia*, edited by Huang Yukon and Alessandro Magnoli Bocchi. Washington D.C.: World Bank, 2009.

He Canfei and J. Wang. "Regional and Sectoral Differences in the Spatial Restructuring of Chinese Manufacturing Industries during the Post-WTO Period". *Geojournal* 77, no. 3 (2012): 361–81.

He Canfei, Yehua Dennis Wei, and XiuzhenXie. "Globalization, Institutional Change, and Industrial Location: Economic Transition and Industrial Concentration in China". *Regional Studies* 42, no. 7 (2008): 923–45.

HKTDC. "China's New, Improved World Electronics Factory; Moving up the Value Chain in the 12th Five-Year Programme". HKTDC Forum, available at <http://www.hktdc.com/info/mi/a/ef/%20en/1X07FRR0/1/>, consulted 2 September 2011. Hong Kong: Hong Kong Trade Development Council, 2011*a*.

———. "Multinational Electronics Firms 'Go West' in China". Trade Watch May 2011. Hong Kong: Hong Kong Trade Development Council, 2011*b*.

———. "Central China the New Location for Electronics Production". Trade Watch May 2011. Hong Kong: Hong Kong Trade Development Council, 2011*c*.

Hong, L. "A Tale of Two Cities. A Comparative Study of Political and Economic Development of Chengdu and Chongqing". In *Cities in China. Recipes for Economic Development in the Reform Era*, edited by Chung J.H. London: Routledge, 1999.

Invest Chengdu. "Status Quo of Electronic and IT Industry of Chengdu", available at <http://www.chengduinvest.gov.cn/EN/htm/detail.asp?id=251> (accessed 1 September 2011).

Jiang, J. "Chengdu: the Highland of Inland and Western Regions in IC Industry". *Chinese Electronics Newspaper*, 20 July 2010, available at <http://www.cena.com.cn/view/2010-07-20/127960991445389.shtml> (accessed 1 September 2011).

Jones Lang Lasalle. "China 40. The Rising Urban Stars". London/Shanghai: Jones Lang Lasalle World Winning Cities Global Foresight Series, 2008.

Kang, Z. "Soft Chengdu vs. Hard Chongqing: A War between Them in Pursuing a Development Pattern of Chicago Model". *Chinese Entrepreneur*, 22 September 2010, available at <http://finance.ifeng.com/city/csjs/20100922/2648547.shtml> (accessed 30 August 2011).

KPMG Advisory. "Offshoring and Outsourcing in Chengdu 2010". Beijing: KPMG, 2010.

Lu, L. and Y.D.Wei. "Domesticating Globalisation, New Economic Spaces and Regional Polarisation in Guangdong Province, China". *Tijdschrift voor Economische en Sociale Geografie* 98, no. 2 (2007): 225–44.

Luo, T. "Taiwanese IT Giants Introduced One Hundred Relative Enterprises into Chengdu". *Huaxi Metropolitan News*, 6 May 2011, available at <http://news.ccidnet.com/art/945/20110506/ 2382185_1.html> (accessed 30 August 2011).

Meyer S., D. Schiller, and J. Revilla Diez. "The Localization of Electronics Manufacturing in the Greater Pearl River Delta, China: Do Global Implants put down Local Roots?" *Applied Geography* 32 (2012): 119–29.

Miu, Q. "Chengdu Building the 'World Office' of Electronics and Information". Chengdu Daily, 15 October 2010, available at <http://ent.ifeng.com/zz/detail_2010_10/15/2788147_0.shtml> (accessed 13 September 2011).

Naughton, B. "The Third Front: Defence Industrialization in the Chinese Interior". *The China Quarterly* 115 (1988): 351–86.

People's Daily. "Labor-intensive Electronics Manufacturers Eye Inland Cities". Published 9 June 2010.

Quan. "General Situation of Chengdu's Software Industry". *Chinese Electronics Newspaper*, 7 April 2011, available at <http://it.cena.com.cn/news/2011-04-27/130388096055612.shtml> (accessed 31 August 2011).

Thun, E. and A. Segal. "Thinking Globally, Acting Locally: Local Governments, Industrial Sectors, and Development in China". *Politics and Society* 29, no. 4 (2001): 557–88.

Thun, E. "'Keeping up with the Jones': Decentralization, Policy Imitation, and Industrial Development in China". *World Development* 32, no. 8 (2004): 1289–308.

———. *Changing Lanes in China: Foreign Direct Investment, Local Governments, and Auto Sector Development.* New York: Cambridge University Press, 2006.

Walcott, S.M. "The Dragon's Tail: Utilizing Chengdu and Chongqing Technology Development Zones to Anchor West China Economic Advancement". *Journal of Chinese Economic and Business Studies* 5, no. 2 (2007): 131–45.

Wang, B. "International IT Giants Settle in Western China". *IT Times*, 10 January 2011, available at <http://tech.163.com/11/0110/22/6Q2P053J000915BD.html> (accessed 30 August 2011).

Wang, C.C. and G.C.S. Lin. "The Growth and Spatial Distribution of China's ICT Industry: New Geography of Clustering and Innovation". *Issues & Studies* 44, no. 2 (2008): 145–92.

Wang, J. and L. Mei. *Dynamics of Labour-intensive Clusters in China: Relying on Low Labour Costs or Cultivating Innovation?* Geneva: International Institute for Labour Studies, 2009.

Wang, J. "Dynamics of Export-Oriented Industrial Districts in China". Presentation at International Workshop "Economic Development and Regional Differentiation in China: Trends and Opportunities". Florence 10 November 2010.

Wang, P. "The Rise of IT Industry and the Rise of Chengdu". *National Business Daily*, 19 July 2011, available at <http://finance.sina.com.cn/roll/20110719/020310167747.shtml> (accessed 2 September 2011).

Wang, X.J. and Meng, F.H. "The Rising of Chengdu's Software Industry with the Support by Local Government". *Chengdu Daily*, 16 April 2007, available at http://news.sohu.com/20070416/n249454985.shtml (accessed 13 September 2011).

Wei, Y.D., W. Li, and C. Wang. "Restructuring Industrial Districts, Scaling Up Regional Development: A Study of the Wenzhou Model, China". *Economic Geography* 83, no. 4 (2007): 421–44.

Xiao, J. "Western Airport New Area in China under Construction". *Huaxi Metropolitan Newspaper*, 3 June 2009, available at <http://news.sina.com.cn/o/2009-06-03/064015725218s.shtml> (accessed 14 September 2011).

Yeung, H.W.C. "Regional Development and the Competitive Dynamics of Global Production Networks: an East Asian Perspective". *Regional Studies* 43, no. 3 (2009): 325–51.

Zhang, J. "Interjurisdictional Competition for FDI: the Case of China's 'Development Zone Fever'". *Regional Science and Urban Economics* 41 (2011): 145–59.

Zhang, S. "Chengdu's Electronics Industry: Flying above the Clouds". *Sichuan Daily*, 25 July 2011, available at <http://news.163.com/11/0725/06/79PPAFGC00014AED.html> (accessed 5 September 2011).

Zhang, Z.H. "Chengdu: Becoming the Fourth Largest Aviation Hub in China". *Chengdu Evening News*, 13 September 2011, available at <http://news.163.com/11/0913/05/7DQCTHAV00014AED.html> (accessed 14 September 2011).

Zeng, D.Z. (ed). *Building Engines for Growth and Competitiveness in China. Experience with Special Economic Zones and Industrial Clusters.* Washington D.C.: World Bank, 2010.

AFGHANISTAN
CHINA
PAKISTAN
NEPAL
New Delhi
INDIA
BHUTAN
BANGLADESH
MYAN
Chennai
TAMIL NADU
Arabian Sea
Andaman Sea
Bay of Bengal
SRI LANKA
INDO
LEGEND
State/Provincial Capital
STATE/PROVINCE
National Capital
COUNTRY
NEIGHBOURING COUNTRY
Geographic Features
National Border Land
National Border Water
State/Province Border
Border Dispute
0 km 150 300
All maps are derived from GADM Data (Global Administrative Areas, www.gadm.org, 2012) edited by maps&more, Hans Hortig / Marcel Jäggi

8

THE ELECTRONICS INDUSTRY IN TAMIL NADU, INDIA: A REGIONAL DEVELOPMENT ANALYSIS

P. Vigneswara Ilavarasan and Francis E. Hutchinson[1]

Once known for its over-regulated and autarkic economy, in recent times India has emerged as a leading exporter of software and IT-enabled services. Following sweeping domestic economic reforms in 1991 and the advent of international outsourcing, India was able to capitalize on its ranks of cost-effective, skilled labour to emerge as a global player in these emerging service industries.

In contrast, India's role as electronics producer is less well-known. Despite its array of state-owned electronics-producers and tradition of electronics production dating back to the 1960s, it has generated only a fraction of the jobs or export earnings of its better-known sibling.

However, this soon may change. The country's growing affluence, urbanization, and IT-literacy, coupled with evolving consumption patterns, mean that India has emerged as a target market of choice for not just consumer electronics, but also other sub-sectors such as office equipment and telecommunications products.

The combination of India's domestic market and its reputation as an attractive site for skill-intensive tasks has seen a large number of electronics-producing market leaders establish a presence in the country. In addition to production facilities, this has also entailed research and design centres geared to understanding the growing and increasingly attractive Indian market. This rapidly expanding market has also offered opportunities for local producer firms to emerge in product spaces such as the manufacture of mobile phones.

The consolidation of local and international market leaders operating in India has, in turn, enabled the country to emerge as a significant electronics exporter, particularly in areas such as: audio/video products; testing and measuring equipment; medical devices; and telecommunication products.

Just as the software sector has changed India's economic landscape, the hardware sector now promises to do the same. New centres of production are emerging, as lead firms and suppliers coalesce in specific areas. While traditional centres of electronics production such as Uttar Pradesh and Karnataka remain relevant for domestic production, Tamil Nadu is emerging as the pre-eminent centre of electronics manufacturing for export.

Over the past two decades, this state of 62 million people in the south of the country has undergone a far-reaching structural transformation. The agricultural sector is making way for the services sector, and Tamil Nadu's industrial sector is being renewed. Traditionally a centre for manufacturing, low-technology activities such as cotton processing and textile fabrication are now making way for the automotive and electronics sectors.

In order to understand why Tamil Nadu has become a new site of electronics production, it is necessary to examine how and what way the state government shaped the local environment for business in general and manufacturing in particular. While it is one of India's more developed states with attractive attributes for skill-intensive manufacturing, these assets do not automatically translate into investment and growth. Thus, while Tamil Nadu possesses a long-standing industrial tradition, it is not the only state with this heritage. And, a review of its economy in the early 1990s found that there were few modern entrepreneurs and the bulk of Tamil Nadu's manufacturing activities were low-value added tasks (Swaminathan 1994). Furthermore, while the state's deep stocks of college graduates undoubtedly played a role in attracting investment into these new, skill-intensive sectors, it is noteworthy that Tamil Nadu was not an early leader in the software sector. Indeed, as late as 1997, the state accounted for a mere 7 per cent of India's revenue from the IT sector (GoTN 1997).

Thus, the answer behind this structural transformation is more likely to be found in exploring how Tamil Nadu acquired or developed a competitive advantage specific to the manufacturing sector and was able to capitalize on it. This chapter will therefore examine how and in what way the state government has attempted to foster the development of its electronics sector. It uses relevant theoretical constructs from a regional development perspective in the analysis.

The chapter is divided into six sections. The first will set out the framework with which the electronics sector will be analysed. The second section will provide a broad outline of the Indian electronics industry, before focussing on its characteristics in Tamil Nadu. The third section will set out the public administration structure in India and highlight the different roles played by the national and state governments. The fourth section will examine how the Tamil Nadu state government has evolved in response to the national and local-level political context. The fifth section will analyse the state government's key policies and how they have shaped the environment within which the electronics industry operates. The final section will conclude.

ANALYZING HIGH-TECHNOLOGY CLUSTERS

It is well-known that firm groups or "clusters" benefit from agglomeration economies. For example, close proximity enables firms in a cluster to benefit from knowledge spillovers and greater aggregate demand facilitates the development of thicker labour markets. However, beyond mere physical proximity, clusters can differ in many ways.

Based on her work on high-technology clusters or "industrial districts" in the United States, Japan, Korea, and Brazil, Markusen (1996) argues that, besides the quintessential Marshallian industrial district — which benefits from the traded interdependencies that arise because of agglomeration — there are other variants. She proposes a typology based on: structure; degree of inter-firm collaboration; local or international orientation; existence of supporting services; and role of local government. Markusen puts forward several types, including: Italianate; Hub-and-Spoke; and Satellite. These different types of clusters offer distinct possibilities for: inter-firm collaboration; the development of a unique local culture; and innovative potential.

The "Italian" type of industrial district is epitomized by the small, artisanal firm clusters found in Northern Italy. In addition to enjoying agglomeration economies, these clusters are characterised by high levels of innovation and dynamism, as well as very active local governments and trade associations. These intermediary organizations: provide shared amenities; attempt to solve collective action problems; and market the region externally. Firm groupings with these characteristics have good long-term prospects if they can generate and sustain sufficient levels of innovation and dynamism. The depth and variety of traded and untraded interdependencies between firms make economic activity highly location-specific and unlikely to relocate.

The Hub-and-Spoke industrial district is sustained by one or more lead firms which generate demand for an array of supplier and supporting services locally and elsewhere. Clusters of this type are characterized by long-term and stable relationships between firms along individual value-chains, but not between lead firms. Lead firms may provide technical support and inputs to supplier firms but, as a result, the clusters themselves do not tend to generate shared facilities or engage in collective learning. Local government activities tend to focus on catering to the needs of lead firms, rather than to those lower down the value chain.

Over time, a regionally-specific culture may develop as firms and workers come to identify with the local context, although workers will tend to identify with lead firms first rather than the district in question. As regards long-term prospects, these clusters depend largely on the fate of the lead firm or firms. That said, if these anchor firms generate enough externalities, they may attract additional firms, which can then help the local economy diversify.

Satellite platforms are a group of branch plants of firms headquartered elsewhere. The branches undertake tasks that may reach a substantial level of sophistication, but follow key decisions made at global or regional centres located outside the cluster. As a result, there is relatively little inter-firm collaboration within the district, and business and inter-personal links are almost all external. Labour mobility, technical support, and communication take place between branches of the same firm and, as a result, there is little development of a locally-rooted culture. While local governments may supply infrastructure and tax incentives, they do not contribute to technical inter-change or collective efforts among firms. Although Satellite Platforms do generate jobs and income, the absence of a unique institutional context or locally-rooted culture means that investment can be easily lured away by competitors.

The firm clusters set out above are ideal types and, as a result, a specific industrial district may possess a combination of traits. In addition, firm groupings and local institutional contexts are dynamic and can evolve over time. Therefore, while there is no certainty that a firm cluster with, say, the characteristics of a satellite platform, will become a hub-and-spoke cluster, the second type of industrial district certainly holds more potential in the way of stability and value-added. Thus, policy-makers need to attract a range of firm types, as opposed to solely branch plants, and implement measures that can improve their territory's potential for the development of a local work culture, more inter-firm exchange, and more value-added activities.

Having put forward a framework for analyzing various types of clusters, the next section will describe the structure and recent growth of the Indian electronics industry, and the subsequent section will apply Markusen's framework to the Tamil Nadu electronics cluster.

THE ELECTRONICS INDUSTRY IN INDIA AND TAMIL NADU

The growth of the Indian information technology (IT) industry is the story of modern India. From its incipience in the 1980s, it has become one of the pillars of the country's economy. In 2010, the IT sector, which includes both hardware and software, was estimated to constitute 6.1 per cent of GDP and directly employ some 2.3 million people (NASSCOM 2011). However, much of this growth has been concentrated in the software sector. According to Electronics and Computer Software Promotion Council (ESC 2011), the hardware sector accounts for only 27 per cent of the total IT sector production of US$103 billion.

However, while starting from a low base, the electronics sector is growing. According to the ESC, in 2010–11, the country's production of electronic hardware was some US$27.6 billion, up from US$12.8 billion in 2005–06. Indeed, over that five year period, production grew at 17 per cent per year. However, in global terms, this is still very small, with India accounting for a mere 1.5 per cent of total production of electronics hardware.

Fuelled by high demand for consumer electronics as well as the increasing penetration of personal computers, production is largely oriented to the domestic market. Thus, in 2007–08, only 23.6 per cent of total production was for export. That said, exports are significant in absolute terms, accounting for some US$8.8 billion in 2010–11. Key exports are electronics components, followed by telecommunications equipment. Other significant items include solar cells, telephone components, PC parts, and television reception apparatus (ESC 2011, pp. 52–53).

Joseph (2004; 2007) argues that the development of the Indian electronics industry can be divided into three phases: the pre-1980s period; the 1980s; and the post-1991 period. The first phase was the import substitution period, where the "thrust was on self-reliant growth in tune with the general industrial/technology policy framework" (2004, p. 101). Thus, the country was largely closed to foreign direct investment, and sought to develop its own domestic technological capabilities through national "champions". The second phase was characterized by controlled liberalization, where "the government implemented a number of initiatives in the direction of a more liberal and open electronics policy" (2004, p. 103). Prior to this, electronics production was almost entirely state-run and these policy changes allowed a small domestic private sector to emerge. The third phase reflected increasing globalization, where there was "abolition of industrial licensing in almost all the electrical components and equipment manufacture except for aerospace and defence equipment (strategic electronics) and consumer electronics" (2004, p. 104).

Recent policies have deepened the liberalization process. FDI is actively courted by the Indian government, and restrictions have been reduced. For example, government permission or licensing for manufacturing activities has been eliminated, with the exception of aerospace and defence equipment. And, subject to approval, private sector firms are allowed to compete in the defence sector. Operations can be set up anywhere in the country, and require minimal paperwork with the Ministry of Industry (ESC 2011, p. 17).

How has the electronics sector fared under the new, more open paradigm? Research by Joseph (2004; 2007) argues that there has been increasing investment in research and development and significant collaborations with foreign firms for technology transfer. However, an assessment of the hardware industry over the 1994–2004 period by Majumdar (2010) argues that, despite liberalization policies, the performance of the industry in terms of total factor productivity is not significant, suggesting a concentration of low-end technology products in the industry.

An analysis of the composition of electronics sector can provide some indication of its relative skill- and capital-intensity. Table 8.1 sets out the educational qualifications across the various electronics sub-sectors. Table 8.2 sets out the production of electronics goods in India by sub-sector over the period 2006–11 in monetary terms, and Figure 8.1 depicts their relative importance.

Table 8.1
Breakdown of Educational Qualifications across the Electronics Sector in India
(in percentage terms)

	PhDs/Research	Engineers	MBA	Other Grads	Diploma/Voc	Secondary
Consumer Electronics	3	9	22	27	17	22
Computer Hardware	4	20	6	19	35	16
Telecommunications Equipment	5	40	6	15	27	22
Strategic Electronics	6	37	5	10	32	10
Components	6	28	2	3	29	32
Manufacturing	1	20	2	2	35	40
Design	25	60	1	8	5	1

Source: MAIT 2008; Primary survey of 22 firms.

Table 8.2
Evolution of Electronics Production in India
(USD billion)

	2006–07	2007–08	2008–09	2009–10	2010–11
Consumer Electronics	4.6	5.6	5.6	6.4	7.3
Instruments, Office Equipment, Medical Equipment	3.5	4.4	4.3	4.3	5.7
Components	2.0	2.4	2.4	2.8	4.2
Telcom Equipment	2.2	4.6	5.4	6.6	7.1
Computer Hardware	3.0	3.9	2.9	3.0	3.3
Total	15.3	20.9	20.6	23.1	27.6

Source: ESC 2011.

Figure 8.1
Composition of the Electronics Sector in India

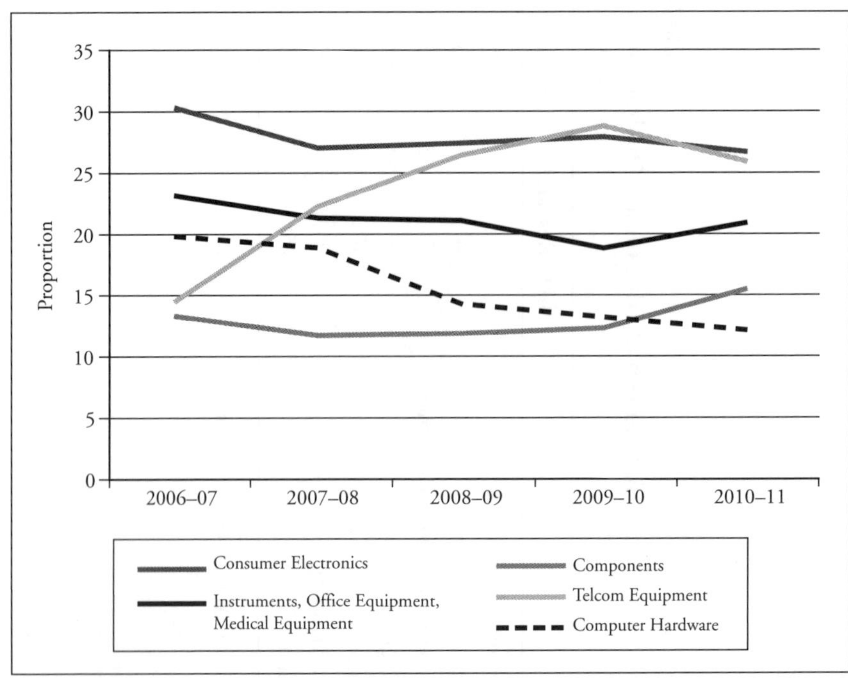

Source: ESC 2011.

As can be seen, consumer electronics is a relatively low-skill sub-sector, requiring comparatively few research staff or engineers. A significant proportion of staff are involved in marketing and administration, as well as low-end production work (17 per cent diploma/vocational and 22 per cent secondary education). Computer hardware, telecommunications equipment, and strategic electronics are more skill-intensive, requiring more research capacity and substantially greater engineering capacity. Conversely, a relatively smaller proportion of the workforce is involved in low-end manufacturing operations. The components subsector overall is quite research-intensive and technical. However, it is comprised of two very distinct activities. The manufacturing aspect of component production requires virtually no research capacity and only modest engineering capacity, with the bulk of the workforce comprised of relatively unqualified production operators. However, the design aspect is highly research-intensive and technical in nature, with almost no low-end labour tasks.

Turning now to the relative performance of the sub-sectors, the two biggest — consumer electronics and telecommunications equipment — cater directly to India's large domestic market. Consumer electronics, consisting largely of televisions, DVD players, and watches, was worth US$7.3 billion dollars in 2010–11. Telecommunications equipment was worth US$7.1 billion the same year and was driven by demand for mobile phones, telephone sets, and television receivers. This sub-sector has seen the most growth over the past five years, and will most likely become the biggest in the near future, driven by the country's burgeoning mobile phone ownership rates.

The other more capital-intensive sectors have also grown, but depend less on the mass market. Thus, instruments, office equipment and medical components (such as X-ray equipment, rectifiers, and medical devices) as well as components (including printed circuits, solar cells, and distribution panels) have more or less doubled over this period. Interestingly, computer hardware production has stayed steady, despite reports of increasing PC penetration. Indeed, in relative terms, this sub-sector is now the least important.

Thus, over 2005–11, the electronics sector has expanded significantly — almost doubling in only five years. While all sub-sectors have expanded, the relatively less sophisticated sub-sectors of consumer electronics and telecommunications equipment have increased from 44 per cent of the total to 52 per cent over this period. In relative terms, components production has increased only slightly and computer hardware production has actually gone down.

While more research is necessary before arriving at definitive conclusions as to the evolution in the technological sophistication of the sector, it is possible that Joseph's (2007) and Majumdar's (2010) arguments are not incompatible. There may be more R&D operations and technology transfer agreements between MNCs and local firms, but these are in relatively less technologically-intensive areas that cater to the domestic market and where local players have an edge in understanding local tastes and needs.

The Geographic Distribution of the Electronics Sector

At present, the electronics sector is comprised of more than 3,500 firms employing more than 780,000 workers across India. Traditionally, production has clustered in the west and north, particularly in Maharashtra and Noida in Uttar Pradesh (MAIT 2010). At present, there are three large concentrations in the country: in the north around the National Capital Region, which represents some 37 per cent of production; in the west, around Mumbai and Pune, which produces approximately 25 per cent; and in the south, in Karnataka, Tamil Nadu, and Andhra Pradesh, which accounts for 32 per cent (ESC 2011, p. 51).

Table 8.3 shows the value added by sub-sector in the key electronics producing states of Tamil Nadu, Karnataka, Uttar Pradesh, and Maharashtra for the period 2000–10. Looking at the production of components, Tamil Nadu has emerged to become, along with Uttar Pradesh, one of the top two components producers in the country. However, Tamil Nadu began from a much lower base, producing only INR1.2 billion in 2000–01 compared to Uttar Pradesh's INR28 billion. In the other sub-sector, comprising consumer electronics, computers, and communication equipment, Tamil Nadu is roughly on a par with Karnataka and Uttar Pradesh, and far behind the prime producer, Maharashtra.

In terms of firm numbers and employment, Tamil Nadu has some 150 firms which employ 25,000 workers. This is substantially below Maharashtra's 220 firms and 34,000 workers, but higher in employment terms than either Uttar Pradesh or Karnataka. On average, firms in Tamil Nadu and Uttar Pradesh employ more people per firm than do those in Maharashtra or Karnataka.

Thus in terms of electronics production, Tamil Nadu is an important centre particularly for electronics components, but it is not the most

Table 8.3
Value Added by Sub-Sector in Tamil Nadu and Selected States (2000–10)
(Rs m)

Components	**2000–01**	**2001–02**	**2002–03**	**2003–04**	**2004–05**	**2005–06**	**2006–07**	**2007–08**	**2008–09**	**2009–10**
Tamil Nadu	1,250	9,580	10,140	88,590	55,720	54,060	47,940	60,450	308,290	572,260
Maharashtra	19,2730	133,510	28,100	62,910	136,750	77,020	54,580	152,630	28,210	75,650
Karnataka	91,150	120,920	95,650	171,370	203,450	269,490	279,160	10,220	–36,600	16,090
Uttar Pradesh	281,860	410,450	835,300	949,320	530,320	469,940	860,240	675,810	696,040	602,680
Computers, Communication Equipment, Consumer Electronics	**2000–01**	**2001–02**	**2002–03**	**2003–04**	**2004–05**	**2005–06**	**2006–07**	**2007–08**	**2008–09**	**2009–10**
Tamil Nadu	217,070	96,000	96,570	52,910	141,970	174,200	117,810	715,300	481,850	1,373,260
Maharashtra	940,660	464,540	1,136,220	1,030,990	1,041,550	1,381,430	1,538,460	2159,390	2,839,800	3,268,260
Karnataka	425,210	561,860	617,040	712170	714,940	969,820	787,720	2,052,570	812,860	1,119,960
Uttar Pradesh	650,300	758,900	985,940	1,051,310	937,490	1,416,570	327,640	658,170	574,710	1,681,310

Source: Authors' compilation from *Annual Survey of Industries*, Govt. of India, various years.

Table 8.4
Firms and Workers in the Electronics Sector in Tamil Nadu and Selected States (2000–10)

Firms	**2000–01**	**2001–02**	**2002–03**	**2003–04**	**2004–05**	**2005–06**	**2006–07**	**2007–08**	**2008–09**	**2009–10**
Tamil Nadu	138	149	95	97	103	110	101	104	166	147
Maharashtra	286	234	228	224	222	223	224	232	267	223
Karnataka	235	234	207	201	211	173	183	221	193	208
Uttar Pradesh	163	108	120	133	128	143	149	129	132	118
Workers	**2000–01**	**2001–02**	**2002–03**	**2003–04**	**2004–05**	**2005–06**	**2006–07**	**2007–08**	**2008–09**	**2009–10**
Tamil Nadu	4,694	6,022	4,836	6,297	7,264	6,049	5,616	6,662	17,772	25,429
Maharashtra	25,477	20,098	19,538	21,556	21,842	23,773	26,743	29,032	25,167	33,956
Karnataka	13,320	11,664	11,453	11,485	11,846	9,904	11,963	25,074	10,239	12,638
Uttar Pradesh	12,569	11,739	13,565	14,032	15,854	20,332	17,641	22,294	20,573	21,283

Source: Authors' compilation from *Annual Survey of Industries*, Govt. of India, various years.

important in employment or value-added terms. However, the picture changes substantially with regards to production for export. Rather than the industry being distributed across different regions of the country, it is clearly grouped in the south, which accounts for 66 per cent of all electronics hardware exports. Tamil Nadu alone accounts for 46 per cent of the total, and other major exporting states include: Karnataka (17 per cent); Uttar Pradesh (14 per cent); and Maharashtra (12 per cent) (ESC 2011, p. 57).

The Electronics Sector in Tamil Nadu

Tamil Nadu's economy has grown at more than 5 per cent p.a. for the last ten years. In 1999, the state's per capita GRP was INR19,432 and in 2009, it was INR30,652. The state's economy has grown roughly at the same pace as the national average, allowing it to remain one of the country's wealthier states. Roughly on a par with Gujarat, Tamil Nadu is substantially wealthier than Andhra Pradesh or Karnataka, but not as wealthy as the fertile states of Haryana and Punjab (GoTN 2009).

As with the national growth trend, this overall economic expansion at the state level has been driven by a structural transformation, as the economy has shifted towards more complex and high value-added activities. Over the last two decades, the contribution of the agricultural sector to Tamil Nadu's GRP has decreased from approximately one-fifth of the total to 12.5 per cent. The service sector, for its part has grown dramatically, from 40 per cent in 1990 to 62 per cent in 2010. This has been driven, in large part, by the burgeoning software and IT-enabled services industries. In contrast, the manufacturing sector has shrunk, from 40 per cent to 26 per cent over

Table 8.5
Composition of Tamil Nadu's Gross Regional Product

	1980–81	1990–91	1999–2000	2010–11
Primary	25.9	20.1	19.7	12.5
Secondary	38.1	39.6	29.9	25.8
Tertiary	36.0	40.3	50.4	61.6

Source: Data from Ministry of Statistics and Programme Implementation, Government of India. www.indiastats.com.

the same period. However, this shrinkage belies important changes within the sector itself.

Manufacturing in Tamil Nadu is well-established, dating back to its central role as a port and trading centre during the colonial period. Much of this manufacturing has centred on the textile industry which, even today, is the state's biggest employer, following agriculture. Other important traditional activities include apparel, leather goods, and paper products (Maclay 2001; GoTN 2009). The state also has firms engaged in precision engineering, petrochemical operations, and instrument fabrication.

However, over the past fifteen years, Tamil Nadu has seen the emergence of two new industries. In 1995 and 1996, the state received large-scale investments from Ford, Hyundai, and Mitsubishi to manufacture autos for the local market. Since that time, Tamil Nadu has built on these investments and is now the second most important auto-producer in India. And, as mentioned above, Tamil Nadu has emerged as the pre-eminent centre for electronics export in the country.

As mentioned, in 2010, the electronics sector was comprised of some 150 firms and 25,000 workers. Of these, some 30 are major electronics manufacturers, which in turn have been joined by an important number of supplier firms (interview, Tamil Nadu Industrial Guidance and Export Promotion Bureau, 30 June 2012). The state is also home to a significant number of production facilities for large local manufacturers.

The Tamil Nadu electronics sector can be broken down into the following sub-sectors:

Components. There are a number of important semiconductor firms in the state, including: Texas Instruments, which set up its second R&D facility in the country there; Cypress Semiconductor; SMSC; and Identive. Firms in the testing, packing, and design fields include: TAPP Semiconductor; SPEL Semiconductor; and Suriya Technology Design Services. There are also a number of solar cell manufacturers including Moser Baer and Signet Solar.

Telecommunications. Nokia has its largest facility in the world in Chennai, employing over 11,000 people and investing US$300 million in its own industrial park. The rest of this park is filled by its supplier firms, such as Salcomp, Perlos, Wintek, and Laird Technologies. Other lead firms include Motorola, Nokia-Siemens, and Sony-Ericsson, who are joined by components suppliers such as BYD, Polymatech, and Shinetsu. Huawei has established a US$500 million telecommunications equipment facility in the state; as has NEC, who has established a centre of excellence for mobile networks; and Spice Mobile, a local mobile phone manufacturer, also has operations there.

Contract Electronics Manufacturers (CEM). Drawn by the telecommunications and semiconductor firms, CEM majors such as Flextronics, Jabil, and Foxconn have established operations in the state, producing components for mobile handsets as well as other hardware items. Sanmina-SCI also employs 1,500 people in Tamil Nadu and provides printed circuit board assembly services, among others.

Computers. Dell has a facility that can produce 2 million personal computers a year; Accel manufactures hard disk drives; and HCL Infosystems makes memory cards.

Consumer electronics. Samsung produces LCD televisions and colour monitors; the local firm Videocon makes LCD and plasma televisions, as well as home theatre systems; Akai Electronics produces a range of audio and video products; and Panasonic has its affiliate for manufacturing small appliances headquartered in the state.

Sources: Guidance Bureau 2011; ESC 2011; and trade reports.

At present, the bulk of the firms mentioned above are clustered around Chennai. Most are located in special economic zones (SEZs), which are approved by the central government and supported by the state government. There are 69 approved SEZ across the state, with the majority in and around Chennai.

However, the state's second city, Coimbatore, also has a strong manufacturing tradition and is home to a growing number of firms. For example, Dell has production facilities in both Chennai and Coimbatore. Other firms in the city and their products are: Robert Bosch — industrial equipment; Indus Electronics — industrial equipment; V-Tech Engineering — modems; UMS Technologies — medical devices and computer components; and Sunmac Enterprises — solar panels and lighting (company websites).

What can be said about the nature of the electronics sector in Tamil Nadu? Its overwhelming export orientation, predominance of large firms, and large number of branch plants would strongly suggest that it resembles a Satellite Platform. While there are a number of large local firms in the state, many of them — such as HCL and Videocon — are headquartered elsewhere. While generating jobs and income, these clusters are highly vulnerable to competition, as their competitive advantage is not territorially-rooted and is easily replicable. Their outward orientation means that it is hard for inter-firm relationships to be established and a local culture to emerge.

However, the electronics sector in Tamil Nadu also displays characteristics reminiscent of a Hub-and-Spoke cluster, especially with regard to mobile phone manufacture. There are a number of lead firms, the most obvious being Nokia, that have close relationships with a number of supplier firms that are also present in the state. This is complemented by inter-relationships between lead firms in the telecommunications and component sectors and contract electronics manufacturers. While much of this is between MNCs and international suppliers, this still implies a significant amount of inter-firm interaction that takes place locally and would be more difficult for competitor regions to replicate. Furthermore, there is greater potential for the development of a unique local work culture in the future that would add to the sector's rootedness in the state.

Thus, the electronics sector in Tamil Nadu presents a combination of characteristics. Given that it is a relatively new industry led by electronics multinationals that has developed very quickly, it is understandable that it resemble a Satellite Platform. In addition, structural characteristics of the Indian economy and national innovation system are likely to exacerbate this. The country has relatively low R&D spending, a limited level of technology absorption, and limited facilities for standards and testing (EXIM Bank 2011). Despite this, the Tamil Nadu electronics sector also demonstrates some characteristics of a Hub-and-Spoke cluster that has the potential of generating deeper firm links and an indigenous firm culture.

The next sections will seek to understand why the electronics sector in Tamil Nadu has developed so quickly and what role, if any, the state government played in this transformation. This will be supplemented by comparisons of key indicators with other electronics-producing states to see how Tamil Nadu fares. However, before that, the next section will set out the overall context within which state governments operate.

INDIA AND ITS STATES

India is a Sovereign Socialist Secular Democratic Republic with a parliamentary system of government. It has a federal governance structure of 28 states and seven union territories. Each state is administered by a Governor, a Chief Minister and Council of Ministers. The real executive power lies in the hands of the Chief Minister and their Ministers. They are answerable to the Legislative Assembly, whose representatives are elected directly by the people.

The 1950 Constitution sets out central and state government responsibilities, in addition to a list of concurrent responsibilities that are shared. The central government is charged with: defence; international relations; currency management; banks; and insurance. The state governments' list includes: agriculture; infrastructure; water supply; health; pre-university education; justice administration; local government; and urban development (Frankel 2005; Reddy 1988). The concurrent list includes: higher education; power development; and economic and social planning (Guhan 1995).

Until the early 1990s, the central government maintained a complex apparatus of economic regulation, particularly through the industrial licensing system, which gave it the power to decide the location and production capacities of individual firms. However, the 1991 measures removed this layer of central control, exposing the underlying regulatory structure at the state level. Beyond their prerogative over land, infrastructure, and urban planning, state governments have re-regulated some aspects of the policy framework, seeking to establish greater control over the economy or assert their own regional autonomy (Sinha 2004). In addition, the demise of the Congress Party's "one-party dominance" in the 1980s has led to the proliferation of locally-based regional parties, lending increasing importance to state-level agendas and issues.

The relationship between the central and state governments has thus changed. The central government's support in terms of finances to the state government has decreased and state governments are asked to "effect economic growth in their respective states" (Rudolph and Rudolph 2001, p. 1541). However, the central government has moved to increase its regulatory control over the states through, for example, limiting the amount of debt state governments can incur. In addition, it has also tried to impose a layer of federal executive agencies overseeing state government prerogatives in areas such as police investigation, security, and the administration of special economic zones (Oxford Analytica 2009).

The relations between state governments have changed markedly in this new context. Prior to 1991, state governments competed with each other for a finite quantity of resources from the central government — what Sinha terms "vertical competition". Since 1991, they have begun to compete with each other to attract investment from a greater range of actors. In addition to courting foreign investment from large multinationals, state governments are now negotiating loans and financing directly with international financial institutions such as the World Bank — what Sinha terms "horizontal competition" (2004). This competition has seen almost all states, regardless of the ideology of their political parties, embrace market reforms and foreign direct investment.

As regards income, state governments are dependent on a mixture of sources. This includes taxes raised locally, state government debt, as well as central government grants and shares of taxation income. Transfers from the central government via the Finance Commission constitute the largest source of income, followed by taxes levied by the states themselves (Oxford Analytica 2009). In the past, the central government used to see its role as equalizing disparities in revenue between states. However, this has changed, with the central government providing additional financing to those states that perform well economically.

THE TAMIL NADU STATE GOVERNMENT

This section will look very briefly at the local-level political context and how this has influenced the approach and outlook of the Tamil Nadu state government.

Recent reviews of Tamil Nadu's overall investment climate have been very positive. For example, Oxford Analytica's 2009 survey of 28 states and two union territories lists Tamil Nadu as the premier state for investing, lauding its infrastructure, human capital base, and quality of governance. The Tamil Nadu state government receives high marks for capacity and fiscal prudence, and its economy is held to be productive, innovative, and outward-looking. Its political context is taken to be stable, underpinned by a highly literate population that participates in periodic elections, leading to a regular turnover of administrations with little turbulence.

It is easy to attribute the development of the electronics sector to these characteristics. However, for much of Tamil Nadu's recent history, its political context meant that economic issues were largely overlooked and relations between the state government and private sector were tenuous.

Historically, the state is associated with Dravidianism, a distinct culture centring on the Tamil language. Starting as a social reform movement in the 1950s, the Dravidian movement converted Tamil language into a representation of identity and a source of self-respect. Since 1967, political power in the state has alternated between two regional political parties that grew out of the Dravidian movement: Dravida Munnetra Kazhagam (DMK); and All India Anna Dravida Munnetra Kazhagam (AIADMK).

The centrality of the Dravidian movement to Tamil Nadu's politics led to the emergence of what Sinha (2005) terms "cultural subnationalism". This has had several effects on the state's political life. The first was the predominance of cultural policy over other concerns and the growth of personality-driven politics. The second was the "institutionalization" of political rhetoric that was anti-northern India and against the central government. The last was the growth of "populist protectionism" in the state, seeking to prevent various interest groups from outside competition.

This cultural subnationalism had a number of effects on the Tamil Nadu economy. First, the state government's economic policy swung between welfare schemes and populist measures, omitting serious investment in political or organizational terms for industry. Second, a series of short-term, political appointments at the apex of the state bureaucracy led to a decline in capacity. Last, the anti-north and anti-central government platform discouraged investors from outside the state (2005, pp. 203–4).

Consequently, despite Tamil Nadu's strong entrepreneurial tradition, the relationship between the government and local private sector was not close. Rather than seeking to stimulate its local business community, state government policy consisted of lobbying the central government for investment from state-owned enterprises. This was particularly prevalent from the late 1970s onwards, when the Chief Minister of the time, M. Gopala Ramchandran, allied with the premier national party, Congress (2005, p. 108). In comparison, during this time and despite the centrally-administered industrial licensing scheme, Gujarat was very active in working with its private sector to stimulate investment, thus consolidating its lead as the pre-eminent site for industry.

This analysis is borne out by Swaminathan (1994), who compared Tamil Nadu's manufacturing output and value-added with that of Maharashtra and Gujarat — the country's other leading industrial states. She found that, in the first state, the private sector consisted of the established business houses, who concentrated on traditional labour-intensive activities such as textiles and apparels. Maharasthra and Gujarat, for their part, had been able to nurture more sophisticated activities and cultivate the emergence of a new business class which, in turn, attracted more capital.

As regards Tamil Nadu's relative lack of dynamism, Swaminathan attributes this to: a risk-averse local business class that was not able to diversify into new areas; and the fact that the state government had not been able to create an environment suitable for the growth of new businesses. Of particular importance is the lack of communication between the state and the private sector. The state had "alienated" the private sector by rapid policy shifts due, in part, to the tendency to appoint politicians to senior technical posts, plus frequent rotation of personnel (1994).

Yet, a number of important changes were to result in a dramatic reshuffling of priorities in the state government which would, over the coming years, signal a significant improvement in the local business environment.

The first significant change was wrought by the 1991 liberalization measures. As the centre's financial might waned and centralized planning became less prevalent, India's "Nehruvian" economy was replaced by a federal market economy, where the centre assumed the role of regulator and states were bestowed an unheralded degree of autonomy (Rudolph and Rudolph 2001, p. 1546).

More than reducing tariffs on imports, these measures removed central regulatory control on investment and production such as licenses for operation and restrictions on foreign ownership in specific sectors. The repeal of this layer of control meant that state governments gained additional responsibilities. As Jenkins argues, this "set off an intense competition among state governments to attract investment, resulting in a proliferation of tax-incentive schemes and promises of speedy administrative procedures, the expedition of land acquisition for industrial uses, and efforts to maintain a "conducive" industrial-relations climate" (2003, p. 144).

Research from Brazil, China, and the United States has shown the galvanizing effect inter-provincial competition can have. In these countries, individual states or provinces have under-taken "state-building" drives in order to revive declining economies and growing income disparities with more industrialized neighbours. In certain cases, these drives can have positive effects for the quality of a state's administration, particularly regarding lead agencies charged with economic development, but also in other areas such as tax collection and infrastructure (Tendler 2002; Cobb 1993).

Tamil Nadu came under considerable pressure from its neighbouring states, who responded quickly to this new form of competition. Andhra Pradesh, under the administration of Chandrababu Naidu, epitomized the new independence and agency offered to state governments in this new context. Actively courting international opinion, Naidu used the media intensively to market his state as an attractive investment destination and to publicize far-reaching public sector reforms. This was coupled with a number of high-profile flagship investments, such as Microsoft's decision to establish its first R&D facility outside the United States (Kennedy 2004). Tamil Nadu was also placed under pressure from Karnataka who —

benefitting from central government investments in scientific and military research institutes — was able to emerge as the pre-eminent destination for the software industry (Heitzman 2004; Parthasarathy 2004).

Secondly, the Indian political scene witnessed significant changes during the late 1980s and early 1990s. Congress was less able to retain nation-wide support and regional parties became more prevalent. Similar developments took place within Tamil Nadu. Thus, social changes — some associated with liberalization — made party affiliation more complicated. Established loyalties and interest groups dissolved and were reconfigured. Congress saw its traditional voter base within the state erode. Of prime importance, neither of the two parties, DMK or AIADMK, were able to include certain previously marginalized groups such as the Dalits, who had become increasing mobilized (Kennedy 2004).

Since 1991, these two parties have continued to alternate in power every five years, each providing the same Chief Minister, Mr. Karunanidhi for the DMK and Ms. Jayalalithaa for the AIADMK, respectively. However, their power base has been reduced, thus requiring a more consensual approach and an awareness of the demands of middle-lower castes and Dalits, who have traditionally resisted liberalization.

Thus, while aware of the need for reform, the Tamil Nadu government has pursued it slowly. In contrast to Andhra Pradesh, it has avoided publicly pursuing liberalization. However, it has moved on a number of issues, including: rolling back support for state-owned enterprises; reducing subsidies to small firms; and seeking to attract investment (Kennedy 2004). While more slow-moving, progress has been steady, continuing across party lines. According to a senior bureaucrat "in Tamil Nadu, things go smoothly irrespective of the government change or elections. ... They fight and cannot see eye to eye, but the business continues as usual. This is not the case with the other states" (interview, former Central Government Official, 10 January 2012).

The subsequent section will look at what key policy frameworks the state government has developed with regard to the industrial sector. Where relevant, the discussion will be complemented by comparisons to the other principal electronics-producing states using key indicators.

POLICY FRAMEWORKS

Industrialization and engagement with the private sector were not the Tamil Nadu state government's foremost priorities in the pre-1991 era.

However, driven by a belief in public sector-led development, during the 1960s and 1970s, the state government created a stable of organizations, which were to later play a role in fostering the electronics sector. They included:

Tamil Nadu Industrial Investment Corporation (TIIC): established in 1949, with the aim of diversifying the industrial sector, it was the first state financial corporation to be established in India. It has played an important role in the development of the sugar, cement, textile, and aluminium industries in the state.

Tamil Nadu Industrial Corporation (TIDCO): incorporated in 1965 as a public limited company to identify and promote the establishment of medium and large enterprises in the state in association with the private sector.

State Promotional Corporation of Tamil Nadu (SIPCOT): established in 1971, one of the main responsibilities of SIPCOT was to set up industrial complexes or special economic zones by procuring land.

Electronics Corporation of Tamil Nadu (ELCOT): created in 1977 to promote electronic industries in the state, it manufactured electronic goods in the 1980s. While not particularly successful, it did generate a nucleus of technical personnel familiar with the sector.

In addition, the state government had a number of concessions (tax incentives, subsidies, and financing) for target sectors, as well as land and industrial parks in place (Sinha 2005).

How did this compare to what was offered by other states? Sinha made a comparison of the organizational and policy frameworks regarding industrialization of Gujarat, Maharashtra, West Bengal and Tamil Nadu for the period 1960–90. Gujarat and Maharashtra were the most proactive, with established investment promotion services, industrial estate laws, entrepreneur development programmes, and joint-venture initiatives. In contrast, Tamil Nadu had the least extensive framework in place, with no investor promotion service or entrepreneur development programme (2005, p. 139). Much as did West Bengal, both depended to a large extent on investment from the central government.

Following the liberalization reforms of 1991, Tamil Nadu changed tack rapidly, moving from depending on centrally-provided resources to seeking to attract private sector investment. The most important organizational innovation was the establishment in 1992 of the Tamil Nadu Industrial Guidance and Export Promotion Bureau, commonly called "Guidance". Inspired by a similar agency in Gujarat, TIDCO established this subsidiary to centralize investor-related functions and information in one agency. Its tasks include: promoting investments and industrial projects; providing information services to investors; providing all investment related information and documentation in a single window, and facilitating projects during start-up (Sinha 2005; Guidance Bureau 2011). In addition, the state government moved quickly to rationalize its investment procedures, propose new tax incentives and subsidies, and establish industrial estates (Kennedy 2004, p. 35).

These measures were helpful in attracting the first large-scale international investments in manufacturing to the state. In 1995, these, coupled with personal assurances from the state's leadership and additional tax incentives, were pivotal in attracting a large manufacturing investment from Ford. This was complemented by another large investment by Hyundai in 1996, and another by a large glass-maker (Kennedy 2004).

Over the remainder of the 1990s, the state government's priorities centred on improving infrastructure and developing the state's human resource base. Tamil Nadu was the first state to formulate an Information Technology policy, which it did in 1997. The Policy's main objectives were to foster the growth of the hardware and software sectors, in their own right and as a source of income for the state. In order to make this possible, the state government committed to: providing infrastructure — largely through industrial parks; developing the human resource base; enabling consumer access to IT; and promoting the use of IT in government (GoTN 1997).

Much of the infrastructure was provided by the state government and its subsidiaries, such as a number of IT parks in and around Chennai. However, key provisions also allowed the private sector to build industrial parks by speeding up permits for operation and extending financial incentives to firms tenanted in privately-owned parks (GoTN 1997). By 2006, in addition to government-owned industrial parks, more than 140 privately-owned industrial parks were in operation (Kumar 2007). A number of related provisions freed up land for industrial use and exempted IT firms from a number of controls and restrictions. And, the state government committed to providing the industry an uninterrupted supply of electricity.

As part of its commitments to expanding consumer access to IT, in the 1990s, Tamil Nadu became the first state in the country to provide access to computers to secondary school students. Services for providing the computers were contracted out to private operators who could then provide classes to the public in the evening, further bolstering access (Maclay 2001).

Tamil Nadu also built up its human resource base by allowing the private sector to open more engineering colleges. The state had liberalized its private higher education system in 1984, making it among the first in India to do so (Arora and Bagde 2010). In addition, it had the only Indian Institute of Technology (IIT) in the south of the country, which produced an elite cadre of technical personnel. However, further liberalization of this sector allowed the state to boost its production of engineering graduates substantially. As can be seen in the table below, capacity rose from 273,000 engineers in 1998 to more than 700,000 in 2002, making Tamil Nadu the premier source of skilled labour in the country.

Figure 8.2
Sanctioned Capacity of Engineers in Tamil Nadu and Selected States, 1990–2003

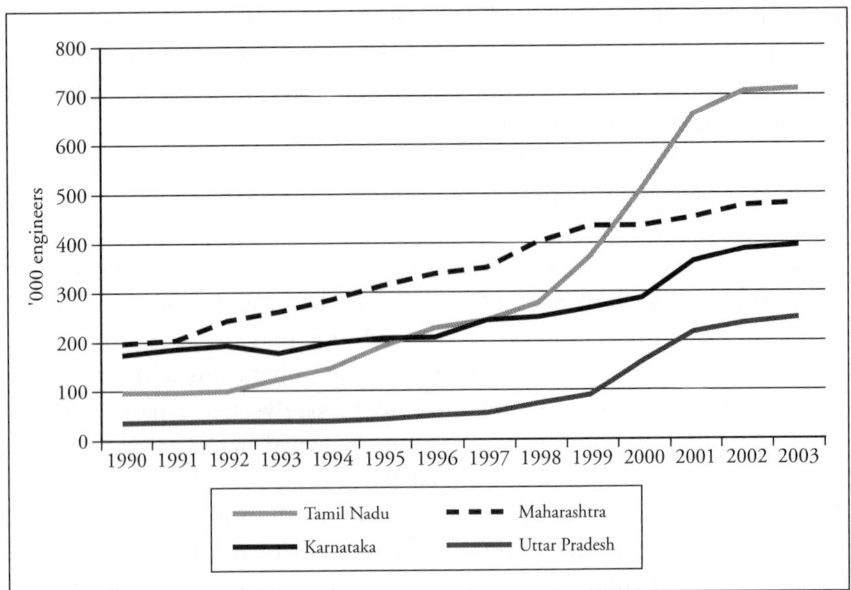

Source: Arora and Bagde 2010.

The state government also tried to foster research and development in the state by founding the Information Technology Institute of Tamil Nadu, with an investment of INR700 million. In addition to developing appropriate curricula for colleges across the state, it was to foster centres of excellence in universities and support research programmes. However, the Institute did not succeed and was subsequently folded into an existing state government-owned university (Kumar 2007).

In addition, the state government established an agreement with the Harvard Institute for International Development to procure high-quality inputs for policy formulation in areas related to trade liberalization, policy reform, and the development of specific sectors (for example, Tewari 2000). This was bolstered by the creation of: a state-level task-force on industry issues, which incorporated senior-level bureaucrats, private sector representatives, and academics; as well as the Department of IT in the state government to provide specialist planning and expertise (Kumar 2007).

After 2000, the state government moved to capitalize on its investments in infrastructure and human resources by marketing the state and seeking to make it more attractive to multinational corporations. Key to this initiative was a series of marketing conferences organized in cooperation with industry associations and interested local enterprises. Small local firms were encouraged to participate in these events in order to showcase local capabilities and establish contacts (Kumar 2007).

However, the sustainability of these reforms was put into question in the late 1990s, when Tamil Nadu experienced a serious fiscal crisis. Its consolidated fiscal deficit rose from 2.2 per cent in 1997–98 to 6.7 per cent in 1999/2000, and 7.2 per cent in 2001/02. Its stock of debt also rose from 16 per cent to 25 per cent of GDP over the same period, and reached a full 30 per cent if guarantees for state-owned enterprises are included (Ianchovichina et al. 2006). Government expenditure on interest payments reached 20 per cent of revenue, and borrowing was increasingly resorted to in order to cover deficits in revenue (GoTN 2004).

The Tamil Nadu state government addressed this issue by implementing a series of fiscal reforms. In 2003, Tamil Nadu became the third state to sign the Fiscal Responsibility Act, which involved establishing targets to bring the fiscal deficit down to 3 per cent of GRP by 2008. Measures were taken to increase tax revenue and government spending was reduced by streamlining the bureaucracy — including closing ten departments and identifying 85,000 surplus positions — and reforming the pension system (GoTN 2004; Ianchovichina et al. 2006). Despite a roll-back of some reforms after 2004, the fiscal deficit was down to 3.3 per cent in 2003/04, and debt also came down quickly — driven in part by increased tax receipts and fewer guarantees of state owned enterprises (Ianchovichina et al. 2006).

As a result, Tamil Nadu was able to improve its financial situation and thus fund additional investments in infrastructure, specialist training schemes, and tax incentives to attract manufacturing firms. Table 8.6 compares Tamil Nadu's fiscal position with the other principal electronics-producing states. Relative to the other states, it is able to raise more of its own money, has a lower state government deficit or gross debt burden relative to GRP, and spends more of its revenue on development projects — rather than current costs. On all these measures, Tamil Nadu is substantially ahead of the national average and compares favourably with the other leading industrialized states of Karnataka and Maharashtra. The only area where Tamil Nadu lags is the number of loss-making state-owned enterprises that it operates.

This favourable fiscal position has enabled additional policy measures to support industry. Thus, in 2003, key policies such as the New Industrial Policy followed. Principal amendments included: the recapitalization of the state financial corporation to fund the upgrading of the technological capabilities of local industries; the decentralization of single window approvals to the district level; and the establishment of an SME centre to promote innovation and the acquisition of technological capabilities (GoTN 2003).

Other policy pronouncements since then have sought to increase the state's attractiveness for investment. In particular, the most recent, such as Tamil Nadu's Electronic Hardware Manufacturing Policy of 2011 have explicitly focused on the hardware sector, as opposed to the IT sector as a whole. Key goals include: doubling the production of electronics hardware by 2016; attracting supporting industries to reduce imports; facilitating downstream linkages and doubling value-added from 15 per cent to 30 per cent; and industrializing the southern part of the state (Guidance Bureau 2011). This policy framework continues previous thrusts of freeing up land for industrial use, streamlining permits and approvals, and pledging to improve infrastructure. In addition, the Policy proposes subsidies for: workplace training; industry-university interactions; vouchers for skills upgrading at approved local institutes; and centres of excellence in priority areas such as chip and electronic hardware design (Guidance Bureau 2011).

Having set out the state's policy frameworks in the pre-1991 period, during the 1990s, and post-2000, the next sections will seek to evaluate the relative attraction of Tamil Nadu by comparing it with other electronics-producing states. Key axes of comparison will be: the business environment; hard and soft infrastructure; and market complementing measures.

Table 8.6

Government Finances in Tamil Nadu and Selected States

No.	Description	Tamil Nadu	Karnataka	Maharashtra	Uttar Pradesh	India Average
1	Proportion of total state revenue that is comprised of transfers from the central government in percentage (2006–07)	22.2	25.4	21.4	48.2	46.9
2	State Government Deficit as a percentage of SGDP (2006–07)	1.51	2.49	2.27	3.08	2.57
3	Government Gross Debt as a percentage of SGDP (2009)	23.9	26.6	26.9	48.1	43.2
4	Proportion of development expenditure of State Government to total expenditure as a percentage (2008–09)	58.0	66.2	57.4	62.8	62.0
5	Average number of industrial units registered for sickness with the govt in 2004 & 2005	28.0	13.0	65.0	8.5	10.0

Source: Authors' compilation from *State Finances: A study of Budgets of 2008–09*, Reserve Bank of India, Govt. of India.

Business Environment

So how does Tamil Nadu's business environment rate in comparison to that of other electronics producing states? Table 8.7 compares the state with the other electronics-producers across a number of key areas.

With the exception of Karnataka, Tamil Nadu has a relatively higher tax burden than the other states. However, as discussed above, the state's solid financial position has allowed it to support the industrial sector as well as develop its competitive advantage. Beyond this, while Tamil Nadu requires a slightly higher number of permits to begin business than the national average, it still scores relatively well in comparison to Maharashtra and Uttar Pradesh. It also has a relatively low registration cost for setting up a business, roughly on a par with Karnataka and Uttar Pradesh, and substantially below the national average or Maharashtra. Of all states, it has the lowest rental price for industrial land, substantially undercutting Karnataka, Maharashtra, and even Uttar Pradesh — which is substantially larger. The only area where Tamil Nadu fares worse than the national average and the other states is with regard to labour relations, as it has a very strong trade-union movement linked to its political parties.

Hard and Soft Infrastructure

Tamil Nadu has managed to construct a formidable competitive advantage based on its hard and soft infrastructure.

With regard to hard infrastructure, Tamil Nadu has a well-developed road network, with some 70 per cent of the total being paved. This is on a par with Maharashtra, and considerably above Karnataka as well as the national average. Relative to all states except Maharashtra, Tamil Nadu also has an extensive number of seaports, and an international airport. While most states in India struggle to provide an adequate supply of electricity, Tamil Nadu fares better than its industrialized competitors. According to the World Bank, Tamil Nadu has one of the better state power utility boards in India, with good collection rates and a lower distribution loss (World Bank 2004). Thus, its shortfall of 4,000 mega units is substantially lower than that of Karnataka, Uttar Pradesh, and especially Maharashtra. Of these states, electricity is clearly the most expensive in Tamil Nadu, but presumably the smaller shortfall in supply translates into a more reliable supply. The state has a relatively high degree of teledensity, with 75 per cent of households owning a telephone. This reflects investment in physical infrastructure for fixed line telephones as well as market penetration by mobile telephone

Table 8.7

Business Environment in Tamil Nadu and Selected States

No.	Description	Tamil Nadu	Karnataka	Maharashtra	Uttar Pradesh	India Average
1	Tax levied by the State as a % of SGDP in percentage in 2006–07	10.6	12.4	7.8	7.4	6.8
2	Number of permits required to start the business (Land, Power, Environment, Water, and Labour) in 2010	9	8	14	15	8.1
3	Cost of registering business in 2010 in INR.	2,638	2,913	8,376	2,418	5,314
4	Rent Price for land per factory in INR in 2005–06.	291,000	590,338	535,188	321,537	352,358
5	Level of Government Finance made available to local private firms (2005–06)	3,949,656	2,425,725	6,344,915	2,414,890	1,311,726
6	Number of labour strikes in 2006	50	13	7	11	11.6

Sources: 1 — Authors' compilation from *State Finances: A study of Budgets of 2008–09*, Reserve Bank of India, Govt. of India.
2 — Authors' compilation from state reports of Indian Brand Equity Foundation.
3 — Authors' compilation of selected variables like company registration, procuring company identification number, tax account number etc.
4 & 5 — *Statistical Abstract India 2007*, 43 Edition, Ministry of Statistics and Programme Implementation, Govt. of India.
6 — *Annual Surrey of Industries 2005–06*, Vol. II. GoI. http://labourbureau.nic.in/Ind_dis_clo_2k6%20Table2(II)%20(d).htm.

operators. Relative to the other electronics-producing states, Tamil Nadu has a significantly higher degree of teledensity.

Tamil Nadu has also managed to allocate a considerable amount of land and special economic zones for industries. In 2009, it had 69 special economic zones in operation, placing it second behind Maharashtra, but considerably ahead of Karnataka. Many of these SEZs are sophisticated, high-end facilities run by international operators, such as Ascendas of Singapore, as well as three specialized SEZs for electronics hardware.

Beyond the number of the industrial parks, Tamil Nadu stands out for the effective way that it has been able to acquire agricultural land and convert to industrial use. In contrast, the development of SEZs has been resisted by farmers in states such as Maharashtra, Gujarat, and West Bengal (Menon 2010; Rediff 2010). Vijayabaskar (2010) credits this to a number of factors. First, the state's long tradition of anti-caste politics and relatively good access to education has allowed many to move from rural areas to cities, freeing up land in the process. Second, land acquired by state government corporations like SIPCOT and TIDCO during the 1990s proved useful in creating the first generation of SEZs. Last, compensation agreements have been relatively generous.

Due to its established tradition of literacy and high rate of urbanization, as well as its decision to allow private sector participation in higher education, Tamil Nadu has a very well-developed human resource base that is deeper and broader than that of either Maharashtra or Karnataka.

While Maharashtra has more university students per capita than Tamil Nadu, the latter state has more students in engineering or technical institutions per capita than any of the comparator states. Tamil Nadu also has substantially more students enrolled in undergraduate courses in science and engineering than either Karnataka or Maharashtra, and considerably more than either Uttar Pradesh or the average for India. The state has more than 450 engineering colleges as well as a group of leading technical institutions, including IIT-Madras, the National Institute of Technology in Trichy, as well as Anna University, which is also a highly-regarded technical university. Tamil Nadu also has a well-developed vocational education system that is surpassed only by Karnataka's. This advantage in qualifications is further extended by having a relatively high female labour-force participation rate that is similar to Karnataka and Maharashtra and higher than the national average.

As a result of its well-developed education system, Tamil Nadu enjoys an unparalleled cost advantage for skilled labour when compared to the other industrialized states. The daily cost for skilled labour in Tamil Nadu is slightly more than half of the comparable rate in Maharashtra and some

Table 8.8
Hard Infrastructure in Tamil Nadu and Selected States

No.	Description	Tamil Nadu	Karnataka	Maharashtra	Uttar Pradesh	India Average
1	Proportion of surfaced roads to total roads in the state, 2002	0.70	0.60	0.70	0.60	0.58
2	Number of ports, 2004–05	15	10	53	0	5.2
3	Number of international airports, 2009	1	1	1	0	0.3
4	Shortfall or excess in electricity production in mega units, 2006–07	–4,006	–9,722	–23,514	–15,514	–2,893
5	Cost of electricity for business, 2008–09 (INR/kW)	5.10	4.34	0.05	4.34	4.30
6	Formally approved SEZs, 2009	69	52	111	34	19.3
7	Households with telephone, 2011	74.9	71.6	69.1	66.9	63.2

Sources: 1 — *Statistical Abstract India 2007*, 43rd Edition, Ministry of Statistics and Programme Implementation, Govt. of India.
2 — "Port Statistics", Ministry of Shipping, Govt. of India (http://shipping.gov.in/index1.php?lang=1&level=0&linkid=46&lid=4).
3 — Brien, O.D. *Penguin Yearbook 2009*, New Delhi: Penguin, 2009.
4 — "Power Supply Position at the end of 10th Plan", Central Power Authority, Ministry of Power.
5 — Authors' compilation from state reports of Indian Brand Equity Foundation.
6 — "Statewise Distribution, Special Economic Zones in India", Ministry of Commerce and Industry, Government of India, 2011.
7 — "Mode of Communication: 2001–11", Census of India, Government of India, 2011.

Table 8.9
Soft Infrastructure in Tamil Nadu and Selected States

No.	Description	Tamil Nadu	Karnataka	Maharashtra	Uttar Pradesh	India Average
1	Number of students in enrolled in UG science and engineering courses per 1000, 2009–10	5.8	4.3	4.6	1.9	3.4
2	Number of students in vocational education per 1000 population, 2004–05	2.2	2.5	1.7	0.5	1.3
3	Number of universities recognized by Govt. of India per million, 2010–11	0.75	0.57	0.25	0.36	0.43
4	State-wise funds allocated to scientific research and development institutes of Ministry of Science and Technology in India in USD Million, 2007–11	9,450	23,920	19,024	12,441	NA
5	Female labour/total number of females, 2001	31.5	32.0	30.8	16.5	26.9

6	Average cost of unskilled labour in INR, 2007	102	83	100	81	83
7	Average cost of skilled labour in INR, 2003–04	239	337	438	263	273

Sources: 1 — *Annual Report 2005–06*. University Grants Commission, Govt. of India.
2 — *Statistical Abstract India 2007*, 43rd Edition, Ministry of Statistics and Programme Implementation, Govt. of India.
3 — *Annual Report 2010–11*, University Grants Commission.
4 — from www.indiastats.com on the basis of data Lok Sabha Unstarred Question No. 3468, dated on 07.12.2010.
5 — *Statistical Abstract India 2007*, 43rd Edition, Ministry of Statistics and Programme Implementation, Govt. of India.
6 — *Annual Report 2007–08*, Ministry of Labour, Govt. of India.
7 — *Annual Survey of Industries, 2003–04*, Vol. I, GI. http://www.labourbureau.nic.in/ASI2K4ELC%20Table%203.1.1.htm. (Total of wages, bonus, PF and welfare expenses of all joint, public and private enterprises and divided by the respective number of mandays worked).

30 per cent cheaper than in Karnataka. It is even cheaper than in Uttar Pradesh, which has a relatively under-developed higher education system.

Market Complementing Measures

Due to India's former state-led growth model, the Tamil Nadu state government developed a range of state-owned corporations and has an extensive range of schemes to support strategic sectors as well as small and medium enterprises.

However, at present much of the state-owned sector is oriented to providing specialized infrastructure. While a great deal of this is to cater to present demand, a significant portion of these investments are strategic in nature. A long-term goal is to open up new areas of the state for investment and to distribute the fruits of growth more equitably. As a result, SIPCOT, SIDCO, and TIDCO will construct four new electronics industrial parks in secondary cities such as Madurai, Trichy and Coimbatore. And, these corporations pooled resources to invest INR3.7 billion in a high-end technology park in Coimbatore (GoTN 2009). This has the additional benefit of reducing diseconomies of scale such as poaching and attrition, and allows skilled labour in other areas of the state to be tapped.

In addition, these corporations establish joint ventures with private sector operators to provide infrastructure facilities in key areas, such as ports and industrial corridors. In some cases, these joint ventures are more technical and commercial in nature. TIDCO has joint ventures in a number of strategic sectors, such as: watches, auto components, textiles, fertilizers, chemical, pharmaceuticals, and processed food (Guidance Bureau 2011). Furthermore, agencies such as the Tamil Nadu Industrial Investment Corporation have an array of investment and support schemes for local firms to acquire land, machinery, and buildings. An estimated 90 per cent of TIDCO's funds go to small and medium firms, of whom 40 per cent are first generation entrepreneurs (TIDCO 2012).

Perhaps, the most interesting market-complementing initiative undertaken by the state government has been the Tamil Nadu Technology Development Promotion Centre, in partnership with the state branch of the Confederation of Indian Industries, the foremost business association in the country. Offering project evaluation, feasibility studies, advisory services, business coaching, and networking services, the centre aims to be a one-stop shop for SMEs in the state to improve their technological capabilities. The state government has contributed 100 per cent of the capital cost, but the Confederation provides all technical and managerial support (TNTDPC 2012).

However, beyond the supply-side, the most direct way the Tamil Nadu state government can shape market outcomes is through its demand-side policies. Under the National Electronics Governance Plan, Tamil Nadu is investing heavily in the use of IT for governance (DIT 2010). Aside from its use of IT, the government also constitutes an important source of demand by purchasing large amounts of laptops and colour televisions to be distributed to its citizens. This is done through ELCOT, which has, since 2006, purchased 7.3 million TV sets (ELCOT 2009). The agency also procures IT equipment for state government bodies and plans to distribute some 6.8 million laptops over the next five years (Indo-Asian News Service 2011). The size of these orders means that successful bidding firms can see their market position can change (*Financial Express* 2012). In the long term, such a large source of demand may help the electronics sector in the state become more rooted, as more firms seek to be closer to the ultimate source of demand.

CONCLUSIONS

Tamil Nadu's industrial sector has moved from low-technology manufacturing activities to export-oriented electronics production in the space of a few years. Despite the state's well-developed human resource base, high rate of urbanization, and manufacturing tradition, this chapter has argued that the state government has played a key role in the sector's emergence and consolidation.

Prior to 1991, like many states in India, the state government lobbied the central government for resources, and did not prioritize entrepreneurship. While a relatively wealthy state, Tamil Nadu's industrial sector was languishing, with little diversification or the emergence of new entrepreneurs.

However, the 1991 reforms changed this and, prompted by the rapid development of its neighbours, the Tamil Nadu state government embarked on a far-reaching — albeit low-key — raft of reforms. Understanding that the central government was no longer the sole or most promising source of investment, successive administrations looked outside the state for technology and funds. Despite changes in the ruling party, a consist investor-friendly stance was maintained that was pivotal in attracting and retaining the first large-scale investments.

Underpinned by fiscal prudence, the state government was able to invest considerable amounts in upgrading its infrastructure, rolling out specialist facilities for large electronics and software firms, and encouraging private sector investment. Streamlined processes and centralized investor liaison services also improved the local business climate. Further liberalization of education provision also enabled Tamil Nadu to emerge as the leading state for science and technology in quantity and cost terms.

These policy measures have enabled Tamil Nadu to emerge as the leading state for electronics production. However, despite its large size in employment terms, the sector is relatively young and still displays characteristics of a satellite platform. It would appear that, despite considerable resources for SMEs and incipient attempts to foster the acquisition of capabilities by local firms, there is little progress in this area. Notwithstanding that, Tamil Nadu's importance as an end-market and the creation of inter-firm linkages — albeit among foreign firms — bode well for the creation of a territorially-rooted competitive advantage.

Note

1 We are grateful to Ms Neha Gupta of IIT-Delhi for her research assistance.

References

Arora, A. and Bagde, S.K. "Human Capital and the Indian Software Industry". NBER Working Paper 16167. Cambridge, MA: National Bureau of Economic Research, 2010.

Bajpai, N. and R. Radjou. "Raising Global Competitiveness of Tamil Nadu's IT Industry". *Economic and Political Weekly* 35, no. 6 (2000): 449–65.

Brien, O.D. *Penguin Yearbook 2009*, New Delhi: Penguin, 2009.

Chandran, S. "From Separatism to Coalition: Variants in Language Politics and Leadership Pattern in Dravidian Movement". *World Academy of Science, Engineering and Technology* 75 (2011): 601–05 <http://ssrn.com/abstract=1859917> (accessed 10 May 2012).

Chandrasekhar, C.P., and J. Ghosh. "How Big is IT?" In *Business Line*, March 2008, available at <http://www.thehindubusinessline.com/2008/03/11/stories/2008031150170900.htm> (accessed 15 Dec 2008).

Cobb, James C. *The Selling of the South: the Southern Crusade for Industrial Development 1936–1990.* Urbana: University of Illinois Press, 1993.

Cooke, P., M.G. Uranga, and G. Etxebarria. "Regional Innovation Systems: Institutional and Organisational Dimensions". *Research Policy* 26, nos. 4–5 (1997): 475–91.

Dataquest."Overview. DQ Top 20". In *Dataquest*, August 2010, available at <http://dqindia.ciol.com/dqtop20/2010/IndustryAnalyses/default.asp> (accessed 10 Sept 2011).

DIT. "India: e-Readiness Assessment Report 2008 for States and Union Territories". Department of Information Technology, Government of India, New Delhi. 2010.

Doloreux, D and Parto, S. "Regional Innovation Systems: Current Discourse and Unresolved Issues". *Technology in Society* 27, no. 2 (2005): 133–53.

Dutta, M. "Nokia SEZ: Public Price of Success". *Economic and Political Weekly* 64, no. 3 (2009): 23–25.

ELCOT. "Thirty Second Annual Report 2008–09". Chennai: Electronics Corporation of Tamil Nadu Limited, 2009.

ESC. "Statistical Yearbook 2010–11". New Delhi, Electronics and Computer Software Export Promotion Council, 2011.

EXIM Bank. "Indian Electronics Industry: Perspectives and Strategies". Research Brief No. 61, Export Import Bank of India, April 2011.

Financial Express. "Delphi Begins Construction of New Electronics Manufacturing Facility in Chennai". In *Financial Express*, 10 November 2008, available at <http://www.financialexpress.com/printer/news/383810/> (accessed 10 August 2011).

Financial Express. "India Top Location for Hardware Mfg: Nokia CEO". In *Financial Express*, 23 August 2007, available at <http://www.financialexpress.com/printer/news/212150/> (accessed 10 August 2011).

Financial Express. "A State-sponsored PC Boost", 4 June 2012.

Frankel, F.R. *India's Political Economy 1947–2004*. New Delhi: Oxford University Press, 2005.

GoI. *Annual Survey of Industries*. Kolkata: Central Statistics Office, Government of India, various years.

———. "Port Statistics". New Delhi: Ministry of Shipping, Government of India, 2010, availabe at <http://shipping.gov.in/index1.php?kabg=1&level=0&linkid=46&lid=47> (accessed 1 August 2010).

———. *Statistical Abstract India 2007*, 43rd Edition. New Delhi: Ministry of Statistics and Programme Implementation, Government of India, 2008.

GoTN. "Tamil Nadu Secretariat — Brief History". Government of Tamil Nadu, 2012, available at <http://www.tn.gov.in/documents/histn.htm> (accessed 10 June 2012).

———. "Demand No. 31", Policy Note, Chennai: Information Technology Department, Govt. of Tamil Nadu, Chennai, 2010*a*.

———. "Electronic Hardware". Chennai: Industries Department, Govt. of Tamil Nadu, 2010*b*, available at <http://investinginTamil Nadu.com/Tamil Nadu/opportunities/electronic_hardware.php> (accessed 16 August 2011).

———. *An Economic Appraisal 2008–09*. Chennai: Department of Evaluation and Applied Research, Government of Tamil Nadu, 2009.

———. *Economic Growth and Poverty Alleviation in Tamil Nadu*. Chennai: Government of Tamil Nadu, 2004.

———. *Tamil Nadu New Industrial Policy, 2003*. Chennai: Government of Tamil Nadu, 2003.

———. *1997 Information Technology Policy*. Chennai: Government of Tamil Nadu, 1997.

Guhan, S. "Centre and the States in the Reform Process", in *India: The Future of Economic Reform*, edited by R. Cassen and V. Joshi. Delhi: Oxford University Press, 1995.

Guidance Bureau. "Tamil Nadu Electronics Hardware Manufacturing Policy 2011". Chennai: Government of Tamil Nadu, 2011.

Heitzman, J. *Network City: Planning the Information Society in Bangalore.* Delhi: Oxford University Press, 2004.

Hutchinson, F.E. and P.V. Ilavarasan. "The IT/ITES Sector and Economic Policy at the Sub-National Level in India". *Economic and Political Weekly* 43, no. 46 (2008): 64–70.

Ianchovichina, E., L. Liu, and M. Nagarajan. "Subnational Fiscal Sustainability Analysis: What can we learn from Tamil Nadu?" World Bank Policy Research Working Paper 3947, Washington D.C.: World Bank, 2006.

IBEF. "Tamil Nadu". New Delhi, India Brand Equity Foundation, 2011.

Ilavarasan, P.V. "Center-for-Global" or "Local-for-Global"? An Examination of FDI in R&D Centers of IT MNEs in India". Report submitted to the Technology Information, Forecasting and Assessment Council (TIFAC), New Delhi, Department of Science and Technology, 2011.

Indo-Asian News Service. "Free Laptops will have Tamil Nadu State Logo Burnt on Chip", 18 June 2011.

Jenkins, R. "India's States and the Making of Foreign Economic Policy: The Limits of Constituent Diplomacy Paradigm". *Publius* 33, no. 4 (2003): 63–81.

———. *Democratic Politics and Economic Reform in India.* Cambridge: Cambridge University Press, 1999.

Joseph, K.J. "The Electronics Industry". In *Structure of Indian Industry*, edited by S. Gokarn, A. Sen, and R.R. Vaidya. New Delhi: Oxford University Press, 2004.

———. "The Electronics Industry". In *International Competitiveness and Knowledge-based Industries in India*, edited by K. Nagesh and K.J. Joseph. New Delhi: Oxford University Press, 2007.

Kennedy, L. "The Political Determinants of Reform Packaging". In *Regional Reflections: Comparing Politics across India's states*, edited by R. Jenkins. New Delhi: Oxford University Press, 2004.

Kumar, Rajendra. "Rise of Regions After Reforms: Late Development Strategies for the Software Industry in Tamil Nadu, Andhra Pradesh, and Kerala in India". Ph.D. Dissertation, Department of Urban Studies and Planning. Cambridge MA: Massachusetts Institute of Technology, 2007.

Maclay, C. "Readiness for the Networked World: A Quiet Information Revolution in Tamil Nadu". Research and Policy Paper for the Tamil Nadu State Government, Cambridge MA: Centre for International Development, Harvard University, 2001.

MAIT. "Report on Mapping the Manpower Skills in the IT Hardware and Electronics Manufacturing Industry — A Study for the Department of Information Technology, Government of India". Manufacturers' Association for Information Technology, 2008.

Majumdar, R. "Indian Electronics Hardware Industry: Growth and Productivity (1993–2004)". *Economic and Political Weekly* 45, no. 14 (2010): 72–77.

Malik, P. and P.V. Ilavarasan. "Trends in Public and Private Investments in ICT R&D in India". Institute for Prospective Technological Studies, Joint Research Centre, European Commission, 2011.

Markusen, A. "Interaction Between Regional and Industrial Policies: Evidence from Four Countries". Proceedings of the World Bank Annual Conference on Development Economics 1994. Washington: The International Bank for Reconstruction and Development, World Bank, 2005.

———. "Sticky Places in Slippery Space: A Typology of Industrial Districts". *Economic Geography* 72, no. 3 (1996): 293–313.

Meena, M. "First public audit of SEZs in Maharashtra". In *The Hindu*, September 2009, available at <http://www.thehindu.com/news/states/other-states/article20868.ece?service=mobile> (accessed June 2011).

Menon, R. "Special Economic Zones: Boon or Disaster". In <www.Boloji.com> 3 March 2010 <http://www.boloji.com/index.cfm?md=Content&sd=Articles&ArticleID=466> (accessed 10 September 2011).

MHRD. "Report of High Power Committee for Faculty Development in Technical Institutions". Ministry of Human Resources Development, Government of India, New Delhi, 2006.

Ministry of Statistics and Programme Implementation. "Value Addition and Employment Generation in the ICT Sector in India". New Delhi: Ministry of Statistics, Govt. of India, 2010.

MoCl. *Annual Report 2009–10*. New Delhi: Ministry of Commerce and Industry, Government of India, 2011, available at <http://commerce.nic.in/publications/annualreport_chapter6-2009-10.asp> (accessed August 2011).

MOLE. *Annual Report 2007–08*. New Delhi: Ministry of Labour and Employment, Government of India, 2009.

MoP. "Power Supply Position at the End of 10th Plan". New Delhi: Central Power Authority, Ministry of Power, 2010, available at <http://cea.nic.in/planning/Power%20Supply%20Position%20at%20the%20end%20of%2010th%20Plan.htm> (accessed 5 August 2010).

NASSCOM. "Executive Summary — The IT-BPO Sector in India — Strategic Review 2010". New Delhi: National Association of Software and Service Companies, 2011.

Okada, A. "Skills Development and Interfirm Learning Linkages under Globalization: Lessons from the Indian Automobile Industry". *World Development* 32, no. 7 (2004): 1265–1288.

Oxford Analytica. *India Deconstructed: Risks and Opportunities at State Level*, Oxford: Oxford Analytica, 2009.

Parthasarathi, A. and K.J. Joseph. "Limits to Innovation Set by Strong Export Orientation: The Experience of India's Information and Communication Technology Sector". *Science, Technology and Society* 7, no. 10 (2002): 13–49.

Parthasarathy, B., and P.V. Ilavarasan. "Small ICT Firms in Developing Countries: An Analytical Case Study of India". Research Project Report. Bangalore: International Institute of Information Technology, 2007.

Parthasarathy, B., "Globalisation and Agglomeration in Newly Industrializing Countries: The State and the Information Technology Industry in Bangalore, India". Ph.D. Dissertation, Berkeley: University of California, Berkeley, 2000.

Ranganathan, M. "Television in Tamil Nadu Politics". *Economic and Political Weekly* 41, no. 48 (2006): 4947–951.

RBI. "State Finances: A Study of Budgets of 2008–09". Mumbai: Reserve Bank of India, Government of India, 2010.

Rediff. "The Ugly Side of Land Acquisition in India". In <*www.Redfiff.com*>, August 2010, available at <http://business.rediff.com/slide-show/2010/may/31/slide-show-1-ugly-side-of-land-acquisition-in-india.htm> (accessed 10 June 2012).

Reddy, K.N. "Indian Federal Fiscal Devolutions: A Critique". In *Indian Federalism and Autonomy*, edited by S. Chandrasekhar. New Delhi: B.R. Publishing Corporation, 1988.

Rudolph, L.I. and S.H. Rodolph. "Iconisation of Chandrababu: Sharing Sovereignty in India's Federal Market Economy". *Economic and Political Weekly* 36, no. 18 (2001): 1541–1552.

Sinha, A. *The Regional Roots of Developmental Politics in India: A Divided Leviathan*. Bloomington and Indianapolis: Indiana University Press, 2005.

———. "The Changing Political Economy of Federalism in India: A Historical Institutionalist Approach". *India Review* 3, no. 1 (2004): 25–63.

Swaminathan, P. "Where are the Entrepreneurs? What the Data Reveal for Tamil Nadu". *Economic and Political Weekly*, vol. 29, no. 22 (1994): M64–74.

Tendler, J. "The Economic Wars Between the States". Paper Presented at the OECD/State Government of Ceara Meeting on Foreign Direct Investment and Regional Development, Fortaleza, 12 December 2002.

TIDCO website, available at <http://www.tidco.com> (accessed 8 August 2012).

TNTDPC website, available at <http://www.tntdpc.com/> (accessed 8 August 2012).

Tiwari, M. "Toward a New Production Sensibility: The Impact of Economic Liberalization on Regional Industry: the Case of Tamil Nadu, India". Paper prepared for the Harvard India Program. Cambridge MA: Center for International Development, Harvard University, 2000.

UGC. *Annual Report 2010–11*. New Delhi: University Grants Commission, Government of India, 2010, available at <http://www.ugc.ac.in/oldpdf/pub/annualreport_english1011.pdf> (accessed 2 August 2012).

Vijayabaskar, M. "Saving Agricultural Labour from Agriculture: SEZs and the Politics of Silence in Tamil Nadu". *Economic and Political Weekly* 36, no. 6 (2010): 36–43.

Viswanathan, S. "Dravidian Power".In *Frontline*, April 2004. Available at <http://www.frontlineonnet.com/fl2108/stories/20040423007701500.htm> (accessed 10 June 2012).

World Bank. "Tamil Nadu: Improving Investment Climate". SASPR Report. Washington D.C.: World Bank, 2004.

SECTION IV

Cases from Industrialized Countries

South China Sea
MALAYSIA
(Johor State)
Straits of
Malacca
SINGAPORE
Straits of Singapore
INDONESIA
(Riau Islands Province)
LEGEND
COUNTRY
NEIGHBOURING COUNTRY
Geographic Features
National Border Water
0 km 5 10 15
All maps are derived from GADM Data
N

9

THE DEVELOPMENT OF SINGAPORE'S ELECTRONICS SECTOR

Toh Mun Heng

INTRODUCTION

Explaining the development of the electronics sector in Singapore inevitably involves a discussion of the country's economic history. In many aspects, the country's industrialization experience highlights the sector's role in creating national economic growth. Two themes underscore the evolution of the electronics sector in Singapore. First, the industry was transformed over a relatively short period of 40 years. Secondly, the government has played a major role in fostering this change.

Unlike the experience of developed economies, where industrialization originated with textiles and apparels, Singapore's development was driven by more capital-intensive industries such as electronics. Its concerted effort in development coincided with the period where multinational corporations (MNCs) from developed economies were seeking production bases abroad to lower costs. This timing was fortuitous, as the microelectronic revolution had just begun. Hence, the manufacturing of a whole stream of electronic products, from simpler peripherals such as diskettes to more sophisticated items like disk drives, provided opportunities for local enterprises to flourish. Singapore rode on this "electronic wave" to nurture and develop a strong base for the electronics sector.

Today, the electronics sector remains a major pillar of Singapore's growth. The country houses 14 silicon wafer fabrication plants, 20 semiconductor assembly and test operations, 40 integrated circuit (IC) design centres, and the world's top three wafer foundries. Furthermore, about 80 per cent of the world's enterprise hard disk drives and 40 per cent of the world's hard disk media volume is produced in Singapore.[1]

Many of the investments described above are the result of a series of policies implemented from 1965 to 2010. Pragmatic management of the economy by the state has kept the electronics sector relevant through encouraging higher use of capital and technology. Throughout Singapore's development, government policies have been characterized by their coherence and consistency towards the single objective of promoting economic growth; and responsiveness towards changing business conditions. Today, the government's plans for the sector include transforming the industry into an innovation-driven electronics hub providing technological, manufacturing and business solutions. This will develop new growth areas essential for the industry's future prospects.

The following section provides an overview of the performance of the electronics sector in Singapore. The subsequent section sets out the state's role in fostering growth within the sector over the past four decades. A final section concludes the chapter.

OVERVIEW

Singapore's electronics sector is comprised of MNC subsidiaries (such as Broadcom Singapore), a number of large indigenous firms (Chartered Semiconductor, Creative Technology, and Venture Corp.), as well as a range of competitive small-cap firms supplying components to the major producers. In addition, a number of joint ventures have also been started by Singaporean government-linked companies, with MNCs such as Texas Instruments and Hewlett-Packard from the U.S. and Canon from Japan. There were about 270 electronics-related establishments in 1999. This increased to more than 310 establishments in 2010, which constituted more than 3 per cent of total establishments in the manufacturing sector.

Many of the nimbler small suppliers have followed their MNC clients by moving production overseas. To secure new investment in the sector, the Government has taken equity stakes in most large semiconductor wafer fabrication projects in Singapore, while maintaining large stakes in local firms like Chartered Industries or Singapore Test and Assembly Services.

Table 9.1, along with Tables A1 to A4 in the appendix, detail the performance measures for the sector for each five year period beginning in 1991. Two points are discernible from the data: the importance of the sector to the economy; and a slight decline in its recent performance.

Output expanded to S$78.8 billion per annum during the period 2006–2010. This amounts to 31.5 per cent of total manufacturing output or 30 per cent in terms of value added. The sector continues to employ one fifth of the workforce, while Fixed Asset Investments (FAI) account for 37 per cent of economy-wide FAI. The latest figures on foreign direct investment (FDI) in the manufacturing sector show that 28.5 per cent of the FDI in 2009 went to the electronics products and components industries. The bulk of electronics FDI comes from U.S. and Japan.

However, the share of the industry's output to that of the total manufacturing sector peaked from 1996 to 2000, when it reached 50 per cent. This figure has declined ever since. Other measures of performance mirror

Table 9.1
Performance Indicators for the Singaporean Electronics Sector, 1991–2010

	% Share of Total Manufacturing			
	1991–95	**1996–2000**	**2001–05**	**2006–10**
Output (S$m)	45.1	50.1	40.0	31.5
Value-Added (S$m)	35.0	40.4	33.2	29.6
Employment (Thou)	33.8	32.3	26.1	21.3
Fixed Asset Investment (S$m)	–	43.5	52.8	37.7
Domestic Export at end of period (S$m)	60.3	54.7	36.3	26.1
Productivity	**Average Annual Growth Rate (%)**			
Real VA per Wkr ($'000 per wkr)	17.5	19.0	11.3	8.9

Source: Ministry of Trade and Industry of Singapore, *Economic Survey of Singapore,* various issues; Spring Singapore 2011 Electronic Industry Background and Statistics.

that of output. The sector's share of value-added relative to that of total manufacturing peaked in 1996–2000. The latest five years saw this number drop by 9 percentage points. Total employment in the sector has also declined quite substantially over the last two decades. This number fell from 122,000 in the first half of the 1990s to about 88,000 during the second half of the 2000s.

Two reasons explain the aforementioned decline in performance. First, the rapid growth of other sectors has decreased the contribution of the electronics sector to total output somewhat. Closer inspection of Table A1 reveals that while the percentage contribution of electronics to total manufacturing has fallen, the absolute size of output and value added have *increased.*

Second, the drop in the absolute number of workers employed by the sector also reflects the shedding of labour-intensive activities and the growing importance of activities that yield higher value-added. Indeed, growth in labour productivity for electronics is the highest among all clusters in the manufacturing sector. Labour productivity soared at an annual rate of 18 per cent during the 1990s, tapering to approximately 9 per cent over 2006–10.

High productivity also explains the continued inflow of investments into the sector. Productivity translates into higher returns for investors and induces investment. Figure 9.1 illustrates the correlation between fixed asset investments in the electronics industry and the expected value-added to be generated by the investment committed in the current years. The electronic sector currently receives the largest amount of Fixed Asset Investment by any sector in manufacturing.

A separate point should be made about electronics exports. The small size of Singapore's domestic market means that the bulk of electronic goods produced are exported. Exports of electronic products constitute 26 per cent of the country's total exports in 2010. Fifteen years ago, the corresponding statistic was 60 per cent. The decline in percentage share of electronic goods can be attributed to the increased share of non-electronic products such as oil, pharmaceuticals, and industrial equipment in recent years. This diversification in exports reflects the growing maturity of Singapore's economy.

Currently, the sector manufactures a range of products such as: integrated circuits (ICs); semiconductors; hard disk drives; electronics modules; personal computers (PCs); and peripherals. However, the composition of these products is changing. Table 9.2 details the domestic export of major electronic items from 1995 to 2010. IC and PC parts account for

Figure 9.1
Fixed Asset Investment and Value-Added in the Electronics Sector

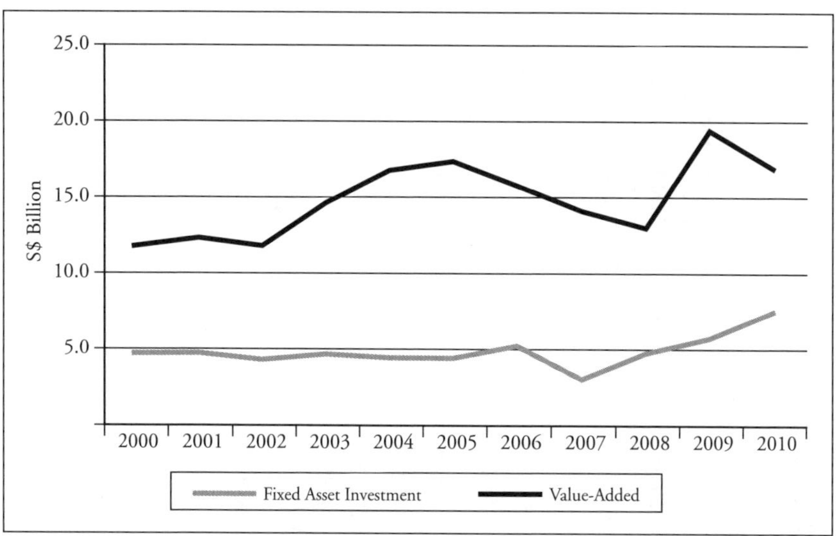

Source: Economic Development Board of Singapore, *Annual Report,* various issues.

more than 60 per cent of domestic exports of electronics in 2010. In fact, with the exception of these two, all other items registered negative growth during the last five years of the 2000s. Singapore's electronics sector is increasingly losing competitiveness in the production and export of electronic hardware.

ICs and semiconductors are the high value products stemming the decline of domestic exports for electronic goods. In recent years, a revolution in digital technology has increased the demand for items such as digital TVs, digital cameras and communication devices. The proliferation of these consumer devices has directly boosted the demand and production for electronic parts like semiconductors and ICs.

The electronics sector has been able to maintain its importance due to its ability to switch over to electronic manufacturing services (EMS). EMS companies conduct activities outsourced by smaller electronic firms that do not enjoy economies of scale necessary to compete in today's economy. These activities include the design and testing, production, provision of return/repair services for electronic components, and assemblies for original equipment manufacturers. They have helped the electronics sector to remain a key sector in the economy.

Table 9.2
Domestic Exports of Electronic Products and Components, 1995–2010

Domestic Exports	Value in S$ Million				% Share of Electronic Sector			
	1995	**2000**	**2005**	**2010**	**1995**	**2000**	**2005**	**2010**
Electronics Products	59,342	74,393	75,386	65,000	100.0	100.0	100.0	100.0
Annual Growth **(%)**		***4.5%***	***0.3%***	***–3.0%***				
Disk Drives	13,509	16,013	13,839	5,473	22.8	21.5	18.4	8.4
Annual Growth **(%)**		***3.4%***	***–2.9%***	***–18.6%***				
PCB Assembled	10,317	13,067	16,310	12,885	17.4	17.6	21.6	19.8
Annual Growth **(%)**		***4.7%***	***4.4%***	***– 4.7%***				
Integrated Circuits	10,500	20,006	21,884	27,187	17.7	26.9	29.0	41.8
Annual Growth **(%)**		***12.9%***	***1.8%***	***4.3%***				
Personal Computers	5,322	3,652	1,067	1,658	9.0	4.9	1.4	2.6
Annual Growth **(%)**		***–7.5%***	***–24.6%***	***8.8%***				
Printers	2,410	3,198	1,619	62	4.1	4.3	2.1	0.1
Annual Growth **(%)**		***5.7%***	***–13.6%***	–				
Telecom Equipment	–	3,167	5,460	1,665	0.0	4.3	7.2	2.6
Annual Growth **(%)**		–	***10.9%***	–				

Source: Department of Statistics, *Yearbook of Statistics 2011*, Singapore.

DEVELOPMENT POLICIES FOR THE ELECTRONICS SECTOR IN SINGAPORE

General Business Environment

In the early 1960s, Singapore embarked on an industrialization drive to jumpstart its economy, which at that time had a high unemployment rate and was mired in uncertainty regarding its future as an entrepôt to the region. Faced with a lack of local industrial entrepreneurs and a small domestic market, the country had to rely on foreign capital, technology, and management techniques as well as access to the world market to move into an alternative economic niche.

Inviting foreign enterprises to invest in Singapore become an important means of promoting its industrialization programme. To this end, providing a conducive business environment to potential investors was deemed vital. Singapore has been governed by People's Action Party since independence in 1965 and has had good economic managers among her ruling elites. A stable political environment, cooperative labour unions and a slew of pro-business, pro-market legislation engendered an economy that can absorb investments, technology, and managerial competencies to produce products that are demanded, and generate jobs that provide a decent standard of living for her citizens.

Singapore's high level of transparency regarding government procedures and effective enforcement of corruption control provides a safe environment for business and investment. Based on Transparency International's Corruption Perceptions Index 2010, Singapore is listed as the least corrupt nation in Asia, and ranked 1st worldwide. In addition, efficiency in handling business procedures such as registering companies, processing permits, getting credits, enforcing contracts, and trading across borders has also made doing business in Singapore easy. In the 2011 World Bank annual survey on Ease of Doing Business, Singapore tops the rankings among all countries surveyed.

Due to its multi-ethnic society, Singapore has adopted four official languages. Among them, English and Mandarin are the most common languages spoken, which facilitates easy communication among local inhabitants as well as international businesses operating in the country. Being a former British colony, the country has inherited several good institutions that help its economic development and nation-building. The country's legal system comes complete with a commercial law system that includes the Law of Contract and Companies Act. Later, the legal system was modified to include the Employment Act, Industrial Arbitration Action, Intellectual Property

Law, and Competition Act so that the interests of investors, innovators, employers and employees are protected.

In addition, Singapore is equipped with world-class infrastructure that enables investments and business ventures to gain a competitive edge. Other than the usual ancillary infrastructures needed for business operations such as electricity, water, and a telecommunications network, Singapore has developed world-class sea and air ports. Changi Airport has won many accolades as a top air passenger and air cargo terminus, providing linkages to more than 200 destinations in some 60 countries. The sea-port is also the world's top container port that provides excellent connectivity to more than 600 ports in over 120 countries. In the field of internet connection, the "next generation" broadband network has been launched and the speed and volume of transmission of data and messages are expected to increase significantly. Concurrently, the development and installation of upgraded accessibility is expected to generate new business opportunities in the digital-related field as well as synergistic linkages to the electronic industries.

As a small nation, heavily dependent on foreign trade and investment for its economic survival, it is in Singapore's best interest to have good relationships with other countries (Toh 2006; Rajan and Thangavelu 2009). Thus, membership in international organizations like United Nations, WTO, and World Bank is important. Singapore is a member of the Association of Southeast Asian Nations (ASEAN), a regional grouping of 10 nations: Brunei Darussalam; Cambodia; Indonesia; Laos; Malaysia; Myanmar; Philippines; Thailand; and Vietnam. Membership of ASEAN confers some benefits to Singapore as a launching pad for investors interested in expanding their businesses to other members in ASEAN. This is especially so when there is an ASEAN Free Trade Agreement (AFTA), which lowers intra-regional tariffs through the Common Effective Preferential Tariff Scheme. Furthermore, ASEAN as a group has also forged several FTAs with major trading nations like, China, India, Japan, South Korea, Australia and New Zealand. At the time of writing, Singapore on its own has a network of 18 regional and bilateral FTAs with 24 trading partners.

Other than signing FTAs, Singapore has endorsed double taxation agreements with more than 60 countries. These provide multinational companies flexibility in managing their fiscal responsibilities when making decisions regarding the import and export of goods, and in remitting their revenue from other countries. It also encourages more exporting and importing since the agreements protect investors from being taxed twice.

A conducive environment is just one of many ingredients for a successful electronics sector. What are delineated above are conditions applicable to all type of industries. In the subsequent sections, we shall dwell in more specific issues, problems and government interventions that shaped the electronics sector over the past four decades.

Development in the early 1960s to late 1970s: Export-Oriented Industrialization

When Singapore was separated from Malaysia in 1965, its industrialization programme became even more important for generating economic growth. Unlike Hong Kong, which absorbed Chinese industrialists and bankers fleeing from mainland China, Singapore did not have a ready-made group of entrepreneurs (Lee 2000). Once it was decided that the world would be the end-market and foreign capital was essential for building production capacity, the government went about providing conditions that would be welcomed by foreign investors.

At that time, industrial relations and trade union activities were not aligned to the country's economic development objectives, and strikes were common. The government managed to rein in the various workers' unions, which were amalgamated in the National Trade Union Congress. In addition, the Employment Act was passed by the Parliament. Under this Act, industrial arbitration is compulsory when there is an industrial dispute and the outcome is mandatory. On the other hand, workers' rights and benefits are also clearly specified. The cooperation and compliance of the union is awarded by having the Secretary-General of the Trade Union as a Minister without portfolio in the Cabinet.

Concurrently, the nation's main designated investment promotion agency, the Economic Development Board (EDB) crafted a list of incentives, ranging from accelerated depreciation allowance to rental rebates and tax holidays.[2] To minimize the duration from signing of agreement to production, industrial estates and pre-fabricated factory buildings supported with reliable utilities and transportation links to air and sea ports were built and furnished for quick occupation and set-up. These factories were leased to companies usually for a period of 30 years. In addition, EDB established offices in major cities like London, Washington DC, New York, Paris and Tokyo to coordinate investment promotion efforts aimed at inviting companies in strategic sectors to invest in Singapore.

Another area of importance for legislative change was to amend the Pioneer Industries Ordinance and the Industrial Expansion Ordinance, introduced before 1960, to include export incentives. And, tariffs that had been introduced during the short era of import substitution implementation when Singapore was part of Malaysia were speedily removed. In the early 1970s, another tax incentive, the Economic Expansion Incentives Amendment Act, which provides tax incentives for capital- and technology-intensive pioneer industries, was passed. In addition, the amount of capital necessary for obtaining pioneer status for companies was increased. For new firms, the capital floor was raised to S$1 million while the requirement for expansion of existing companies was raised to S$10 million. To encourage export activities, pioneer firms are able to enjoy tax holidays for up to 10 years if their total exports exceed S$100,000 and are at least 20 per cent of their total sales. To facilitate more investments, firms with fixed assets worth S$150 million and above were entitled to tax exemptions for up to 15 years. Electronics firms, which produce items that are upgradable in technology and capital intensity, benefited from these incentives and ploughed back their profits into expanding their facilities in the country.

EDB functions as a "one-stop" shop that caters to all needs of potential investors, thus eliminating red tape. Singapore's EDB worked closely and maintained direct linkages with other government bodies such as: International Enterprise Singapore; Agency for Science, Technology and Research (A*STAR); the Standards, Productivity and Innovation Board; and the Jurong Town Corporation. The close relationship between the EDB and these agencies that continues to this day helps foster efficiency and information-sharing.

The careful planning of Singapore industrial estates was an important factor in attracting foreign investment. The concentration of industries in one location made provision of utilities, transport linkages and power more cost-effective and also addressed concerns of environmental pollution. Firms in these estates could also enjoy cost-savings when they transacted with each other on intermediate goods and supplies. Other than facilitating the development of infrastructure, financial and consulting services were provided by the EDB as incentives. Financial services included making medium and long term loans; and consulting services included feasibility and market studies, industrial and market research, as well as industrial training in some instances.

In addition, the EDB moved to ensure the availability of workers with specific skill sets. Thus, the Technical Education Department was established to encourage training to improve productivity. The EDB also introduced and invested in industrial training schemes. Other than working with local educational institutions, the EDB collaborated with MNCs to set up training institutes to produce technicians and skilled workers demanded by firms. Examples of such institutes include the German-Singapore Training

Institute; Franco-Singapore Technical Institute, and Japan-Singapore Institute of Software Technology.

By the early 1970s, the labour market had become tight and wage rates were rising quite rapidly. To meet the demands of industry, workers from neighbouring countries were allowed to be hired based on a work-permit system. The rising wage cost was reckoned to diminish Singapore's attraction as an investment destination and put a dampener on the country's competitiveness. A tripartite body consisting of representatives from the employers, trade unions and government, called the National Wage Council was formed in 1972 to provide guidelines annually for orderly wage changes across all industries in the economy. The Council's recommendations need not be uniform for all industries, as long as the long term competitiveness of the economy and welfare of workers are taken into account. The government plays a leading role in adopting the recommendations for paying workers in the Civil Service, and the private sector has been following the government.

By the mid 1970s, there were already a number of multinational electronics companies with subsidiaries in Singapore. These MNCs were mainly from the United States, followed by those from Europe and Japan (Chia 1997). This included: Texas Instruments; Hewlett Packard; Fairchild; National Semiconductor; General Electric; Toshiba; Sanyo; Philips; and Siemens. The incoming investment from these MNCs also constituted the demand for locally-owned supporting firms to cater to.

Development in the 1980s

By the end of the 1970s, Singapore was determined to move from labour-intensive manufacturing to more capital- and technology-intensive manufacturing activities. Working through the National Wage Council, wage rates were recommended to be increased by 20 per cent annually for two consecutive years prior to 1981. At the same time, it was determined to reduce dependence on imported labour from non-traditional sources. However, it appears that such changes were implemented a bit too hastily. The escalating cost together with slackening international demand after the second oil crisis brought about negative growth in exports and GDP in 1985. This was the first recession since Singapore's independence in 1965. The government quickly put in place an Economic Committee to deliberate on the causes of the slowdown and to make recommendations for restoring growth.

However, on the electronics front, there were many exciting developments which would have a bearing on Singapore. In the West, the miniaturization of the computer and the invention of micro-processors had promulgated the development of personal computers. Steadily improving microprocessors triggered a related explosion of other peripherals: printers; modems; disk drives; interlinked networks; as well as equipment for building chips, video games, and computer-assisted design. In a short time, many of manufactures of electronic products in the U.S. were seeking offshore production bases to remain competitive in their domestic markets.

For Singapore, there was a large inflow of foreign companies setting up subsidiaries and plants to produce electronic products. In 1981, Apple Computer established its first manufacturing plant for the production of personal computers in Singapore. Many leading hard disk drive producers such as Maxtor, Conner Peripherals, Seagate, and JVC also set up factories in Singapore. At one point in time, Singapore produced as much as half of the world's five-and-a-quarter inch Winchester drives.

In response to the changing environment in the 1980s, Singaporean policymakers embraced modern business concepts like competitive advantage, clustering, business eco-system, agglomeration economies, as well as industrial planning and management. The EDB organized the manufacturing sector into five major clusters, with the electronics cluster being one of them. Dedicated programmes and schemes for the development for each cluster follow the precepts of the business eco-system and nurturing synergies and agglomeration economies where possible. The restructuring of the manufacturing sector, especially after the 1985 recession, to move the electronics cluster towards capital- and technology-intensive manufacturing activities, was able to attract new investments by Texas Instruments and SGS-Ates in wafer fabrication. In addition, Hewlett-Packard opened an IC Design centre in Singapore and also upgraded its manufacturing operations to include IC wafer fabrication.

Singapore's recovery from the mid 1980s recession was hastened by the Plaza Agreement, signed between the governments of U.S., UK, France, Germany and Japan in 1985. The Agreement resulted in substantial revaluation of the Japanese and major European currencies against the U.S. dollar. This provided the impetus for many companies in Japan and Europe to invest overseas, especially in Southeast Asia.

While Singapore was one of the main beneficiaries, it also experienced increasing competition from neighbouring Malaysia, Thailand, and Indonesia. Endowed with more land and population than Singapore, they offered significantly lower production costs. This engendered new competition and made it even more imperative to move away from labour-intensive, low-value activities to more high value-added ones. Singapore took the opportunity to reposition itself as an ideal place for MNCs to establish their regional headquarters and other facilities. For instance, electronics manufacturing companies not only established factories in Singapore, but also their marketing, procurement, and R&D departments.

Other than repositioning its locational focus, Singapore came up with the Growth Triangle concept, which involved the neighbouring countries of Malaysia and Indonesia. This was envisioned to be a key component of the Singapore regionalization scheme of the 1980s and 1990s. Singapore's industries, in particular those from the electronics sector, could benefit by relocating their land- and labour-intensive industries to neighbouring places such as the Malaysian state of Johor and the island of Batam in the nearby Indonesian Riau Islands Province, thus retaining higher-value added activities (such as headquarters activities) in Singapore.

To facilitate the movement up the chain into higher value-added activities, Singapore made the first move in the region to attract technologies, management know-how and foreign investment from developed countries. To attract and encourage more knowledge-based activities, Singapore passed its first Copyrights Law in 1986. Immigration rules and regulations were liberalized to enable importation of skilled workers and foreign talents. Singapore's education system was also revamped in the 1980s in order to meet the demand of skilled staff and engineers.

Local enterprises, which are largely small in scale and operations, have not been neglected. The EDB set up the Small Enterprise Bureau in 1986 to assist small local enterprises in improving and modernizing their plants and technology, product design, management skills, and marketing capabilities. Financial support was also provided in order to enable more small and medium enterprises to emerge and grow (Lim 2007).

In addition, to improve the competitiveness of local enterprises and to foster technology transfer, the EDB introduced the Local Industry Upgrading Programme (LIUP) in 1986.[3] MNCs act as mentors to local companies which have the potential and capabilities to be their suppliers of parts and components. Participating MNCS assist local firms to attain know-how and capabilities in producing products that meet their stringent standard and quality needs. There were two examples of local firms that have benefited from the LIUP, and each has become an international company as well as integral part of the Singapore electronics sector. Advanced Systems Automation (ASA), founded by a Singaporean, is an engineering and electronics company that is currently listed on the Singapore Stock Exchange. It expanded its business by providing services to multinational companies and participated in joint ventures to enable technology transfer. It is now an international manufacturer of automated back-end equipment for the semiconductor assembly process. Manufacturing Integrated Technology is also a supplier of semiconductor testing equipments company that grew from a small local firm that was built in Singapore in the 1980s (Mathews 1999).

The Government's effort to reposition Singapore paid off for the electronics sector when Hewlett Packett, which already had a manufacturing plant in the country, established an Asia-Pacific distribution centre and a network software development centre in the country. STMicroelectronics, a European MNC, soon followed, basing its regional management activities in Singapore, in addition to a manufacturing plant. This created spillover effects for other activities. Dr Goh Keng Swee, the founding father of the EDB as well as Singapore's modern economy, notes that "as large corporations locate their headquarters and production facilities in or near cities, the cities act as magnets for talents of all kinds." The result is that these cities become "brain centres ... allowing quality education and research to flourish" (Goh 1995).[4]

Development in the 1990s

With the fall of the Berlin Wall in 1989, the 1990s was heralded as a decade of triumph for capitalism. In the context of Singapore, the unfolding of the international events following the end of the Cold War was quite overwhelming, especially for the electronics sector. Competition ratcheted up due to processes within the electronics sector itself as well as the entrance of new, large players.

A case in point is the hard disk drive (HDD) industry. Technological innovation and progress is generally rapid in the electronic sector, and this is amply true in the HDD industries. The capacity and size of the disk drive are constantly improving and it is no guarantee that if one model is produced in Singapore, all other more advanced models will do the same. If Singapore is simply satisfied with companies doing assembly of HDD, the future of the HDD sub-sector in the country will be bleak. There are many alternative locations in the world that can do assembly at a much lower cost than Singapore. Singapore remains relevant by partnering the companies in research and development at both the technological and marketing fronts.[5] For example, Seagate is a world-renowned HDD manufacturer. Its presence in Singapore began in the early years of industrialization and its business operations have evolved over time from low-value activities to high-value activities in designing, customization and marketing. Most of the big players, Conner Peripherals, Maxtor, and Micropolis from the U.S. now have establishments in Singapore (Hobday 1994).

The reliance on production of electronic hardware for export cannot be expected to be long term and expansive in scope. The rising cost in Singapore can only be buffered to some extent by excellent infrastructure and service efficiency. In fact, whenever an idea or service becomes "commoditized", it can be easily copied and the competitive edge enjoyed by the originator will be eroded quite rapidly. As a countervailing measure, there is a need for continuous innovation and production of knowledge-based goods and service which are not easily imitated.

Thus, in 1990, the Singapore government signalled its intention to transform the country into a knowledge-driven economy providing more value-added services to manufacturing. The EDB, in conjunction with other government agencies, introduced Industry 21. The aim is to develop Singapore into a global hub for knowledge-driven activities, both manufacturing and services. Therefore, strong emphasis was put on technology, innovation and research and development to accomplish the mission. Greater effort is placed in attracting investments for knowledge-driven activities to be carried out in Singapore. Foreign MNCs are invited to set up their regional headquarters and to adjust their manufacturing activities in Singapore towards knowledge-based activities, especially research and development, design and testing, product development, high technology production, and value-added logistics.

The aspiration of a knowledge based industrial system has to be backed by investments in research and development, supported by capable research scientists and engineers. The National Science and Technology Board, a government statutory board in charge of science and technology development, launched its first 5-year National Technology Plan in 1991, followed by a second plan in 1996.[6] In total, S$6 billion were set aside to: establish research institutes; strengthen manpower training; as well as provide support for industry research and development (R&D) and indigenous R&D capability. Of particular relevance to the electronics sector is the establishment of the Institute of Microelectronics in 1991. The Institute is ISO-9001 certified, and its mission is "to add value to Singapore's semiconductor industry by developing strategic competencies, innovative technologies and intellectual property; enabling enterprises to be technologically competitive; and cultivating a technology talent pool to inject new knowledge to the industry".[7]

The Singapore Institute of Manufacturing Technology (SIMTech) was formed to develop high value manufacturing technology and human capital. It aims to enhance the competitiveness, technological edge and indigenous capabilities of the electronics sector in Singapore. In addition, the Agency for Science, Technology and Research (A*STAR) collaborated with local universities to set up research centres. This includes VIRTUS, which is known as the Singapore's IC Design Centre of Excellence, as well as the establishment of Data Storage Institute to perform R&D for both the hardware and software aspects of electronic data storage as well as related data security issues.

By the end of the fourth 5-year Science & Technology Plan started in 2006, Singapore's Gross Expenditure on R&D as a percentage of GDP has increased from 0.85 per cent in 1990 to 2.3 per cent in 2009. The number of Research Scientists and Engineers in Singapore has grown significantly, from 28 in 1990 to 101 per 10,000 workers in 2009.

In tandem with moving Singapore towards a knowledge-based economy, the EDB also aims to develop Singapore into a world-class electronics hub where both foreign and local enterprises work in unison to cater to global demand. The products that were focused on are those electronics components, modules and systems with higher value-added, such as integrated circuits, water fabrication plants and storage drives. To stimulate local enterprises and entrepreneurship, the EDB introduced an initiative, named Technopreneurship-21 which encourages entrepreneurs to commercialize technology. In addition, the Technopreneurship Investment Fund was set up to assist innovators and high technology start-up companies such as those in the electronics sector.

To encourage foreign companies to perform R&D in Singapore, or set up research subsidiaries in Singapore, tax concessions are granted on royalty income received from approved inventions or innovations in Singapore. This initiative has had some success. A notable example is the case of the firm International Semiconductor Products, which responded by building a higher-end technology plant for wafer polishing in Singapore.

EDB also established a S$1 billion Cluster Development Fund in 1994, and expanded it to S$2 billion by the late 1990s to support the investments in wafer fabrication parks which are deemed critical for advanced semiconductor production and R&D activities. In addition to the four wafer fabrication parks, there is the Advanced Display Park jointly set up by the Jurong Town Corporation, Toshiba and Matsushita.[8] It is dedicated to supporting the advanced display industry and is home to AFPD, which is fully-owned by Toshiba and operates Singapore's first Thin Film Transistor — Liquid Crystal Display plant.[9] These initiatives have obtained favourable results.

By 1998, no fewer than 11 advanced wafer fabrication facilities including SGS, Chartered Fab I, SGS-Thomson, Chartered Silicon Partners and

Hitachi/Nippon Steel were in operation or fully committed in Singapore (Mathews 1999). By 2010, the electronics industry of Singapore is home to 40 integrated circuit design companies, 20 chip assembly and test plants, 8 Micro Electro Mechanical Systems (MEMS), Compound, Epitaxy Wafer Fabrication plants and 14 silicon wafer fabrication facilities. This has helped to maintain the sector's relevance in the global market and its contribution to the Singapore economy.

Development in the 2000s

The decade of 2000s was marked by international terrorism, epidemics (SARS and avian flu), and financial crises. Despite the turbulence, the international economic system (trade and GDP) continued to expand. World GDP, on average, expanded by 3.0 per cent per annum and world exports grew by 5.1 per cent (Figure 9.2). Between 2000 and 2009, the world export of electronics products grew by 3.1 per cent per annum. However, there is great disparity in performance across countries. For instance, China's export of electronics products grew by 21 per cent per annum during the

Figure 9.2
World GDP and Merchandise Exports, 2000–10

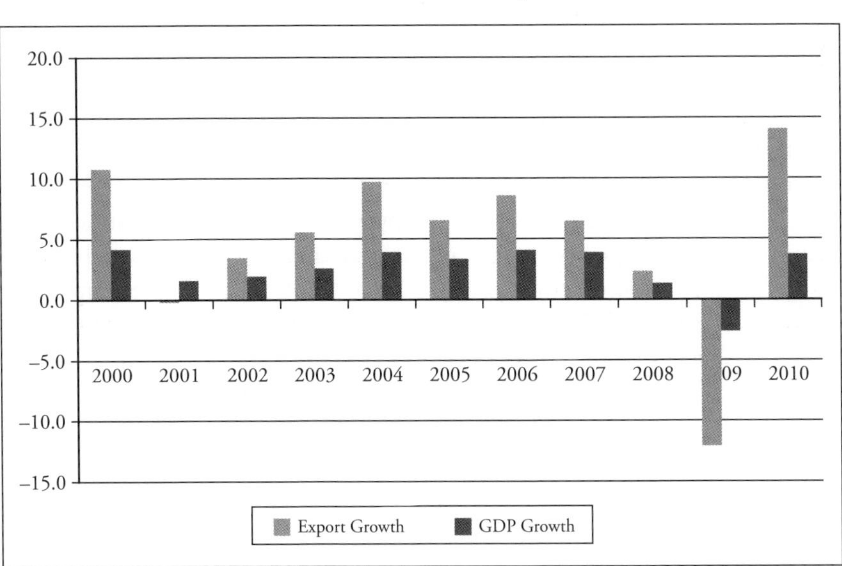

Source: World Trade Organization; URL: <http://www.wto.org/>

same period while Singapore could only manage a modest growth rate of 2 per cent.

As globalization and fragmentation of production became pervasive in the mid-1990s, companies offering electronic manufacturing services (EMS) began to proliferate in Singapore. EMS is a term used for companies that design, test, manufacture, distribute, and provide return/repair services for electronic components and assemblies for original equipment manufacturers (OEMs). EMS activities generally require good spread of technological support and efficient logistic connection. With a short product life cycle, being first to market is vital for a company's survival. Consumers increasingly demand a highly-customized, high-quality product to be delivered quickly at a competitive price. The Dell make-to-order model has become industry standard. The top EMS providers such as Solectron and Flextronics are a new breed of super contractors that command dozens of factories and supply networks around the world. Increasingly, they also manage their customers' entire product lines, offering an array of services from design to inventory management to delivery and after-sales service (Barnes et al. 2000).

On the technology, research and training fronts, there was no slackening of effort in getting the basic ingredients for a top grade electronic hub. The EDB also continued playing a role in expanding the electronic sectors and invested in programs that increased the quality of graduates in the research sector. In 2007, the EDB also invested S$8 million into a Wafer Fabrication Specialist Manpower Programme and co-operated with local universities to incubate undergraduates who specialise in Electrical, Electronics, Mechanical, Materials and Chemical Engineering. In June 2009, the EDB announced that it would invest S$16 million in the Integrated Circuit Design Postgraduate Scholarship. Scholarships will be provided to 150 IC design postgraduates students over 5 years.[10] This investment will help nurture high quality designs of integrated circuits and research in the future. The EDB also sponsors some other scholarships. For example, the EDB also co-funded 70 per cent of the Specialist Manpower Programme which is specially granted to final year undergraduates who focus in IC design. In addition, Singapore has proceeded to establish the fourth public funded university, Singapore University of Technology and Design which will have their first intake in 2012. The university is established in collaboration with the Massachusetts Institute of Technology to develop their science and technology curriculum.

A*STAR continues to provide R&D support to the semiconductor and data storage industry across the entire electronics value chain —

materials, IC design, process technology and packaging technology. Core research capabilities cover areas in spintronics, optical and magnetic materials, recording channel, network storage technologies, integrated circuits design, silicon processing, MEMS and advanced packaging technologies. A*STAR provides funding for several research programmes including Bioelectronics Programme, Nanoelectronics and Photonics Programme, and the Sensor & Actuator Microsystems Programme.

It was noted that Singapore's role as an important player in global HDD industry has diminished in the early 2000s due to intense competition from emerging markets. However, a new related industry emerged — hard disk media manufacturing. Singapore's hard disk media industry expanded rapidly from the 1990s, rising from a global market share of 9.5 per cent in the mid 90s to 41 per cent today. Today, Singapore is a leading location for hard disk media production in the world with the top 3 HDD media producers — Seagate, Showa Denko, and Western Digital having significant media manufacturing operations in Singapore. These companies also have plans to further expand their media manufacturing operations in Singapore. For instance, Showa Denko announced a ¥15.5 billion investment in 2010 to expand its media manufacturing operations in Singapore.

By 2010, semiconductors output successfully accounted for 62 per cent of total electronics manufacturing output in Singapore, and its manufacturing output share of global revenues increased from 11.2 per cent to 13.5 per cent. Nominal growth of 49.8 per cent for Singapore's semiconductor industry outpaced the global growth of semiconductor industry of 32.5 per cent, showing that Singapore's semiconductor industry is growing rapidly. In addition, 15 per cent of the manufacturing output is comprised of data storage devices, in particular hard disk drives recording a 25 per cent increase in production in 2010 compared to 2009. Currently, Singapore is again moving towards more activities with higher technology needs, such as bioelectronics, printed electronics, security, and green electronics that include energy harvesting technologies.

Development Philosophy

The preceding paragraphs clearly show the state's importance in the electronics sector's development. At the very least, the government's consistent push towards capital intensification and knowledge accumulation seem to have instigated the sector to remain relevant in a changing world, prolonging its lifespan.

Two questions relating to development arise in the context of Singapore's electronics industry. The first question concerns the measure by which the government's channelling of foreign investment and resources to the electronics sector has deprived other industries of them. To put it bluntly: did strategic industrial policy benefit the national economy on balance? Goh Keng Swee probably addresses this question better than anyone else:

> What should be stressed is that government policies were market oriented ... There was no target list of industries in the sense that any MNC not in the list would be discouraged from direct investment. Some claim that targeting growth industries amounts to picking winners in a horse race. The reply is that if you bet on every horse in the race, you must pick the winner. (Goh 1995, p. 42)

The second question relates to the specific features of the electronics industry. The industry's inherent nature means that items and production methods in electronics are easily upgradable but require high capital and technological content. Hence, it is of interest as to whether the nature of electronics itself might have made the sector more amenable towards the capital-friendly policies of the time in Singapore. As Segal and Thun (2001) point out, differences in the municipal policies of Shanghai and Beijing meant that IT firms could only thrive in Beijing while industries like automobiles only took off in Shanghai. Curiously, Segal and Thun attribute the success of Beijing's IT sector to its laissez-faire "institutional endowment that lent itself to small innovative firms". In contrast, Shanghai's authoritative set-up led to the "development of large hierarchical corporate groups" that required high amounts of capital instead.

Accepting the premise that Singapore's thriving electronics sector constitutes a success, the similarities between Singapore and Segal's description of Shanghai are interesting. The nascent stage of the sector's development certainly required high amounts of capital input. This was satisfied by the Government's direct approach in attracting foreign investment. Further growth in the sector however will require smaller firms more apt towards creating knowledge, making Beijing's model more attractive. Indeed by 2007, Dr Tony Tan, then Chair of the National Research Foundation, spoke of the need to create an "eco-system of enterprises, big, small, local and foreign" to drive innovation and entrepreneurship — a far cry from the MNC-centric approach taken in the past.

Communication between the government and private businesses takes different forms. It is not uncommon for EDB officers to sit on the boards of MNCs, keeping the government in touch with major companies. Crucially, the government sees foreign investments from a relationship point of view — foreign investors have to be constantly listened to and looked after to ensure they remain invested in Singapore. The sustained interaction with MNCs in particular ensures the country remains responsive to business needs, especially in the dynamic electronics industry. The Association of Electronics Industries of Singapore and the Singapore Manufacturers' Federation are the two organizations working for the development of electronics industry in Singapore. In addition, the Singapore Business Federation represents the wider business community. Companies whose paid-up capital exceeds a particular threshold are automatically federation members, allowing the Federation to present a unified voice for the business community. Lastly, technological scans of various industries are conducted regularly to pinpoint strengths and weaknesses within the industry's value chain.

CONCLUSION

In Singapore, a sovereign state in a single city of 709 square kilometres, there is no difference between central government and sub-national policies. Coordination of policies adopted by different agencies is perhaps less complex than in big countries like Vietnam, China or India. The Cabinet of Ministers provides the overall framework for growth and regulates the overall process, variation and distinction in developmental outcomes is the result of the specific characteristics and abilities of local public agencies like the EDB and others.

The electronics sector continues to remain important to Singapore's manufacturing sector. Nevertheless, its contribution to total manufacturing has declined in recent years. This is due to the natural diversification of economic activity encountered by a country as it matures, and the changing nature of the industry itself. Economic activity within the sector itself has seen a shift away from high employment work such as PC assembly towards production of higher value-added semiconductors.

The government played a crucial role in the sector's development. Policies that affect it are characterized by their coherence towards national objectives. Much effort is taken to ensure that owners of capital, land and labor consistently face incentives well-aligned towards value creation. In addition, the government also ensures that Singapore remains viable as a business destination by taking a long-term view of foreign investment and being responsive to industry demands. As mentioned, foreign investment is seen from a relational point of view, and government officials place importance on cultivating good relations with MNC counterparts to ensure they have reason to remain rooted in Singapore. This differentiates Singapore's strategy from that of countries that focus on building up their indigenous production capabilities rather than luring MNCs.

The electronic sector has undergone several transformations since 1965. The continual change in the domestic industry reflects the government's efforts in initiating such change within the sector as much as the dynamic nature of the industry at large. As a result, economic development of the sector and government policy during the same time period is oft intertwined, each responding to the other.

However, the advent of globalization means that the electronics sector must stay relevant with respect to world development. This is essential in providing MNCs the incentive to remain in Singapore. As shown above, these MNCs transfer technology into the country and provide employment.

R&D and innovation are paramount to Singapore's relevance. This explains the government's commitment to R&D, as S$16.1 billion for 2011–15 have been set aside for investment under the 2015 Research Innovation and Enterprise plan. The government's long-term aim is for Singapore to be among the most research-intensive, innovative and entrepreneurial economies in the world. This should result in high-value production that can create jobs and prosperity for Singapore. The electronic sector will continue to be important for Singapore for a long time.

APPENDIX

Table 9.3

Singapore's Manufacturing Sector: Average Annual Output and Value-Added

	Value in S$ Million				% Share of Total Manufacturing			
OUTPUT	**1991–95**	**1996–00**	**2001–05**	**2006–10**	**1991–95**	**1996–00**	**2001–05**	**2006–10**
ELECTRONICS	41,802	67,910	68,158	78,820	45.1	50.1	40.0	31.5
Semiconductors	6,550	14,954	23,645	42,132	7.1	11.0	13.9	16.8
Computer Peripherals	5,917	11,512	12,172	10,203	6.4	8.5	7.1	4.1
Data Storage	10,087	18,847	16,124	11,992	10.9	13.9	9.5	4.8
Infocomms & Consumer Electronics	16,262	19,015	13,017	12,015	17.5	14.0	7.6	4.8
Other Electronic Components	2,986	3,582	3,200	2,478	3.2	2.6	1.9	1.0
CHEMICALS	16,275	24,541	45,329	80,657	17.6	18.1	26.6	32.2
BIOMEDICAL MANUFACTURING	1,882	4,359	13,011	22,679	2.0	3.2	7.6	9.1
PRECISION ENGRG	12,065	16,284	18,092	23,782	13.0	12.0	10.6	9.5
TRANSPORT	5,876	7,235	11,602	25,635	6.3	5.3	6.8	10.2
GENERAL MANUFACTURING INDUSTRIES	14,793	15,146	14,415	18,912	16.0	11.2	8.4	7.5
Total Manufacturing	92,693	135,476	170,608	250,485	100.0	100.0	100.0	100.0

Table 9.3 *(Cont'd)*

OUTPUT	Value in S$ Million				% Share of Total Manufacturing			
	1991–95	1996–00	2001–05	2006–10	1991–95	1996–00	2001–05	2006–10
ELECTRONICS	7550	12466	13364	15547	35.0	40.4	33.2	29.6
Semiconductors	1436	3552	5740	9227	6.6	11.5	14.3	17.6
Computer Peripherals	1305	1685	2468	2282	6.0	5.5	6.1	4.3
Data Storage	1458	2938	1836	1706	6.8	9.5	4.6	3.2
Infocomms & Consumer Electronics	2491	3116	2175	1555	11.5	10.1	5.4	3.0
Other Electronic Components	859	1176	1145	777	4.0	3.8	2.8	1.5
CHEMICALS	2666	3552	5730	5606	12.3	11.5	14.3	10.7
BIOMEDICAL MANUFACTURING	1353	2556	7087	11457	6.3	8.3	17.6	21.8
PRECISION ENGRG	3676	4985	5640	6817	17.0	16.2	14.0	13.0
TRANSPORT	2010	2543	4123	7795	9.3	8.2	10.3	14.8
GENERAL MANUFACTURING INDUSTRIES	4344	4735	4266	5332	20.1	15.4	10.6	10.1
Total Manufacturing	21598	30837	40209	52555	100.0	100.0	100.0	100.0

Table 9.4

Singapore's Manufacturing Sector: Average Annual Employment and Fixed Assets Investment

	Number in Thousands				% Share of Total Manufacturing			
Employment	**1991–95**	**1996–00**	**2001–05**	**2006–10**	**1991–95**	**1996–00**	**2001–05**	**2006–10**
ELECTRONICS	122.2	114.4	93.1	87.6	33.8	32.3	26.1	21.3
Semiconductors	16.2	23.9	32.5	40.7	4.5	6.8	9.1	9.9
Computer Peripherals	21.8	20.0	15.0	13.4	6.0	5.7	4.2	3.2
Data Storage	29.7	31.8	21.4	14.9	8.2	9.0	6.0	3.6
Infocomms & Consumer Electronics	39.0	23.4	12.9	10.0	10.8	6.6	3.6	2.4
Other Electronic Components	15.5	15.2	11.3	8.7	4.3	4.3	3.2	2.1
CHEMICALS	17.7	21.0	22.7	23.1	4.9	5.9	6.4	5.6
BIOMEDICAL MANUFACTURING	5.4	5.9	8.6	12.4	1.5	1.7	2.4	3.0
PRECISION ENGRG	81.1	90.6	92.2	96.3	22.4	25.6	25.9	23.4
TRANSPORT	37.7	40.7	56.1	103.8	10.4	11.5	15.7	25.2
GENERAL MANUFACTURING INDUSTRIES	97.5	81.5	83.5	88.6	27.0	23.0	23.5	21.5
Total Manufacturing	361.5	354.1	356.2	411.7	100.0	100.0	100.0	100.0

Table 9.4 *(Cont'd)*

Fixed Assets Investment	Value in S$ Million				% Share of Total Manufacturing			
	1991–95	1996–00	2001–05	2006–10	1991–95	1996–00	2001–05	2006–10
ELECTRONICS	–	3,617	4,483	4,629	–	43.5	52.8	37.7
CHEMICALS	–	2,546	1,823	5,491	–	30.6	21.5	44.7
BIOMEDICAL MANUFACTURING	–	423	851	808	–	5.1	10.0	6.6
PRECISION ENGRG	–	908	642	461	–	10.9	7.6	3.8
TRANSPORT	–	375	400	669	–	4.5	4.7	5.4
GENERAL MANUFACTURING INDUSTRIES	–	453	294	230	–	5.4	3.5	1.9
Total Manufacturing	–	8,322	8,494	12,288	–	100	100	100

Source: Department of Statistics, *Yearbook of Statistics*. Singapore: Department of Statistics, various years.

Table 9.5

Singapore's Manufacturing Sector: Average Annual Value-Added per Worker

	Value in S$ Thousand per Worker				Average Annual % Growth			
Productivity	**1991–95**	**1996–00**	**2001–05**	**2006–10**	**1991–95**	**1996–00**	**2001–05**	**2006–10**
ELECTRONICS	32.5	82.2	130.2	207.2	17.5	19.0	11.3	8.9
Semiconductors	45.3	105.3	158.5	263.9	26.9	24.2	10.2	9.7
Computer Peripherals	31.2	63.7	150.6	202.2	11.3	18.7	11.2	9.9
Data Storage	25.5	68.8	76.7	136.2	11.9	12.7	8.7	9.3
Infocomms & Consumer Electronics	33.9	101.8	155.0	174.3	18.6	14.1	16.1	(2.4)
Other Electronic Components	29.2	56.7	93.4	101.4	17.3	11.5	9.3	(0.6)
CHEMICALS	211.1	245.6	321.2	266.6	4.0	3.3	6.0	0.7
BIOMEDICAL MANUFACTURING	350.4	633.9	1,037.0	1,022.6	1.5	6.2	6.3	(12.8)
PRECISION ENGRG	41.8	55.3	66.4	66.2	8.6	1.6	13.9	2.3
TRANSPORT	27.6	45.1	65.4	84.1	4.4	2.9	(2.0)	10.2
GENERAL MANUFACTURING INDUSTRIES	41.6	58.3	55.5	56.1	12.4	7.5	(4.0)	4.8
Total Manufacturing	52.6	91.8	121.4	134.4	(0.8)	14.2	4.6	(3.5)

Source: Department of Statistics, *Yearbook of Statistics*. Singapore: Department of Statistics, various years.

Table 9.6
Singapore's Domestic Exports, 1995–2010

Domestic Exports	Value in S$ Million				% Share of Total Manufacturing			
	1995	2000	2005	2010	1995	2000	2005	2010
TOTAL DOMESTIC EXPORTS	**98,473**	**135,938**	**207,448**	**248,610**	**100.0**	**100.0**	**100.0**	**100.0**
OIL	13,721	22,867	52,798	75,011	13.9	16.8	25.5	30.2
NODX	84,751	113,071	154,650	173,599	86.1	83.2	74.5	69.8
ELECTRONICS	**59,342**	**74,393**	**75,386**	**65,000**	**60.3**	**54.7**	**36.3**	**26.1**
Disk Drives	13,509	16,013	13,839	5,473	13.7	11.8	6.7	2.2
PCB Assembled	10,317	13,067	16,310	12,885	10.5	9.6	7.9	5.2
IC's	10,500	20,006	21,884	27,187	10.7	14.7	10.5	10.9
PC'S	5,322	3,652	1,067	1,658	5.4	2.7	0.5	0.7
Printers	2,410	3,198	1,619	62	2.4	2.4	0.8	0.0
Telecom Equipment	0	3,167	5,460	1,665	0.0	2.3	2.6	0.7
Non-Electronics Products	25,409	38,678	79,263	108,598	25.8	28.5	38.2	43.7

Source: Department of Statistics, *Yearbook of Statistics*. Singapore: Department of Statistics, various years.

Table 9.7
Tax and Financial Incentives in Singapore

EDB Tax Incentives for Investments

Scheme	Benefits	Suitable for
Pioneer (Manufacturing)	Tax exemption on income from qualifying activities	Manufacturing
Pioneer (Services) (also available for IHQ Award)	Tax exemption on income from qualifying activities	• Services • GHQ
Development and Expansion Incentive (also available for IHQ Award)	Reduced tax 5% or 10% on incremental income from qualifying activities	• Manufacturing • Services • RHQ/IHQ • IP Hub
Investment Allowance	Allowance of 30% or 50% of approved fixed capital expenditure on top of normal 100% capital allowance	• Manufacturing
Finance & Treasury Centre Tax Incentive	Reduced tax 5% or 10% on fees, interest, dividends and gains from qualifying services/activities	• FTC
	WHT exemption on interest payments on loans from banks and network companies for FTC activities	
Approved Royalties Incentive	Reduced WHT 0% or 5% on royalty payments to access advanced technology and know-how	• Manufacturing • FTC
Approved Foreign Loan	Reduced WHT 0%, 5% or 10% on interest payments on loans taken to purchase productive equipment	• Manufacturing

Table 9.7 *(Cont'd)*

Scheme	Benefits	Suitable for
S19B writing-down allowances for IP acquisition	Automatic 5-year write-down if legal and econ IPR are acquired EDB's approval is required if only econ IPR is acquired	• IP Hub
S19C writing-down allowances for R&D cost-sharing	1-year write-down for R&D cost-sharing payments	• Manufacturing • IP Hub

Financial Incentives

Scheme	Benefits	Suitable for
Research Incentive Scheme for Companies (RISC)	Co-funding to support the set-up of R&D centres, and/or the development of in-house R&D capabilities in strategic areas of technology Supportable project costs include expenditure in the following: • Manpower • Equipment and Materials • Professional Services • Intellectual Property Rights	Singapore-registered business entities undertaking R&D activities
Initiatives in New Technology (INTECH)	Co-funding to support the manpower development in the application of new technologies, industrial R&D and professional know-how	Singapore-registered business entities introducing or developing new capabilities

Source: Economic Development Board, Singapore. <http://www.sedb.com/edb/sg/en_uk/index/why_singapore/Guide_to_Investing_in_Singapore/financial_assistance.html>

Notes

1 Available at URL: <http://www.sedb.com>.

2 A list of fiscal and financial incentives managed by EDB is included in the Appendix. For a detailed rendition of the formation of the Economic Development Board and the programmes implemented before 1990 and investment promotion experiences, please see Low, Toh, Soon and Tan (1992) and EDB (2002).

3 LIUP aims to raise the efficiency, reliability and competiveness of participating firms in 3 phases: improvement of operational efficiency; transfer of new product and processes to local enterprises; joint production and process R&D with foreign affiliates as partners. EDB contributes to the salary of the foreign representative seconded to a local supplier. Typically, MNC partners choose which local suppliers to work with in consultation with EDB. According to UNCTAD (2001), in 1999, about 30 investors with 11 large local enterprises and government agencies were partnering with 670 local suppliers.

4 When Dr Goh became Singapore's first Finance Minister in 1959, he put his attention to jump-starting the stagnant economy. He initiated the setting up of the Economic Development Board which was established in August 1961 to attract foreign multinational corporations to invest in Singapore.

5 For a vivid study of the dynamics of the hard disk drive industry showing how and why U.S. firms developed complementary clusters in the U.S. and Southeast Asia, see McKendrick, Doner and Haggard (2000).

6 The National Science and Technology Board was later renamed as Agency for Science, Technology and Research (A*STAR) with the passing of the A*STAR Act on 27 August 2002.

7 More detail on the activities of IME is available at URL: <http://www.ime.a-star.edu.sg/html/co_a.html>.

8 JTC (16 September 2010). Wafer Fabrication and Advanced Display Park: <http://www.jtc.gov.sg/industrycluster/Electronics/Pages/index.aspx>.

9 In July 2010, AU Optronics Corp of Taiwan acquired the entire share capital of AFPD Pte Ltd, a manufacturer of TFT liquid crystal displays, from Toshiba Corp's Toshiba Mobile Display Co Ltd.

10 Economic Development Board. (2010). *Revolutionising Electronics — IC Design in Singapore.* <http://www.sedb.com/edb/sg/en_uk/index/news/articles/electronics_industry_briefing.html>.

References

Barnes E., J. Dai, S. Deng, D. Down, M. Goh, H.C. Lau, and M. Sharafali. "Electronics Manufacturing Service Industry". White Paper, The Logistics Institute — Asia Pacific, National University of Singapore, Singapore, 2000.

Chia, S.Y. "Singapore: Advanced Production Base and Smart Hub of the Electronics Industry". In *Multinational and East Asian Integration*, edited by S.Y. Chia, & W. Dobson, Canada and Singapore: International Development Research Centre and Institute of Southeast Asian Studies, 1997.

Department of Statistics. *Yearbook of Statistics 2011*. Singapore: Department of Statistics available at <http://www.singstat.gov.sg/pubn/reference/yos11/yos2011.pdf>.

Economic Development Board. *Heart Work: Stories of How EDB Steered the Singapore Economy from 1961 into the 21st Century*. Singapore: Singapore Economic Development Board/EDB Society, 2002.

Goh Keng Swee. *Wealth of East Asian Nations*. Singapore: Federal Publications, 1995.

Hobday, M. "Technological Learning in Singapore: A Test Case of Leapfrogging". *Journal of Development Studies* 30, issue 4 (1994): 831–58.

Lee Kuan Yew. *From Third World to First: The Singapore Story, 1965–2000*, London: HarperCollins Publishers, 2000.

Lim, H. "ASEAN SMEs and Globalization". *Research Report No. 5*. Jakarta: Economic Research Institute of ASEAN and East Asia, 2007.

Low, L., Toh M.H., Soon T.W., and Tan K.Y. *Challenge and Response: 30th Anniversary of the EDB*. Times Academic press, Singapore, 1992.

Mathews, J.A. "A Silicon Island of the East: Creating a Semiconductor Industry in Singapore". *California Management Review* 41 (1999): 55–78.

McKendrick, D.G., R.F. Doner, and Stephan Haggard. *From Silicon Valley to Singapore: Location and Competitive Advantage in the Hard Disk Drive Industry*. Palo Alto: Stanford Business Books, 2000.

Rajan, R., and Thangavelu, S.M. "Regional Integration Strategies: Singapore". In *National Strategies for Regional Integration: South and East Asian Case Studies*, edited by J. Francois, P.B. Rana, and G. Wignaraja. Manila: Asian Development Bank, 2009.

Segal, A. and E. Thun. "Thinking Globally, Acting Locally". *Politics and Society* 29, no. 4 (2001): 557–88.

Spring Singapore. *Electronics Industry Background and Statistics*. Singapore: SPRING, 2011, available at <http://www.spring.gov.sg/enterpriseindustry/elc/pages/industry-background-statistics.aspx>.

Tan, K.Y. "Innovation and Entrepreneurship in a Knowledge-Based Economy". Speech given at the Marriott Hotel, 27 November 2007.

Toh M.H. "Singapore's Perspectives on the Proliferation of RTAs in East Asia and Beyond". *Global Economic Review* 35, no. 3 (2006): 259–84.

UNCTAD. *World Investment Report 2001*. New York and Geneva: United Nations, 2001.

U.S. Embassy. *Singapore's Electronics Industry — Facing Challenges, But First Mover Advantage*. White Paper, Singapore: United States Embassy, 2004.

NORTH KOREA
Seoul
SOUTH KOREA
Gumi City
GYONGSANGBUK
East China Sea
JAPAN
LEGEND
State/Provincial Capital
STATE/PROVINCE
National Capital
COUNTRY
NEIGHBOURING COUNTRY
Geographic Features
National Border Land
National Border Water
State/Province Border
0 km 25 50
All maps are derived from GADM Data
(Global Administrative Areas, www.gadm.org, 2012)
edited by maps&more, Hans Hortig / Marcel Jäggi

10

THE EVOLUTION OF AN INDUSTRIAL CLUSTER AND ITS POLICY FRAMEWORK: THE CASE OF GUMI CITY, KOREA[1]

Sam Ock Park and Do Chai Chung

INTRODUCTION

Korea has achieved remarkable economic growth over the past five decades. After the devastation of the Korean War (1950–53), the country was among the poorest in the world. From a GNP per capita below US$100 in 1960 (in 1996 USD), the country's GNP reached US$20,000 in 2007. Such a remarkable level of economic growth is closely tied to the successful implementation of the government's strategies for export-oriented and sector-specific industrial development, human resources, and innovation since the launch of the First Five-Year Economic Development Plan in 1962.

Because of the development of industrial clusters, dynamic spatial patterns and processes have occurred in Korea's "space economy". In the country's early industrialization phase, the spatial disparity of economic activities increased with the bipolar concentration of industries. This spatial disparity has been persistent, accompanied by the continuous concentration of population, which has created a new spatial division of labour between the Capital Region and the rest of the country as well as path-dependent trends of industrial development.

However, a conscious state-led strategy of supporting industrial clusters in the non-Capital Region has mitigated this somewhat. Despite the persistent spatial disparity of economic activities, the intra-regional disparity of per capita GRP has been considerably decreased over the last two decades. Along with the decrease of the spatial disparity of per capita GRP, a new path creation trend is evolving in cities of provincial areas and rural areas. The development of high tech industries such as electronics in the non-Capital Region and the development of ICT have influenced the spatial dynamics of the Korean economy (Park 2009).

Gumi city in Gyungsangbuk province is a good example of economic development in the non-Capital Region that has contributed to lowering intra-regional disparities as well as the progress of the dynamics of Korea's economic space. Over the last four decades, Gumi city has emerged as a leading electronics cluster and, in 2011, the city's export value reached US$33 billion. Gumi's industrial structure has been transformed, evolving away from the branch plant agglomeration characterised by the co-location of many branch plants in the electronics sector to an industrial cluster with evolving local networks.

This chapter will analyse the case of Gumi city. This chapter will first put forward national industrial and regional policies. From there, Gumi's history of industrial development and the evolution of its electronics cluster will be analysed. The final section will summarize the chapter's principle conclusions.

KOREA'S INDUSTRIAL POLICIES AND THEIR EVOLUTION

Over the past five decades, the national government's industrial policy frameworks have had a substantial impact on Korea's industrial development. However, these frameworks have, themselves, evolved over time from a concentration on the development of industrial parks in the early period to a focus on innovation in more recent years.

Industrial Policy Frameworks until 2002

Since the first Five Year Economic Development Plan was launched in 1962, the national government has taken a leading role in the promotion of sectoral and spatial industrial policies. Export-oriented industrialization has been the major strategy in place since the early 1960s, and the strategy was fashioned to promote the most promising industries at specific points in time. Labour-intensive industries such as textile and apparel were the target sectors before the mid-1970s, while heavy and chemical industries such as petrochemicals, shipbuilding, automobiles, and consumer electronics were the leading industries for export expansion in the late 1970s and early 1980s.

Typical supporting measures for target sectors included allowing access to foreign capital markets and providing investment incentives. These policies geared to the heavy industry and chemical sectors contributed to the development of the *jaebol* system (conglomeration) in the Korean economy (Markusen and Park 1995).

Since the mid 1980s, high-technology industries such as semiconductors have been increasingly favoured. Due to the shift in sectoral industrial policies, Korea's industrial structure shifted from a labour-intensive to a technology-intensive industrial structure. Labour intensive industries took a dominant role during the rapid industrialization in the 1960s and early 1970s, but showed a continuous decrease of their share since the late 1970s. Conversely, assembly-type industries, which include high-tech industries, have continuously increased since the late 1970s (Park 2009).

Since the mid-1990s, especially following the foreign exchange crisis in November 1997, the Korean government has made great efforts to: promote the development of knowledge-intensive industries and services; open the country fully to trade and capital movements; restructure the economy including the financial sector; and make the labour market flexible.

In the 1960s and 1970s, along with sectoral industrial policies, the national government promoted spatial policy with the establishment of several large industrial estates especially in the southeastern part of the country in order to decentralize industries. The major new industrial cities or production complexes of Ulsan, Changwon, Pohang, Gumi, Gwangyang, and Ansan were the result of these policies. However, the role of FDI from foreign countries was relatively insignificant in the 1970s and 1980s, compared to imported technology and foreign capital borrowing. Instead, *Jaebols* significantly contributed to the development and growth of these industrial cities by establishing large branch plants with imported technology and borrowed foreign capital. The industrial policy focusing on the development of industrial parks can be regarded as one of the principal strategies to establish production systems in the nation. However, the idea of territorial production systems was not successfully implemented at earlier stages because of poor local industrial linkages. That is, at first, the industrial parks had only limited local inter-firm linkages and were just agglomerations of production activities without significant intra-regional production networks.

Up until the 1980s, the sectoral and spatial industrial policies had a significant impact on the spatial structure of the Korean economy. The government's spatial industrial policy resulted in the spatial division of labour, concentrating headquarters of *Jaebols* in Seoul while decentralizing production functions to the non-Capital Region (Park 1993).

However, since the mid-1980s, the high-technology industrial policy resulted in the re-concentration of industry in the Capital Region, due to its locational advantages such as easy access to skilled labour, knowledge, technology information, and finance. The concentration of high-technology industries and advanced services including R&D activities in the Capital Region has again intensified the spatial division of labour in Korea's production systems and space economy (Park 1993; 2009).

Since the mid-1990s, several local governments have made significant efforts to attract knowledge-based industries. Despite this, knowledge-based industries are still overwhelmingly concentrated in the Capital Region. Thus, even though new industrial parks have developed in the non-Capital Region as a result of the national government's regional industrial policy, regional variation in the development of knowledge-based industries has been persistent in the Korean space economy (Park et al. 2009).

Innovation Policy Frameworks until 2002

In the early phase of industrialization, innovation-related issues were relatively neglected, since the main goal was focused on establishing manufacturing production bases. In the 1970s, science and technology policy focused on the expansion of education in the technical and engineering fields and on the establishment of a number of government-supported research institutes in the field of heavy and chemical industries. This was complemented with the promotion of the heavy industry and chemical sectors. Government-supported research institutes took a leading role in the improvement of industrial technologies during this period. Most of the firms were more interested in technology transfer from industrialized countries rather than their own promotion of R&D activities. However, through the establishment of government-initiated research institutes, the national systems of innovation began to evolve in tandem with industrial development.

However, innovation systems in Korea have significantly changed since the 1980s. This is largely due to the considerable growth of private firms' in-house R&D investments. Firms have themselves begun to emphasize technology development due to increasing competition in international markets. Private firms' share of national total R&D expenditure was 56 per cent in 1981 and reached 81 per cent in 1985 (MOST 1990). In 1980, only 54 firms, most of whom belonged to *Jaebols*, had their own R&D

centres, but the figure increased to 2,226 in 1995 (KITA 1995). In the early 1980s, *Jaebols* aggressively established R&D centres.

Kim (1997) identified three major characteristics of innovation activities of firms in the 1980s: (1) Large firms of Jaebols established strategic alliances with world-wide high tech firms; (2) Large firms, which mostly belonged to Jaebols, were aggressive in establishing foreign R&D centres and labs; (3) Due to the difficulties in getting a license for leading-edge complex technology, large firms were actively involved in the merger/acquisition of high-tech firms in industrialized countries in order to secure original technology.

After the late 1980s, even smaller firms began to establish R&D centres. Accordingly, in the 1990s, beyond the national innovation systems, regional innovation networks began to evolve due to the development of regional clusters of SMEs in technology-intensive sectors. The establishment of science parks and high-tech parks in the non-Capital Region in the 1990s, in addition to Daeduck Science Town, has also contributed to the development of local clustering of innovation networks.

According to Park's (2000) surveys of SMEs, the role of such firms has become important in the development of regional innovation systems in Korea. Based on the survey, SMEs have been more involved in R&D activities during the 1990s, especially as one of the strategies to cope with industrial restructuring. Out of 825 firms replying to the questionnaire, 20 per cent of firms conducted R&D activities in 1993. The ratio of firms conducting R&D activities increased to 34 per cent in 1996. Overall, larger SMEs are more active in R&D than smaller ones. However, among the firms which conduct R&D activities, the smaller SMEs show a higher ratio of R&D expenditure to total sales than do the larger SMEs. Thus, a considerable proportion of smaller SMEs could be regarded as venture businesses, even in the early 1990s.

It should be noted that inter-firm networks between large contract firms and suppliers of SMEs within a local area, and collaborations with other firms and trade associations of the same industry have become an important mechanism for fostering innovation among SMEs in Korea. Large firms belonging to *Jaebols* have a critical role in forming inter-firm networks through establishing cooperative supply arrangements with local SMEs. For some SMEs, collaboration with universities, government-sponsored research institutes, and other public institutes are also important contributors for their technological development and innovation. The importance of inter-firm networks for the innovation of SMEs was supported by the extended survey of 1999 (Park and Nahm 2000). It should also be noted that, since the financial crisis in 1997, even some SMEs have been established as focal points for networks in Silicon Valley, showing that global innovation networks for SMEs are becoming important.

Recent Innovation and Cluster Policies

The Participatory Government of President Roh Moo-Hyun, inaugurated in 2003, set out "democracy with the people", "the balanced national development" and "the Northeast Asian cooperation initiative" as three policy goals. In the regional policy area, the Government strongly promoted balanced national development by emphasizing regional innovation and cluster policies (Park 2007). Major policies for regional innovation are providing the basis for the establishment of regional innovation systems (RIS), by: strengthening the innovation capacity of universities in provinces; promoting science and technology in the provincial regions; and establishing industry-university-research centre networks (PCBND 2007).

The innovative cluster policy was strongly promoted by the Participatory Government at the regional level. Seven innovative clusters were designated as a model of innovative clusters to be accompanied by the reorganization of selected national industrial complexes: electronics and IT in Gumi; machinery in Changwon; automobiles in Ulsan; parts and components in Ahnsan; parts and components of automobile and machinery in Gunsan-Janghang; photonics in Gwangju; and medical instruments in Wonju. Each innovative cluster has three to seven mini-clusters, which focus on a specific technology or product to promote collaboration and solve production problems. The Participatory Government subsequently designated six more innovative clusters and a total of 13 innovative clusters have been supported outside Seoul.

In addition to these innovative clusters, Daeduck special R&D district in Daejun city has been supported under a special law to promote commercialization of R&D and diverse innovations. High tech IT clusters and local culture clusters are also supported to strengthen local innovation through collaboration among the diverse economic actors. The development of clusters by private firms such as Suwon IT cluster by Samsung, Paju's LCD cluster by LG-Phillips, and Pohang's material cluster by POSCO are also promoted.

These regional innovation policies have contributed to the increase of the density of regional innovation networks in the provinces. However, the availability of high quality manpower in the provinces has been an issue. That said, intraregional disparity in terms of per capita income between the

Capital Region and other parts of the country has decreased, even though this has not been accompanied by shifts in population (Park 2007).

Following the beginning of the Lee administration, the Presidential Committee on Regional Development formulated a new regional policy. Because of drastic changes in the global and domestic economic environment, the new administration introduced a new philosophy to its regional development policy (Richardson et al. 2011). Under this new approach, the boundaries of regional development plans are not limited to one administrative area, but rather include several administrative units to make a larger economic region for synergetic effects of cooperation with other regions. The new regional policy aims to build a spatial foundation for each region to attract and develop new industries, while ensuring their effectiveness and competitiveness (Choe 2011). The new policy also emphasizes both localization and decentralization for regional development. The Committee suggested that the new government alter its policy direction from declaratory "balanced regional development" to "regional specialization and competition" that attempts to implement effective self-reliant localization policies (Choe 2011).

The new regional policy comprises five key strategies. First, the new policy has a three-tiered approach based on different geographical scales: (1) the Economic Region (ER) scheme is to regroup upper-tier local government (16 metropolises and provinces) into seven Economic Regions; (2) the Local Area scheme is to provide quality-of-life services and income-earning opportunities for all of the residents in lower-tier local government (163 cities and counties); and the Supra-Economic Region (SER) scheme is to create the SERs along the coastlines and one inland for the promotion of cross-economic region and cross-border cooperation. Second, the government will promote regional development by providing new regional growth engines based on specialization. Two selected industries for each ER will be fostered to enhance global competitiveness. Interregional linkages and cooperation will be boosted by promoting the synergetic effects of strategic industries. Third, to enforce decentralization by delegating power from the central government to local governments, several tasks will be promoted. They are: the devolution of central government's regional agencies; the redistribution of taxes between the central and local governments; the integration of diverse national subsidies into block grants for regional development; and the transfer of development authority to local governments for better planning and implementation (Table 10.1). Fourth, the symbiotic development of the Capital Region and the non-Capital Region is pursued through diverse channels. Some examples are the transfer of capital gains earned by developing land in the

Table 10.1
Local Government Revenue in Absolute Terms and as a Share of National Total Revenue

	1999	**2000**	**2001**	**2002**	**2003**	**2004**	**2005**	**2006**	**2007**	**2008**	**2009**	**2010**
Total amount (Billion KRW)	67,11	82,67	122,89	122,59	149,11	144,69	194,85	204,41	226,24	257,80	250,80	263,46
Proprtion of national total revenue(%)	13.27	15.00	15.00	15.00	15.00	15.00	19.13	19.24	19.24	19.24	19.24	19.24

Source: Korean Statistical Information Service (www.kosis.kr).

Capital Region to local governments in the non-Capital Region; reducing regulations and improving the institutional environment to reinvigorate the regional economy; and a stepwise deregulation of the Capital Region concurrent with regional development in the non-Capital Region. Fifth, existing regional projects inherited from the former administration will be continued. For example, the dispersal programme of government offices and public institutions will be furthered considering its contribution to make growth hubs in the ER; incentive measures such as tax exemptions and low-priced land can be used to attract high-tech business in the non-capital Region (Choe 2011).

Overall, the Korean industrial policy has evolved from promoting industrial complexes in the 1960s and 1970s, to high tech industrial development in the 1980s, to industrial restructuring in the 1990s, to innovation and industrial clusters from 2002 to 2007, and to the current focus on building economic regions through emphasizing interregional networks and cooperation (Figure 10.1). Along with the changes and reshaping of the industrial and innovation policies, Korea's space economy has shown several

Figure 10.1
A Brief History of Korean Industrial Policy

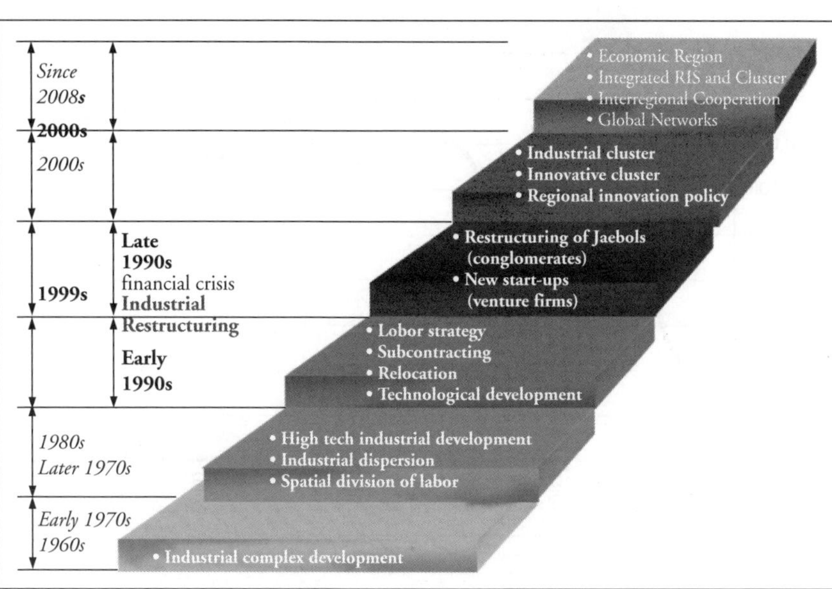

Source: Park 2010.

dynamic characteristics, such as: spatial division of labor; development of ICT and spatial changes with the dominance of Seoul; and virtual innovation networks in peripheral areas (Park 2009). More details will be examined through the case of Gumi, a specialized electronics cluster in Korea.

THE CASE OF GUMI

Gumi is a new industrial district established in 1971 and designed as an electronics district. The city is in Gyongsangbuk province, about 267 km south of Seoul, and is located about half an hour's drive north of Daegu, the third largest city in Korea.

In 1977, the erection of factories and buildings began in the first industrial complex in considerable numbers, and over 240 firms employing more than 70,000 workers resided in the complex by early 1990s. Since the early 1990s, the number of firms has continuously increased and reached at 1,177 in 2009, but the number of employees has fluctuated from 65,000 to 80,000 (Figure 10.2).

Figure 10.2
The Number of Manufacturing Establishments and Employees in Gumi

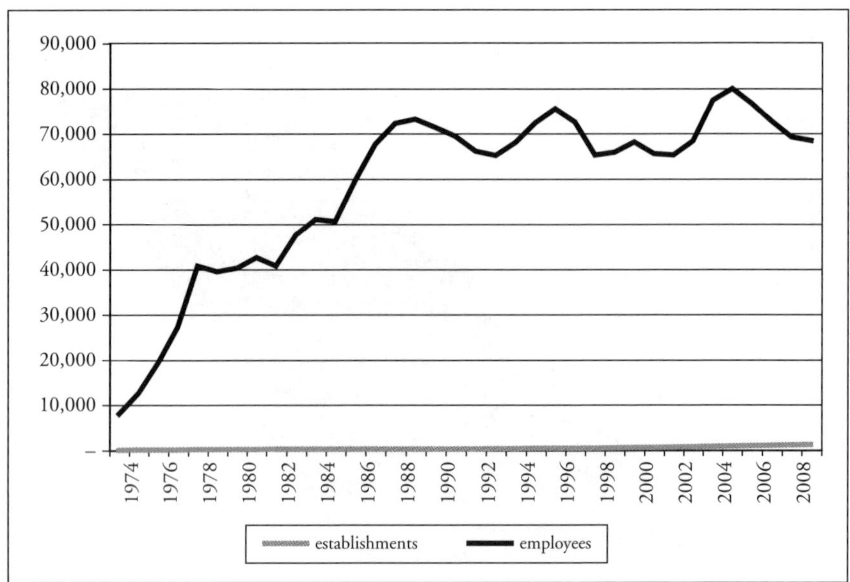

Source: Korea Industrial Complex Corp. 2011.

Growth in industrial output and export has been impressive, with an annual growth rate of more than 18 per cent during the 1980s and continuous growth up to 2005 except the period of financial crisis in 1997 and 2001 as seen in (Figure 10.3). Of course, there has been a fluctuation of the growth during the last half-century. In the changes of the growth rate of production, export, establishment, and employment, the negative growth rate in some of the criteria noticed in a few years, especially during the financial crisis in East Asia in 1997 and global economic crisis in recent years. Most of the years during the last four or five decades, export growth rates were more than 10 per cent and more than half of the years revealed growth rate of more than 20 per cent (Figure 10.4).

Gumi's resident population grew almost four-fold, from 105,000 in 1980 to 404,920 in 2010. Gumi fitted into the Park administration's economic strategy of a new round of export-oriented development in the 1970s, based on electronics in particular. The development of electronics was an obvious route for a resource-poor economy attempting to diversify away from South Korea's dependence upon heavy industry (Markusen and Park 1995). The choice of Gumi owed more to the exercise of discretionary political power than to a commitment to regionally-balanced industrial growth. Gumi was a personal project of President Park, whose military regime targeted

Figure 10.3
Production and Exports from Gumi

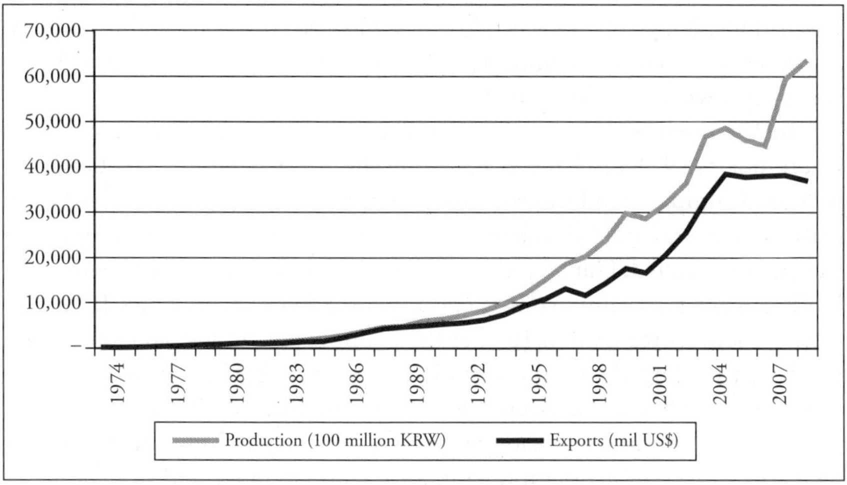

Source: Korea Industrial Complex Corp. 2011.

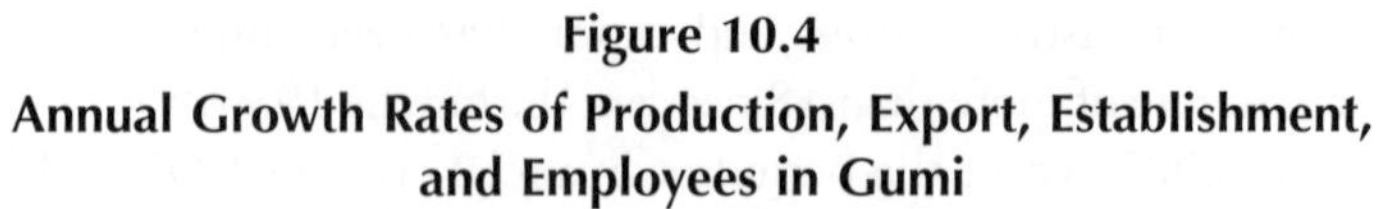

Figure 10.4
Annual Growth Rates of Production, Export, Establishment, and Employees in Gumi

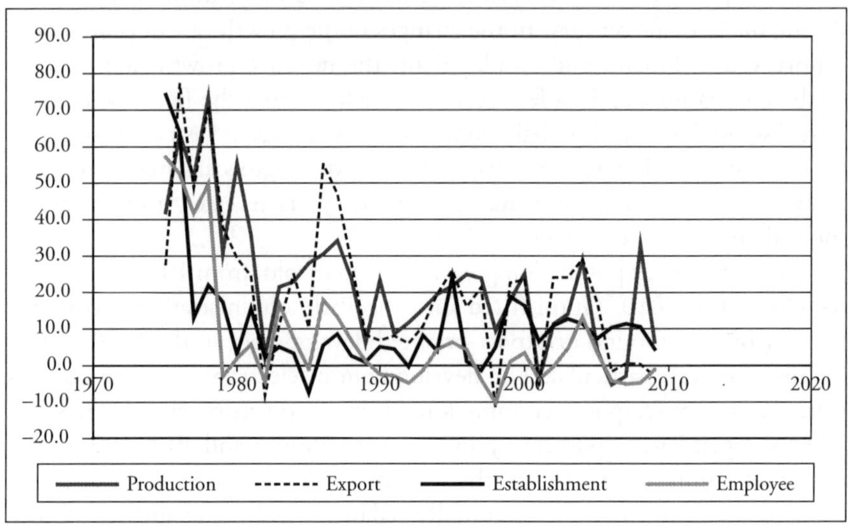

Source: Park 2010.

new investments toward the South-eastern region of Korea. Although it was situated far from the coasts — which had been a consideration that dictated the earlier export-oriented and port-oriented selection of Pohang, Ulsan, and Changwon as industrial cities for steel, chemicals, and machinery, respectively — Gumi's countervailing asset was its good fortune in being the hometown of President Park.

In 1969, when its fate was sealed, Gumi was a small agricultural village, where not much had changed economically for generations. To build the industrial complex, land was reclaimed from the Nakdong River and a 12 km embankment was built to rechannel the river's flow. Factories and dormitories were rapidly erected, filling up the grid lines laid down on planners' maps. The leading agent for Gumi's construction was the Korea Electronics Industrial Corporation, reorganized in 1974 as the Gumi Export Industrial Corporation (KEIC), to acknowledge the fact that the textile business was becoming as important a tenant as electronics in the complex. KEIC was an arm of the national government, belonging to the Ministry of Trade and Industry, with its own special trust fund to finance land clearance and development. Over the years, KEIC's operations have been increasingly funded out of the proceeds of its land and energy activities, but its development policy is still overseen by the national government.

Government subsidies and incentives have played a major role in inducing companies to site plants in Gumi. Land clearance, site preparation, infrastructure, and water and energy supplies are ample and cheap, and efficient transportation links to Seoul and southern seaports are assured. Not all the inducements are physical in nature. Tax breaks, worker education and training programs, and a modest level of business services have been provided. Although a plant is supposed to be "clean" to qualify as a Gumi resident, the government has tolerated water and air pollution, an advantage to firms who find restrictions tightening in Seoul.

At first, the principal input into the production process in Gumi was labour, the target labour force being young women from the 400 villages within 15 km of Gumi. KEIC actively helped to recruit the labour forces by combing the high schools in rural areas. Larger plants were required to set up their own in-house training facilities and technical and vocation schools in Gumi supplemented the labour supply. A major inducement to immigration by very young women was the offer of a high school education in the factory itself, as part of the conditions of employment. Managers and engineers, on the other hand, were drawn from urban centres such as Daegu, especially from the College of Engineering of Gyungbuk National University. Job opportunities within the region have helped to staunch the flow of educated labour from places such as Daegu to the Seoul metropolitan area.

The Evolution of Gumi's Industrial Cluster

The creation of the Gumi industrial complex as an agglomeration of branch plants could transform itself by generating spin-offs, spawning localized supplier networks, and creating governance structures that resemble those of industrial districts. Several questions were raised in the early 1990s studies (Markusen and Park 1995). Have the economies of Gumi continued to exhibit the features of a satellite platform? Over time, are there signs of modernization, local entrepreneurship, vertical disintegration, increased networking among firms, or greater interest in flexible specialization? Or do state management, branch plants, non-place embeddedness, and exogenous decision-making still predominate?

In this section, we examine the changing industrial specialization and industrial and firm organization, as well as the prevalence of indigenous versus exogenous linkages in the context of industrial district's evolution. Pressures to restructure in these districts were intense in the early 1990s. Since the late

1980s, national factor advantages such as cheap and abundant labour and cheap industrial land have almost disappeared in Korea compared with other developing countries. As a result, industry in South Korea had to undergo significant restructuring in order to retain its competitive advantage. Major triggers of this industrial restructuring were a rapid increase in wages, the eruption of labour disputes, currency revaluation, and high financial costs (Park 1993). Corporate strategies and regional characteristics as well as the role of the state have been shown elsewhere to be important factors in the industrial restructuring at the national level (Park 1993). Overall, this study suggests that the Gumi industrial complex originally started as a satellite industrial district. Although it faced some difficulties, it evolved into an industrial cluster with continued diversification under the main trend of specialization by electronics industry.

Changes in Industrial Specialization

Gumi was initially designed to promote a high level of industrial specialization, in order to encourage localization economies and inter-firm networking. Such specialization is a key feature of industrial districts in the New Industrial District literature (Markusen and Park 1995). Gumi started with considerable sectoral specialization. It was originally designed to host electronics manufacturing. In 1990, the electric machinery and electronics sectors accounted for 49.9 per cent of total employment (Markusen and Park 1995) and the share increased during the last two decades to reach 60.2 per cent in 2009 (Table 10.2). Along with the increase of the share of the electronics sector, the share of textiles has continuously decreased, from 38 per cent in 1980, to 28.1 per cent in 1990, and to 6.4 per cent in 2009.

At first, Gumi's success stood on two sectoral pillars, textiles and electronics. In the initial phase, the high incidence of textiles, not part of the original plan, is attributable both to the strength of the industry in domestic and export markets and to early difficulties that the complex had in attracting electronics investment. The nearby older industrial city of Daegu had specialized in the textile industry, and textile firms looking to expand out of the Daegu area were eager to come to Gumi in the early 1970s. Furthermore, when the first Gumi complex was built in the early 1970s, it proved harder than anticipated to attract enough electronics plants to absorb the prepared industrial sites. The management of the complex, despite its official designation as the Gumi Electronic Industrial Corporation, thus lowered its high-tech aspirations and built space for textile factories.

Table 10.2
Changes in Gumi's Industrial Structure

Industry	Leading industries in Gumi Industrial Complex (in %)								
	1999	**2000**	**2002**	**2004**	**2006**	**2007**	**2008**	**2009**	**2010**
Textile & Clothing	16.19	15.93	14.21	11,47	8.51	8.75	6.28	6.44	6.11
Nonmetallic	13.00	14.46	14.94	12.49	11.34	11.84	13.80	9.04	11.38
Machinery	4.03	4.16	3.89	5.24	4.17	4.72	11.97	13.47	13.01
Electric & Electronics	55.43	53.84	55.3	61.44	66.23	64.55	58.7	60.16	59.52
Others	11.35	11.61	11.66	9.36	9.75	10.14	9.25	10.89	9.98

Source: Park 2010.

It should be noticed that there are three important trends in the changes of the specialization in Gumi. First, even though the degree of specialization on electronics has increased continuously, on the one hand, the share of textiles has decreased on the other. Due to the national government's restructuring policy in the 1990s, many textile firms closed or relocated overseas. New sectors such as new energy industries have emerged in Gumi. For example, LG Electronics changed the production line of PDP (Plasma Display Panel) into a solar battery and module production line in 2009; LG InnoTech established solar battery production facilities in 2007; and Seronics began to produce solar battery parts by joint investment with Japanese firm in 2009. It can be recognized as the diversification of industry within the general trend of intensification of specialization in the electronics sector.

Second, within the intensification of specialization on electronics, there has been diversification of production within the electronics sector. In the early development of the district, televisions were the major product. Since the mid 1980s, however, with the construction of additional districts in Gumi, semiconductors, computers, and additional electronics components have been produced.

Third, Gumi's economy has become more diversified with the remarkable increase of service sector. The share of services to total employment increased from 3.7 per cent in 1980, to 10 per cent in 1990, to 34.5 per cent in 1995 and to 46.1 per cent in 2009. Much of this expansion has been due to the swelling of the ranks of SMEs in the city. With the increase in the number of SMEs, the occupation structure of the city has changed with the increase of managerial people living in the city. This, in turn, has led to the demand for more diverse services in the city.

Changes in Firm Size and Industrial Linkages

The remarkable changes of industrial development in Gumi can be found in the distribution of firm size. Small independent and locally headquartered firms are more likely to engage in and benefit from the potential for networking and cooperation in local areas. A proliferation of locally-linked establishments is considered to be an asset for an industrial district in insulating it from volatile external demand; enabling firms to be flexible and sharing the burden of restructuring during adverse times (Markusen and Park 1995).

In Gumi, about three quarters of establishments employed less than 300 workers in 1992, and these plants accounted for about 20 per cent of all employment. Gumi's major plants are all large branch plants headquartered mostly in Seoul, revealing a branch plant economy in the 1980s. It showed some tendency toward the proliferation of smaller firms even in the 1980s. In the period 1981–86, when new construction rates peaked, large plants predominated. Since then, smaller plants have been more prevalent among new additions. Especially, as seen in Figure 10.2, the remarkable increase of the number of establishments since the financial crisis in 1997 is mainly attributable to the new additions of small independent firms. With the increase of the additions of SMEs, the share of the number of SMEs increased to 93.9 per cent in 1999 and to 96 per cent in 2008, while the share of employment increased to 34.8 per cent in 1999 and to 49.4 per cent in 2008 (Chung 2011). Considering the share of employment of SMEs to total employment of Gumi is almost half, it can be regarded that new industrial structure has been evolving from the specialization of large branch plants in Gumi. The changes of industrial structure can be realized from the labour restructuring as seen in the change of occupation structure. The share of managerial and administrative personnel increased from 2.6 per cent in 1987, to 3.5 per cent in 1992, and to 17.2 per cent in 2010, while the share of production workers decreased from 82.3 per cent in 1987, to 75.6 per cent in 1992, and to 17.8 per cent in 2010 (Markusen and Park 1995; Chung 2011).

Considering the remarkable increase of the number of SMEs, it can be expected that growing local linkages between suppliers and buyers, and among competitors, might set off endogenous growth dynamics. In 1988, Daewoo Electronics, which had branch plants in Gumi, Chunan, Incheon, and Gwangju, sourced only 14 per cent of input materials from the Southeastern Region, revealing that branch plants located in Gumi had only limited local linkages (Park 1990). Samsung Electronics demonstrated a similarly low level of linkages in Gumi (Park 2004).

Because of the weak local linkages of large branch plants located in the city, the prospects of Gumi's economy were not optimistic in the 1980s. However, positive indications have emerged in more recent years. According to Chung's 2011 survey, 54 per cent of responding firms procure more than 75 per cent of the total procurement from Gumi, and 55 per cent of responding firms sell more than 75 per cent of their total products in the city, revealing a high level of local input and output linkages. The strong local linkages, of course, are related with the restructuring strategy of large firms. Large branch plants in Gumi encouraged spinning-off by employees in order to reduce the number of employees and establish subcontracting relationships with the spin-offs. The senior workers of large branch plants can utilize their experience and know-how to start new firms in related industries, which is a benefit both to large branch plants and the local economy.

However, a new trend is emerging, where firms establish considerable non-local linkages. Twenty-four per cent of responding firms procure less than 25 per cent of total procurement from Gumi, and 17 per cent of responding firms sell less than 25 per cent of their total products to firms in the city, suggesting that a considerable number of SMEs in Gumi have non-local industrial linkages and are not just subcontractors of large branch plants. It is still true that many SMEs are subcontractors to large branch plants and there is an atmosphere of a subcontracting economy. It should be, however, realized that the trend of economic reshaping is ongoing.

Innovation and Knowledge Networks

One of the most significant changes in Gumi industrial cluster is the dramatic increase in innovation since the early 1990s. The number of registered patents in Gumi was negligible up to the mid-1980s. Since then, the number of registered patents has increased dramatically. The number of patents during the financial crisis in 1997 decreased, but continuously increased after 2004 (Figure 10.5).

The emergence of the innovation cluster started from the early 1990s. About 61 per cent of the responding firms in Markusen and Park's (1995)

Figure 10.5
The Number of Registered Patents in Gumi by Year

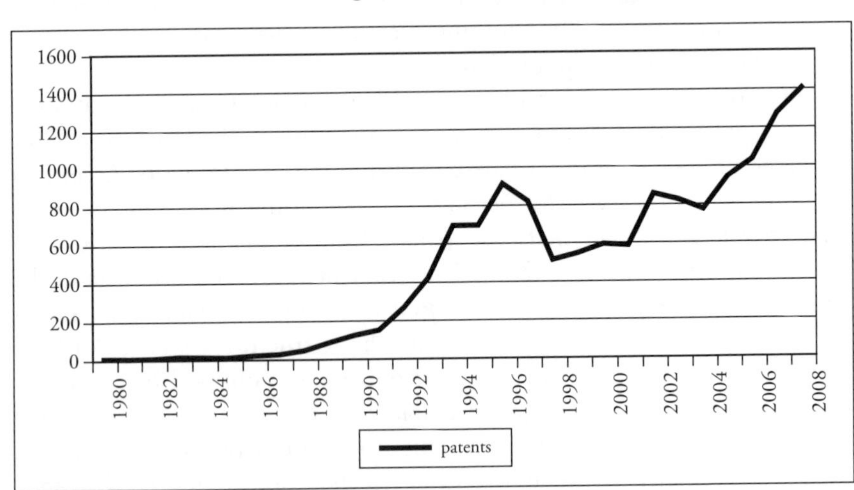

Source: Chung 2011.

survey in 1992 introduced new technology in the last two years, and 43 per cent of the firms responding to the introduction of new technology introduced the new technology in order to produce new products. The most important thing is the fact that the major actors of the innovation are increasingly SMEs as opposed to large firms. Before 2005, large enterprises were responsible for 64 per cent of registered patents and SMEs generated 26 per cent. But the positions were reversed after 2005. In this year, large enterprises were responsible for only 24 per cent, while SMEs generated 57 per cent of the registered patents (Table 10.3). After 2005, the shares of the individual patent and joint patents increased from 6.3 per cent and 3.2 per cent to 8.5 per cent and 8.5 per cent respectively. From Table 10.3, it is clear that the role of SMEs in the innovation activities in Gumi has proliferated in recent years and collaboration with other actors is increasing. Technology alliances with transacted firms and own technology development were regarded most important for the introduction of technology and innovation (Chung 2011). In the process of restructuring and specialization, large firms in Gumi have supported subcontracting SMEs to improve technology and to promote R&D activities in order to improve the quality of supplied parts and components. They regard the technological progress of the subcontracting SMEs are critical for retaining the competitive advantage of their final products in the global market.

Table 10.3
Electronics-Related Patent Applications in Gumi

	Before 2005		Since 2005		Total	
Applicants	**# of Patents**	**%**	**# of Patents**	**%**	**# of Patents**	**%**
Large firms	101	63.9	40	24.2	141	43.7
SMEs	42	26.6	94	57.0	136	42.1
Individuals	10	6.3	14	8.5	24	7.4
Univ. and R&D centres	0	0.0	3	1.8	3	0.9
Joint Patents	5	3.2	14	8.5	19	5.9
Total	158	100.0	165	100.0	323	100.0

Source: Korea Intellectual Property Rights Information Services <www.kipris.or.kr>, accessed 31 August, 2010; Chung 2011, p. 78.

It is true that large customer firms are the most important as original source of innovation information. However, small and medium customer firms, supplier firms, and competitive firms are also regarded as important sources of innovation information. Such knowledge and innovation networks of the firms in Gumi suggest that its industrial complex has now moved from the early satellite platform type of industrial district towards an innovative cluster, although the importance of large branch plants to the overall industrial ecosystem cannot be ignored.

Changes in the Role of Local Government in Recent Years

Along with the changes in central government policies, changes in the role of local government have also considerably contributed to the evolution of Gumi's Industrial Cluster. Since 1995, city mayors and provincial governors have all been directly elected. As a result, local governments have transformed themselves from mere administrators into active planners for their own development (Park 2007; Choe 2011).

Korea has adopted a two-tier local government system. There are sixteen upper-level local governments (seven metropolitan and nine provincial governments) under the central government. There are also 230 lower-level local governments, including: 75 cities; 86 counties (rural local bodies); and 69 urban districts. Local governments have the authority to control all administrative affairs within their own jurisdiction including: policy formulation and implementation; personnel and financial management; and organizational re-engineering. The revenue sources of local governments are divided into two categories: self-generated revenue; and grants from the central government.

Gyongsangbuk province is located in south-eastern Korea, and in 2010 was the third largest province with a population of 2,600,000. The province is divided into 10 cities and 13 counties. The largest city in the province is Pohang, with a population of 517,088. Gumi is the second largest city with a population of 413,446. The GRP of Gyongsangbuk province was KRW79,444 billion (US$71.7 billion) in 2010 and accounted for 6.8 per cent of GDP in Korea. The GRP per capita in Gyongsangbuk province is KRW30,555 thousand, which is higher than the national average of KRW24,190 thousand. The province's annual budget in 2011 is KRW18,054 billion (US$16.3 billion), but only 28.1 per cent is self-generated revenue.

In line with the central government's innovation policies, Gyongsangbuk provincial government and Gumi city government have cooperated to build local innovative capacity in — and cooperative relationships between

— firms in Gumi. While the Gyongsangbuk provincial government has developed a blueprint for local industrial development and tried to attract funds from the central government for industry support programs, the Gumi city government has established R&D institutes and business associations (Chung 2011).

The central government targets key industries for each province considering the opinion of the provincial government as well as national long-term development goals. In 2002, the central government in collaboration with provincial governments selected locally promising strategic industries for each province and fostered them intensively. As for Gyongsangbuk province, IT and new energy were selected as its leading industries. In addition, based on the Korean government's action plan for regional development, Gyongsangbuk province developed the revised "Third Comprehensive Provincial Territorial Plan (2008–20)", this included initiatives to restructure its electronic industry and develop a new energy industry (DGDI 2011; Gyunsangbukdo 2011).

Since the central government set up a general plan for development, the local government focused on increasing local R&D capacity and promoting industry. Thus, in 2002, a complex of R&D facilities for the electronic industry was built up within the Gumi industrial cluster and, in 2007, the Gumi Electronics & Information Technology Research Institute (GERI) was established through a joint investment by the central government as well as the governments of Gyongsangbuk province and Gumi city (DGDI 2010). GERI supports: SME R&D activities; R&D in the electronics sector; training and education; incubation of high tech industry; and the development of the renewable energy industry. Gumi city is in charge of operating GERI, and it is the first case of such an institute being operated by a city government (Chung 2011).

Meanwhile, the governments of Gyongsangbuk province and Gumi city have tried to attract foreign firms to invest in Gumi's industrial cluster. They designated a Foreign Investment Zone within the industrial cluster and attempted to foster agglomerations in new industries such as LED and new energy industry through several subsidies and incentives (DGDI 2011). In addition, they tried to establish business networks between the existing electronic firms and the newly entering energy sector firms and to enhance local R&D capacity, in addition to the relocation of large facilities to Gumi. For example, the photovoltaic devices industry, a kind of new energy industry, needs intermediate goods which are also utilized in semiconductor and display industries. This makes it easy for SMEs in the electronics sector to expand their businesses into these new sectors. Considering this fact, GERI operates a technical assistance team for promoting the photovoltaic devices industry in Gumi. Also, the team supports networking for collaborative R&D among SMEs in the photovoltaic devices and electronic industries (Chung 2011).

In addition, the Gumi city government supports forming cooperative relations between SMEs by supporting local business associations. Previously, cooperative relations in Gumi consisted largely of subcontracting activities from locally-present large branch plants (Markusen and Park 1995). Because of the increase of the proportion of SMEs in the region and the change of government industrial policy towards innovation-oriented policy, the importance of strengthening cooperative relations among firms was emphasized (Chung 2011). The Gumi Small & Medium Business Association was set up through voluntary participation from local SMEs in 2002, which cannot be found in other cities in Korea. The Association plays a role as a channel agency for central and local government support programmes for SMEs such as providing joint research activities for product development and process innovation, business support services, and so forth. Gumi city government furnishes the Association with operational funds. As of 2011, the Association had 758 member SMEs, which represented 65 per cent of the total population (Chung 2011).

CONCLUSION

Up until the 1990s, Gumi was the typical satellite platform type of industrial district. The previous study on Gumi by Park and Markusen concluded as follows:

> In Gumi, large branch plants predominate, surrounded by a growing number of smaller subordinate firms with captive subcontracting relationships. Gumi's branch plants have maintained significant intra-organizational but non-local linkages to other establishments of their parent firms located in other regions. Most establishments, with the exception of captive subcontractors, have extensive relationships with firms headquartered outside the region, even outside the country. In other words, they are non-locally rather than locally embedded, and embedded within the vertically integrated structure of their parent firms, rather than among a set of vertically disintegrated firms. Most continue to operate with mass production processes and little local R&D, with the more customized and innovation-intensive work remaining in Seoul. They fit the depiction of "global Fordism" more closely than they do that of "flexible specialization". (1995, p. 100)

During the last two decades, however, Gumi has experienced considerable changes and has evolved to an industrial cluster with significant local linkages and innovation activities. This is the case, even though the role of branch plants is still significant in shaping Gumi's industrial and economic atmosphere.

The national government's industrial policies have significant impacts on the changes of the Gumi industrial cluster. The state was the developer and locator of business activities within the confines of the Gumi industrial complex, both in origins and in their contemporary operation, as the case of Changwon (Markusen and Park 1993). The state moved to develop the several national industrial complexes in the late 1960s and 1970s such as Ulsan (petrochemical and transportation equipments), Changwon (machinery), and Pohang (steel).

During the Participatory Government (2003–08), the promotion of an innovative cluster policy has contributed to the R&D activities and the formation of local networks of knowledge and technology. Before the 1990s, local universities had not been an important actor in R&D and the formation of industrial clusters, although the region's ability to supply university-educated, engineering labour facilitated Gumi's development. However, in recent years, with the emphasis of innovative clusters, Geumho Institute of Technology located in Gumi has strengthened its role of participating in R&D activities and technology networks with local firms.

Over the last decade, the role of provincial and city governments has been increasingly important in supporting innovation in SMEs. The role of the central government is still important within the broader picture of industrial development, and the large branch plants still account for more than half of the Gumi's production and export. However, the local government, local universities, and SMEs are becoming important agents in reshaping the local economy. Over the past five years, the share of SMEs in the number of registered patents has become greater than that of large firms. Local inter-firm networks among the SMEs for input and output material linkages, information and knowledge networks, and cooperation for technology development with other local actors has also evolved. It should be noticed that, despite the hierarchical relationship between the large firms and SMEs, large branch plants have significantly contributed to the emergence of the local networks of SMEs through technology transfer, management consulting, and collaborative research.

The case of Gumi suggests important aspects of industrial cluster development in industrializing countries. The formation of local networking and cooperation between firms is possible even if the original firm grouping is just an agglomeration of establishments without significant local networks. Indeed, the initial momentum of industrial development was based entirely on exogenous rather than endogenous factors, such as government policy and the investments of large firms. However, an industrial cluster with local innovative networks for information, knowledge, and technology in addition to material linkages has evolved over time. However, in the case of Gumi, more than a quarter century was required for this to occur.

The case of Gumi can be regarded as a successful Korean model of industrial specialization and development. However, future prospects for the electronics cluster may depend on cluster formation dynamics and what they imply for innovation. In addition, the supply of qualified labour and knowledge-intensive advanced services will also be important in the future.

Note

1 This paper is a revised version of an article published in *The Journal of Korean Geographical Society* 47, no. 2 (2012): 226–44.

References

Choe, Sang Chuel. "Introduction: Reshaping Regional Policy in Korea". In *Reshaping Regional Policy*, edited by Harry Richardson, Chong-Hee Christine Bae, and Sang Chuel Choe. Cheltenham: Edward Elgar, 2011.

Chung, Do Chai. "Evolution of Industrial Cluster through Overcoming the Lock-in Effect of Branch Plant Agglomeration". Ph.D. Dissertation. Department of Geography, Seoul National University, 2011.

Daegu Gyungbuk Development Institute. *New Design Daegu Gyungbuk*. Daegu: DGDI, 2011.

Gyungsangbukdo Provincial Government. *The Provincial Administration of Gynsangbukdo in 2011*. Daegu: Gyungsangbukdo Provincial Government, 2011.

Kim, H.S. "Innovation Systems and Science and Technology Policies in Korea". In *Technology Capacity and Competitiveness of Korean Industry*, edited by K. Lee et al. Seoul: Kyungmungsa, 1997.

Korea Industrial Technology Association. *Statistics of Industrial Technology*. Seoul: KITA, 1995.

———. *Directory of Korea Technology Institute 95/96*. Seoul: KITA, 1995.

Markusen, A. and S.O. Park. "The State as Industrial Locator and District Builder: The Case of Changwon, South Korea". *Economic Geography* 69, no. 2 (1993): 157–81.

———. "Generalizing new industrial districts: a theoretical agenda and an application from a non-Western Economy". *Environment and Planning A* 27 (1995): 81–104.

Ministry of Science and Technology. *Science and Technology Annual.* Seoul: MOST, 1991.

Park, Sam Ock. "Corporate Growth and Spatial Organization". In *The Corporate Firm in a Changing World Economy*, edited by M. Smidt and E. Wever. London: Routledge, 1990.

———. "Industrial Restructuring and the Spatial Division of Labor: The Case of Seoul Metropolitan Region, the Republic of Korea". *Environment and Planning A* 25, no. 1 (1993): 81–103.

———. "Innovation Systems, Networks, and the Knowledge-based Economy in Korea". In *Regions, Globalization, and Knowledge-Based Economy*, edited by J.H. Dunning. Oxford: Oxford University Press, 2000.

———. "The Impact of Business-to-business Electronic Commerce on the Dynamics of Metropolitan Spaces". *Urban Geography* 5, no. 4 (2004): 289–314.

———. "Regional Innovation Policies for Maximizing Endogenous Regional Development Capabilities". In *Balanced National Development Policy in Korea: Theory and Practice.* Seoul: Presidential Committee on Balanced National Development, 2007.

———. "A history of the Republic of Korea's Industrial Structural Transformation and Spatial Development". In *Reshaping Economic Geography in East Asia*, edited by Yukon Huang and Alessandro Magnoli Bocchi. Washington D.C.: The World Bank, 2009.

———. "Regional Resilience and Path Dependence: Examples of Four Korean Industrial Clusters". Conference Paper from 57th Annual North American Meeting of Regional Science Association International, Denver, USA, 10–13 November 2010.

———. "Long-term Strategies for Regional Development Policies". In *Reshaping Regional Policy*, edited by Harry Richardson, Chong-Hee Christine Bae, and Sang Chuel Choe. Cheltenham: Edward Elgar, 2011.

Park, Sam Ock and Nahm, Keebom. "Development of Regional Innovation Systems and Industrial Districts for the Promotion of Small and Medium Enterprises". *Journal of Korean Planners Association* 35, no. 3 (2000): 121–40.

Park, Sam Ock, Yang, Seung-Mock, Yoon, Young-Kwan, Lee, Keun, and Lim, Hyun-Chin. *The Sustainable Development Model and Growth Engines for Korea.* Seoul: Seoul National University Press, 2009. (In Korean with English summary)

Presidential Committee on Balanced National Development. *Balanced National Development Policy in Korea: Theory and Practice.* Seoul: Presidential Committee on Balanced National Development, 2007.

Richardson, Harry W., Chong-Hee Christine Bae, and Sang Chuel Choe. *Reshaping Regional Policy.* Cheltenham: Edward Elgar, 2011.

CHINA
East China Sea
Taipei
TAIWAN
Taiwan Straits
Pacific Ozean
Kaohsiung City
KAOHSIUNG
PHILIPPINES
LEGEND
Provincial/Municipal Capital
PROVINCE/
SPECIAL MUNICIPALITY
National Capital
COUNTRY
NEIGHBOURING COUNTRY
Geographic Features
National Border Land
National Border Water
State/Province Border
0 km 25 50
All maps are derived from GADM Data
(Global Administrative Areas, www.gadm.org, 2012)
edited by maps&more, Hans Hortig / Marcel Jäggi

11

THE CASE OF THE ELECTRONICS SECTOR IN KAOHSIUNG MUNICIPALITY, TAIWAN

Ting-Lin Lee and Thanut Tritasavit

INTRODUCTION

As one of the Four Asian Tigers, Taiwan is known for attaining exceptionally high growth rates and rapid industrialization from the 1960s to the present. With a per capita income of US$70 at the end of World War II to US$2,280 by 1980 (Kuo, Ranis and Fei 1981) and US$18,458 in 2010 (International Finance Corporation 2012), the country's growth has been phenomenal. Beyond this, it is one of the few countries to have successfully engineered both structural and sectoral transformation. In addition to Taiwan's economy experiencing a large-scale shift from agricultural to manufacturing activities, its firms have emerged to become market and technology leaders in a range of areas.

Kaohsiung, Taiwan's second city, has played an important role in the country's development. With a total population of 2.77 million people, it is the country's centre for heavy industry, with important steel-making, ship-building, and petrochemical operations (UHDD 2010). Kaohsiung is also the site of the island's biggest harbour and its port used to be the third largest in the world. As an educational city, it ranks second only to Taipei, with 12 universities and colleges in the metropolitan region. Kaohsiung has also been long associated with export-oriented manufacturing, housing the country's first export processing zone (EPZ) and an important cluster of leading electronics majors.

Taiwan has a multi-levelled government structure, with special municipality, provincial city, and county governments. However, while the role of the central government in supporting the country's industrialization is well-documented, relatively little research has been carried out on the role of Taiwan's sub-national governments. While the country's political system has traditionally been centralized, municipal leadership positions were made elective in 1994, and important responsibilities regarding taxation and economic development have progressively been decentralized.

This chapter will therefore examine the electronics sector in Kaohsiung and the role of Taiwan's central and municipal governments in fostering its development. To this end, it is divided into four sections. The first section will provide an overview of Kaohsiung's economy and its electronics sector. The second section will discuss central government policies which have been implemented to promote the development of Taiwan's electronics sector, including those that affect Kaohsiung. Following that, Kaohsiung's local institutions and policies will be examined, seeking to establish how and whether they have played a role in the sector's development. The fourth and final section will conclude.

KAOHSIUNG'S ECONOMY AND ITS ELECTRONICS SECTOR

In 2010, the city of Kaohsiung was merged with the adjacent county of the same name. Prior to the merger, Kaohsiung City had a population of 1.5 million people and most of the area's industrial base. Kaohsiung County had a smaller population of 1.2 million people and a more agriculturally-based economy. In 2009, Kaohsiung's residents had an average disposable income per household of NTD928,214 and NTD789,837 for Kaohsiung City and County, respectively, with the former being substantially higher than the national average of NTD887,605 (UHDD 2010).

Kaohsiung City houses the bulk of the municipality's industrial and service sectors. The secondary industrial sector of the city includes: high-tech industry (electronics); heavy industry (ship-building, ship demolition, steel, and petrochemicals); and other types of manufacturing (sugar, cement, and fertilizers, among others). As can be seen in Table 11.1, the tertiary sector in Kaohsiung Municipality accounted for 59.8 per cent of employment in all of the sectors, with the secondary sector at 36.3 per cent and the primary sector at 3.8 per cent, respectively. However, the financial

Table 11.1
An Overview of Kaohsiung Municipality's Industrial Structure, 2010

Sector	Total production (2006) (million dollars)	Proportion	Employment (2010)	Proportion
Primary industry	27,790	0.9%	48,000	3.8%
Secondary industry	2,299,092	74.5%	458,000	36.3%
Tertiary Industry	760,277	24.6%	753,000	59.8%
Total	3,087,159	100.0%	1,260,000	100.0%

Source: Liu 2011.

output of the secondary sector is approximately three times larger than the tertiary sector.

As of the third quarter of 2011, there were 6,464 registered industrial firms in the city. While the firm base is not that big, the scale of the firms is very large. They are divided into five types of industries: (i) metal production, with 3,120 factories for iron and steel, hardware, and machinery equipment industries; (ii) electronics, with 293 factories for computers, communications, and electronic components; (iii) petrochemicals, with 936 factories for chemical materials and products, rubber, and plastic; (iv) transportation, with 292 factories that are divided into land transportation and shipbuilding; and (v) food, with 609 factories.

As can be seen, the city's firm structure is heavily industrial, with the metal production and petrochemical sectors being the largest. However, in 2008, the two largest sub-sectors in terms of investment were basic metals and electronic components (Liu 2011).[1] In terms of overall research and development (R&D) in all sectors, investment in Kaohsiung has grown consistently. In 2008, the R&D amount was NTD17.1 billion, up by 47.4 per cent when compared to the NTD11.6 billion in 2005. Most of the accounted R&D expenditure was allocated to electronic components, chemical materials, basic metals, and transportation tools (Liu 2011).

With regards to the electronics sector, the electronic parts and components manufacturing sub-sector is the largest, accounting for some 250

Table 11.2

Overview of the Electronics Sector in Kaohsiung

		Nation-wide			(Former) Kaohsiung City			(Former) Kaohsiung County		
Industry	**Year**	**Revenue (hundred million)**	**Numbers of Firms**	**Employment (persons)**	**Revenue (hundred million)**	**Numbers of Firms**	**Employment (persons)**	**Revenue (hundred million)**	**Numbers of Firms**	**Employment (persons)**
Electronic Parts and Components Manufacturing	2007	37,520	3,832	518,000	2,523	134	48,710	267	109	8,129
	2008	34,114	3,837	492,000	2,072	135	42,682	276	114	7,341
	2009	33,514	3,785	499,000	1,540	136	39,043	264	115	7,506
Computers, Electronic and Optical Products Manufacturing	2007	13,045	3,104	205,000	78	83	2,677	81	52	1,826
	2008	12,000	3,017	199,000	90	79	2,151	81	1736	60
	2009	10,238	2,925	190,000	150	78	3,072	85	56	1,620

Source: Department of Statistics, Ministry of Economic Affairs, 2011.

Table 11.3
International Tenants in Kaohsiung's Export Processing Zones

Park	Company	Nationality	Invested money (US$)
Nanzih Park	EBT Technology, Inc.	USA	10,673,958
	Siliconix Electronic Co., Ltd.	USA	2,273,258
	NEC Tokin Electronics (Taiwan) Co., Ltd.	Japan	5,840,596
	INTERNATIONAL Electronics Co., Ltd.	Japan	17,829,174
	KOA Kaohsiung Corporation	Japan	4,481,830
	Sumika Technology Co., Ltd. Kaohsiung Branch	Japan	122,168,414
Kaohsiung Cianjhen processing zone	ITW Electronic Business Asia Co.,Ltd.	USA	2,618,535
	Interpoint Taiwan Corp.	USA	2,896,476
	Kaohsiung Hitachi Electronics Co., Ltd.	Japan	16,270,003
	Hitachi Chemical Co., (Taiwan) Ltd.	Japan	22,994,752
	Nidec Sankyo Taiwan Corp.	Japan	10,740,325
	Okazaki Manufacturing (Taiwan) Co., Ltd.	Japan	2,572,911
	Taiwan Gold Electronics Industries Co., Ltd.	Japan	1,090,036
	Microtips Electronics Co., Ltd.	Japan	4,532,280
	LINTEC Advanced Technologies (Taiwan) Inc. Kaohsiung EPZ Branch	Japan	9,697,750
	Mitsui High-Tec (Taiwan) Co., Ltd.	Japan	45,357,266
	GEM Power Semiconductor Co., Ltd.	EU	4,356,330
Linguang Park	Global Fiberoptics Inc.	USA	12,181,630
	Senju Metal Co., Ltd.	Japan	1,407,570

Note: as of December, 2011.
Source: Authors.

firms and 46,500 workers in 2009. The computers, electronic, and optical products sub-sector accounted for 134 firms and 4,600 workers. Over time, the electronic parts sub-sector has been shrinking in employment terms, although the number of firms has remained steady. Conversely, while the number of firms in the computer, electronic, and optical products sub-sector has also remained steady, the number of workers has increased.

At present, Kaohsiung's manufacturing sector is located in two main areas: one in the urban area of the city; and the other in the EPZs in the outskirts of the city. The bulk of the heavy industries is found in the urban area, with the electronics firms concentrated in three export processing zones, which between them house 222 factories. The Kaohsiung EPZ, the oldest in the country, was established in 1966, with Texas Instruments and Philips investing in plants there (Crook 2010). This EPZ, along with the Linguang EPZ, today house a large number of LCD makers, such as HannStar Display, Arima Display Corporation, and Emerging Display Technologies Corporation, among others. In contrast, the tenants of Nanzih EPZ are predominantly in the semiconductor sub-sector, including Advanced Semiconductor Engineering (ASE), Orient Semiconductor Electronics (OSE), and WUS Printed Circuit, among others. Most of these factories are assembly and testing plants.

An important contingent of the EPZ tenants are foreign-owned MNCs. Table 11.3 shows some of the companies in these EPZs, their country of origin, and the amount of foreign investment for each firm. Despite some important American concerns, as well as one European semiconductor firm, the majority are large Japanese firms.

It is worth asking whether investments in the EPZs have had any effects on the surrounding firm base. Hsu (2009) explored whether the Kaohsiung EPZ brought any positive externalities to the Kaohsiung Area from 1970–2008 by using a Feder model, which is suited to investigating differences in technology between sectors. It concluded that the EPZ did indeed bring beneficial effects to the area through the introduction of foreign technologies, indicating that the establishment of the EPZ helped increase the production technology standard of local firms.

CENTRAL GOVERNMENT POLICIES

Given the concentration of responsibilities at the central level, as well as a long tradition of state intervention in the economy, the central government has played a major role in fostering Taiwan's industrialization process. In order to understand the context within which the country's municipal governments operate, this sector will set out central government policies with regards to: industrialization; Kaohsiung's economic development; and inter-governmental relations.

Central Policies Regarding Industrialization

The electronics sector, and particularly the semiconductor sub-sector, has been at the centre of Taiwan's industrialization drive. In the 1960s, in order to attract electronics-related investment, the central government established a network of export processing zones, under the Ministry of Economic Affairs, in urban centres of the country. Accompanied by low wage costs, tax incentives, and a favourable policy environment, Taiwan was able to attract investment from American and European semiconductor companies. Much of the initial investment was low-skill and labour-intensive assembly work (Hobday 1995).

This laid the foundation for the semiconductor industry and was a major step to accomplishing the central government's goal of turning Taiwan into an "electronics industry centre" (Wade 1990). The country also gained from this through spillover effects from technical cooperation and joint ventures, among others. In addition, domestic firms such as Hua-tai, Wang-pang, Chi-chen, and Huan-yu began to operate in the semiconductor industry. Although there were many difficulties when the industry first started, these firms helped cultivate the future professionals of Taiwan's information electronics industry.

In the 1970s, leadership for the industry was vested in the public sector rather than the private sector through public research organizations and public enterprise spinoffs. A non-profit organization called the Industrial Technology Research Institute (ITRI) was established in Hsinchu, near Taipei, by the central government in 1973. Over the decades, the Institute has played a vital role in the development of Taiwan's economy by fostering local technological capabilities; as well as generating a human resource base over time due to its high turnover in technical staff, many of whom have transferred to the private sector (Hu and Mathews 2005). In the second half of the 2000s, ITRI employed more than 4,000 researchers (Yang et al. 2009).

In 1976, a publicly-owned company called the Electronic Research and Service Organization (ERSO) was also formed by state officials and slotted under the umbrella of ITRI, and its main responsibility was to recruit foreign partners to help develop and commercialize technology. Under ERSO, the country opened its first model shop for wafer fabrication and signed a technology transfer agreement with a U.S. firm, RCA, in integrated circuit design (Wade 1990).

In the 1980s, international trade protectionism, the rise of wages, competition from other industrialized countries moving into the same market, and overstretched physical infrastructure caused many problems for Taiwan; the appreciation of the New Taiwan Dollar against the U.S. dollar did not help matters either. This led to reforms on multiple levels which eventually relaxed the outdated tax and trade systems. At this point, the central government began to promote strategic industries with high-level technology, high value-added, and low energy consumption.

In 1980, the central government established the Hsinchu Science-based Industrial Park (HSIP), with the aims of: providing a conducive environment, accompanied by suitable incentives, to attract FDI and human capital in high-technology sectors; generating links between foreign and local firms in these sectors; and encouraging participation from the local private sector as well as members of the Taiwanese diaspora (Chen and Sewell 1996). It was hoped that this would help spawn new technology-based firms to strengthen the country's innovative potential and domestic capabilities. The Park's six key sectors include semiconductors, computers and peripherals, and precision machinery. Located near Taipei, the Park is close to ITRI as well as three key universities: National Tsing-Hua University, National Chiao-Tung University, and National Central University — all with a focus on science and engineering.

During the 1980s, the government then introduced policies to promote the high-tech and export industries and successfully helped them transform into a major production base for semiconductor IC packaging and testing, LCDs, and flat panels. However, Taiwan's labour-intensive industries faced the pressure of industrial upgrading and transformation in 1987 because of the appreciation of the NTD and the change in the international market division of labor. This posed problems for the industry as many of the factories in the EPZs started to disinvest, closed their plants, and relocated them to mainland China or Southeast Asia. This resulted in structural unemployment problems as the industrial transformation could not keep up with the external changes.

In the 1990s, Taiwan gradually lost its competitiveness in labour-intensive products with low value-added. Therefore, the central government initiated the *Statute for Upgrading Industries*, which encouraged enterprises to engage in investments, research, and development via multiple forms of tax incentives and reductions (Chan 2012). At this stage, firms in the EPZs transformed further into producing technology-intensive and high-tech products with high value-added due to: the adjustments from multiple markets; pressure from industrial offshore migration; and the emergence of new products and technologies, such as integrated circuits (ICs), liquid crystal displays (LCDs), and other electronic and computer components.

The policy prioritization of industrialization began to pay off, as a generation of Taiwanese electronics firms emerged in and around Taipei and

Table 11.4
Principal Generators of Industry-focussed Research in the Electronics Sector in Taiwan

Firm Name	Year Founded	Location
United Microelectronics Corporation	1980	Hsinchu
Taiwan Semiconductor Manufacturing Company	1987	Hsinchu
Unipac Optoelectronics/AU Optronics	1990	Hsinchu
Mosel Vitelic, Inc.	1991	Taipei
Sinotech Engineering Consultants, Inc.	1993	Taipei
Vanguard International Semiconductor Corporation	1994	Taipei
Chunghwa Telecom	1996	Taipei

Source: Based on Choung and Hwang 2000.

Table 11.5
Principal Generators of Patents Filed in the U.S. from Taiwan

Firm Name	Year Founded	Location
Foxconn	1974	Hsinchu
Silitek Corporation/Lite-On Technology Corporation	1975	Taipei
Acer, Inc.	1976	Taipei
United Microelectronics Corporation	1980	Hsinchu
E-Lead Electronics	1983	Changhua
Taiwan Semiconductor Manufacturing Company	1987	Hsinchu
Winbond Electronics Corporation	1987	Hsinchu

Source: Based on Choung and Hwang 2000.

Hsinchu. Beginning in the 1970s with Foxconn and throughout the 1980s and 1990s, a number of soon-to-be successful firms emerged. Over the years, the most successful would attain technological leadership in Taiwan, coming to account for the larger part of research collaborations and patents filed in the U.S.

The deepening of economic globalization along with Taiwan's inclusion into the World Trade Organization (WTO), provided a more liberalized economic environment. In 2002, the central government initiated a plan to turn Taiwan into a "Green Silicon Island", which was its vision for national development in the new century based on a sustainable knowledge-based economy (Government Information Office 2002).

In 2011, the EPZ Industrial Innovation Service Group was established by the Ministry of Economic Affairs (MoEA) to assist manufacturers in EPZs. It combined six cooperation units into one entity, which comprises: the ITRI; Marketech International Corporation; Corporate Synergy Development Center; Plastics Industry Development Centre; and Precision Machinery Research Development Center. With their help, manufacturers are able to overcome technical bottlenecks in areas such as: R&D; process improvement; application of value-added; logistics; and other operational developments.

In terms of soft infrastructure, the central government has been very aggressive in establishing academic institutions and public research organizations. A majority of the universities involved in the dispersion of knowledge and R&D to the electronics sector are public universities which obtain their funding through the Ministry of Education, such as National Cheng Kung University and National Sun Yat-Sen University. To bolster the universities' capacities, the Ministry of Education initiated a plan in 2005 called the *Five Year, Fifty Billion* programme that injected NTD50 billion into Taiwanese research intensive universities. Additionally, the central government has strategically placed science parks near the vicinity of these universities and key transport infrastructure such as airports or seaports. This gives two major advantages for firms that are located in these parks: first, they enjoy many investment incentives due to the park's policies; the second is that the parks increase the potential for learning and inter-firm networks for tenants (Yang et al. 2009). Moreover, being near these universities provides for high quality human capital, arguably the parks' most important intangible assets.

With regards to incentives for investment, the central government has implemented a number of key policies. Taiwan has lowered corporate tax to 17 per cent, which is lower than the 25 per cent in China and 22 per cent in Korea while being similar to the 17 per cent in Singapore and 16.5 per cent in Hong Kong. The *Industrial Innovation Act* provides multiple tax incentives when meeting certain criteria, such as tax credits or a five-year corporate tax reprieve for firms that invest in R&D and talent training, all of which are documented in the *Act for the Establishment of Management of Free Trade Zones*. Loans for promoting industrial research and development are readily available to qualified applicants (the Internet, manufacturing, technical services, and distribution services industries) to encourage service upgrade capabilities or to expand their operational scale, while loans for the revitalization of traditional enterprises exist for applicants in traditional industries aside from the ones categorized as emerging strategic industries.

Additionally, the central government strives for technological innovation, information exchange, and knowledge diffusion using the MOE regional industry university cooperation centre and SME innovative incubation centre platforms. This is reinforced by the central government's proposal for local small business and innovation research plans and local R&D programmes to cooperate with the local government; this method provides R&D subsidies to regionalized SMEs in strategic industries including the electronic field. From 2008–11, the number of applications for related electronic industries to the central government have been five, thirteen, seven, and six, respectively.

Another important contributor to the development of Taiwan's electronics sector is the Taiwan Electrical and Electronic Manufacturers' Association (TEEMA), founded in 1948. The Association is an organization that aims to bridge the gap between industry and the government to promote economic development via promoting mutual interests, improving electrical equipment industrial products, and pushing technological upgrades.

In 1967, TEEMA played a role in discussing the development of the local electronics industry with the government. This eventually led to the formulation of the first long-term development plan for the industry at its request and in response to the oil crisis (Weiss 1998). Kuo (1995) argues that a significant class of indigenous industrialists emerged and expanded only after the intimate cooperation between TEEMA and the central government. As of the end of 2008, TEEMA had 3,762 member firms, with a total of 880,000 employees from upstream, midstream, and downstream operations (TEEMA 2010).

Central Policies and Kaohsiung's Economic Development

The central government influenced Kaohsiung's economic and industrial development. Due to its port and well-developed infrastructure, the government pursued two goals in its policies towards the municipality: fostering the development of its heavy industry, and encouraging export-oriented sectors such as electronics.

From the 1950s onwards, the government implemented the Ten Major Infrastructure Projects, which set up Kaohsiung's heavy industry base. This included the establishment of the state-owned firms of China Steel and China Shipbuilding, which complemented the earlier founding of the Chinese Petroleum Corporation in the city. Over time, these initial investments would grow to comprise important sub-sectors for the municipality's economy.

In the 1960s, the focus of the government was to develop the light industries and EPZs and to continue implementing the *Ten Major Infrastructure Projects*. In order to attract foreign capital and in cooperation with the government's implementation of the *Statute for the Encouragement of Investment* and the *Statute for the Establishment and Administration of the EPZ*, the government set up the Kaohsiung and Nanzih EPZs, under the management of the Ministry of Economic Affairs.

This succeeded in starting a clustering effect which drove Kaohsiung's local economic development. Meanwhile, the industrial imports and exports were thriving and the industry was moving towards a fast-growth period. Steel, petrochemical, shipbuilding, cement, and other heavy industries played a leading role in the industrial development. In 1969, the construction of a second harbour began, accompanied by additional conversion of land to industrial use.

In the 1970s and 1980s, the central government continued to heavily promote the *Heavy and Chemical Industries Import Substitution* policy and the *Ten Major Infrastructure Projects* with the aim of upgrading its industrial structure. The government used its own resources to import raw materials for iron, steel, petrochemicals, machinery, shipbuilding, and other basic industries, transforming Kaohsiung into the core heavy and chemical industrial area in Taiwan.

In the 1990s, keeping Kaohsiung as an important gateway in the south of the country became a priority. As a result, the central government sought to: upgrading the EPZs; foster the emergence of high-tech industries; actively enhancing the industries' competitiveness; and establish a high-quality urban environment. It also began to work with the local branch of TEEMA, which was established in Kaohsiung in 1985.

The central government began to invest in the municipality's human resource base. Thus, over the past three decades, a number of centrally-funded technology-related universities and polytechnics have been established in the municipality. Of particular note are National Kaohsiung First

Figure 11.1
The Evolution of Industries in Kaohsiung Municipality

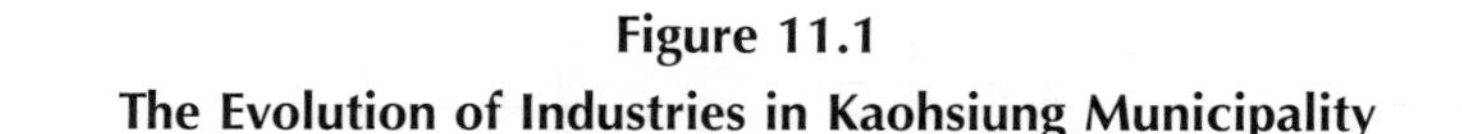

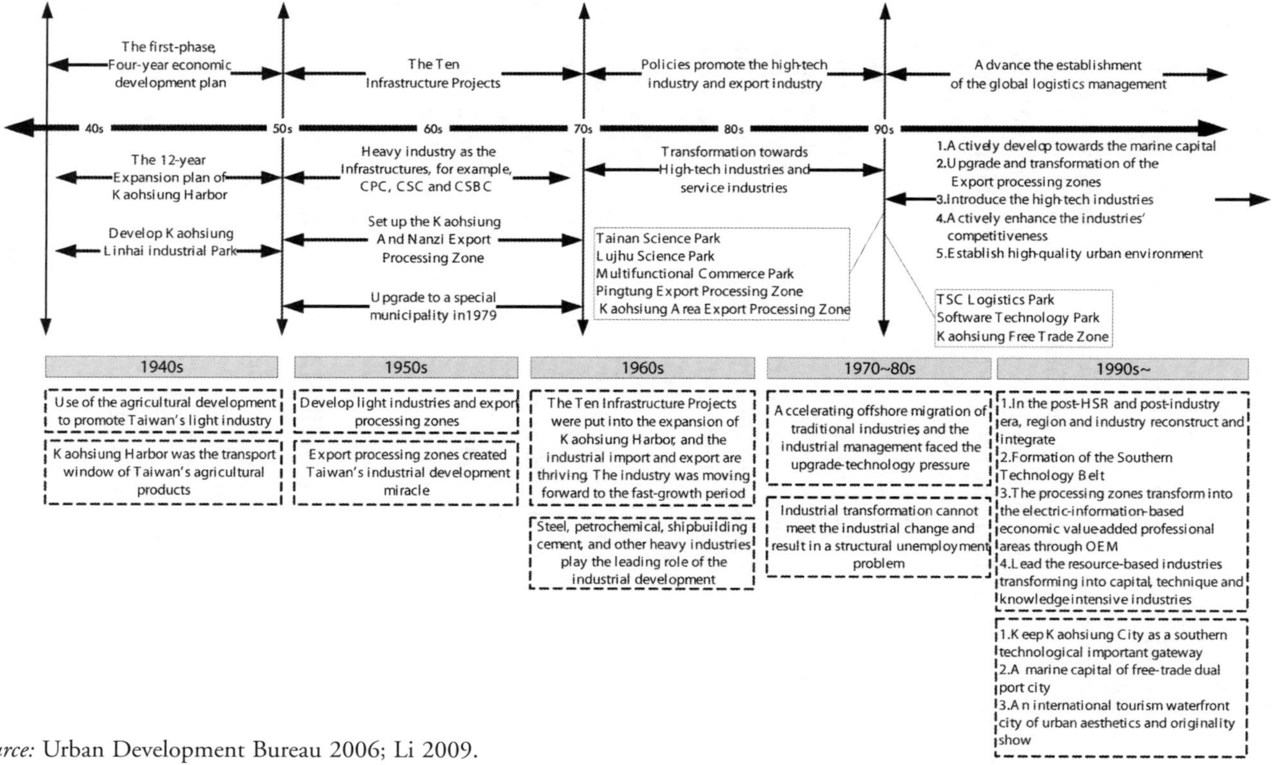

Source: Urban Development Bureau 2006; Li 2009.

University of Science and Technology and National University of Kaohsiung established in 1995 and 2000, respectively.

These have been complemented by three industry-university cooperation centres in Kaohsiung, which work in conjunction with the EPZs to enhance industrial competitiveness. There are also a number of business incubators, most of which are university-based and offer R&D services for enterprises.

Furthermore, the central government established high-end facilities similar to Hsinchu High-Tech Park in order to foster local capabilities. Thus, the Kaohsiung Science Park was established in 2001, with the goals of "forming an IC industry cluster and establishing an optoelectronics technology hub in southern Taiwan". At present, the Park hosts tenants in science-based high-tech industries such as precision machinery, biotechnology, optoelectronics, integrated circuits, and communications. As of September 2010, there were 59 companies, with land allotment having reached 76.5 per cent. Several of the country's leading research institutes, such as Academia Sinica, and the National Nano Device Laboratories have branches there, as do several universities (Ministry of Foreign Affairs 2010). There are also plans to combine this park with a number of nearby EPZs into growth corridors (Chang 2010).

However, the unique advantages of, and synergies between, ITRI, Hsinchu High Tech Park, local universities, and U.S.-based entrepreneurs are difficult to duplicate (Yang et al. 2009). As seen in Table Four, none of the Taiwanese firms that have emerged to become market- or technology-leaders were founded in or around Kaohsiung. In addition, the local branch of TEEMA is made up largely of affiliates of larger firms from Taipei. Only a small number of its members, some 100–150, have been founded locally.[2]

CENTRAL-SUB-NATIONAL RELATIONS

In Taiwan, the central government decides on revenue arrangements between the various levels of government. Charging important regulatory fees and fines are all controlled at the national level; and the central government also makes decisions regarding what is transferred to sub-national governments through a tax allocation and grant system. From the local government's side, its primary financial resources are split into three categories: (i) self-revenue, including local taxes, fees, and overall taxes, among others; (ii) subsidy income, including general planning subsidy; and (iii) debt income, including issuing bonds and bank loans (Tseng

2010). The fund allocation and regular grants that the local governments receive from the central government are used to meet local basic needs and economic development expenditures. However, municipal, provincial and local governments do have some leeway in generating their own revenue through locally-levied taxes and bond issuances, for example.

Prior to the municipal elections of 2010, in which municipal mayors, municipal councillors, and ward chiefs of the five current and newly created special municipalities (Kaohsiung, New Taipei, Taichung, Tainan, and Taipei) were elected, there were three special municipalities, 22 cities or counties, and 319 townships. The ratio of overall tax allocations transferred to these levels of government was: 43 per cent for municipalities; 39 per cent for counties or cities; and 12 per cent for townships. After the elections and government restructuring — which resulted in six special municipals, 16 cities or counties, and 211 townships — there was a need to adjust the ratio of overall tax allocations. Therefore, the figures are currently at 61 per cent for municipalities, 24 per cent for cities or counties, and 9 per cent for townships.

Taiwan's financial structure has been an issue for Kaohsiung as the centralized financial allocation structure has resulted in insufficient local

Figure 11.2
The Fiscal Relationship between Central and Local Government

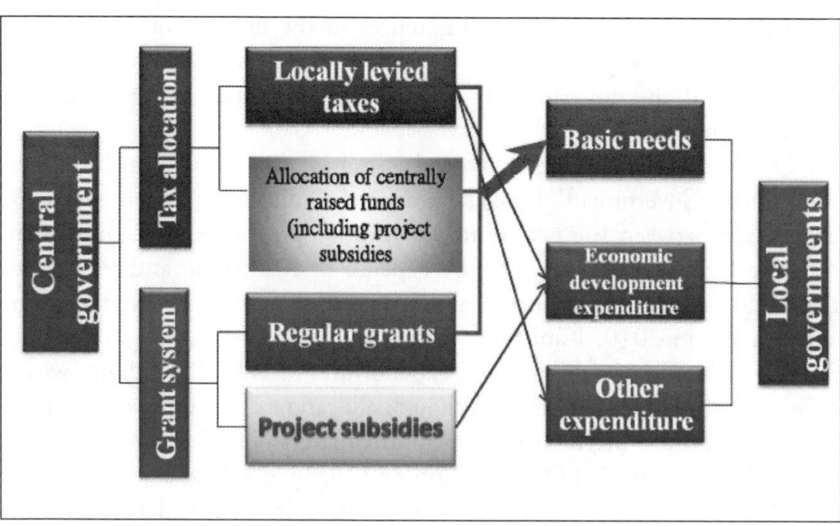

Source: Chang 2010.

Table 11.6
Ratio of Overall Tax Allocation to Each Level of Government

	Ratios of Central Tax Allocation (100 million Euros)			Sharing Counties/Cities	
Local Government	**Old System**	**New System**	**Increase/ Decrease**	**Old System**	**New System**
Special municipalities	43%	61%	8.1	3	6
Cities or counties	39%	24%	–6.8	22	16
Townships	12%	9%	–1.4	319	211
Special funds	6%	6%	–	–	–

Source: Chang 2010.

finance and has limited local construction and development (Chang 2010). The *Act Governing the Allocation of Government Revenues and Expenditures*, which stipulates how powers are demarcated, regulates central and local fiscal relations and how revenues and responsibility for expenditures are divided. Additionally, there are differences in the division of revenue and responsibility for expenditures between the governments of special municipalities and the governments of cities and counties due to different demarcations.

However, a large issue that persists is how this redistribution affects Kaohsiung's Government's budget. Prior to the adjustments, there was some contention between the people in Kaohsiung and the central government over the favouring of Taipei at the expense of Kaohsiung and the rest of the island in the form of budget allocation (Williams 2004). Even after the adjustments in 2010, Kaohsiung City's total budget in 2012 projected a revenue of NTD115.3 billion and expenditures of NTD133 billion, with a difference of NTD17.7 billion. Although allocation of centrally raised funds in 2012 to Kaohsiung increased by approximately NTD2.0 billion, regular grants decreased by NTD 6.3 billion and planned subsidies decreased by NTD8.4 billion. As a result, the local finances are still in deficit and this could affect development plans.

In 2010, the *Industrial Innovation Act* was announced. The Executive Yuan, the executive branch of government in Taiwan, proposed industrial development guidelines that allowed municipal and city governments to formulate industrial development strategies, although consultation with the central government would still have to be conducted. This Act marks a sea-change in the country's industrial policy framework, as it marks the first time that other levels of government are explicitly tasked with this responsibility.

Having set out the overall national policy context, the next section will look at Kaohsiung's local institutional and policy context.

KAOHSIUNG'S LOCAL INSTITUTIONS AND POLICIES

In the early 1940s, Kaohsiung's population was 200,000. By 1975, Kaohsiung City's population had reached one million. Subsequently in 1979, it became the second city in the country, after Taipei, to be designated as a special municipality. However, martial law was enforced nation-wide until 1987, which meant that the central government retained control (Williams 2004).

Following the end of martial law in 1987, municipal and local elections were introduced in 1994. The first municipal mayor of Kaohsiung City, Wu Den-yih, was elected on 3 December 1994. This is of importance as Wu, who questioned the central government's policies towards Kaohsiung and its role within Taiwan, started a number of bold initiatives to transform Kaohsiung into a more livable city and diversify its economy (Williams 2004).

The Kaohsiung municipal government is the top-tier local government administrative body in Kaohsiung and is the fusion of the city and county governments. It is under the jurisdiction of the Executive Yuan. The mayor is elected by citizens for a term of four years and is eligible for consecutive re-elections. The municipal administration congress is the highest decision-making body. There are 30 principal department units under the mayor and the three deputy mayors (including 23 bureaus, four offices, and three commissions).

The current political position of Kaohsiung is balanced between the Pan-Blue Coalition (Kuomintang) and the Pan-Green Coalition (Democratic Progressive Party). Although Kaohsiung City's former mayors were all from the Pan-Blue Coalition, this trend was reversed starting with Frank Hsieh Chang-ting in 1998, a member of the Pan-Green Coalition and the second special municipal mayor. He has been widely credited transforming the city from an industrial sprawl into an attractive modern metropolis.

Subsequently, all of the special municipal mayors have been from the Pan-Green Coalition, with significant postings consisting of Chu-lan Yeh, Taiwan's first female special municipal mayor, and Chu Chen, Taiwan's first elected special municipal mayor. The municipal government is sometimes seen as the political opposite of Taipei, with Kaohsiung leaning towards the Pan-Green Coalition since the early 1990s, while Taipei has traditionally been and still is Kuomintang-dominant.

Policies

The local government has followed the central government's policies regarding economic development due to the lack of autonomy in this area. However, economic development led by the central government has changed over the years, with local leaders gradually becoming more influential as a result of continuing decentralization.

Enabling Businesses

The local government has also implemented its own policies to support local entrepreneurs and increase the municipality's firm base. According to the *Kaohsiung City Self-Government Ordinances for Governing Receipts, Expenditures, Custody and Utilization of the Private Investment Encouragement Fund*, applicants that meet a series of criteria are entitled to apply for loans, rent rebates, and tax subsidies.

In addition, the *Kaohsiung City Government Mini Commercial Loans* programme is designed to help businesses by providing loans and credit guarantees to help them obtain the venues, equipment, and cash needed for operation. The local government also has a programme called *Measures for Rewarding Convention and Exhibition Activities in Kaohsiung City*, which essentially rewards promotional activities held in Kaohsiung City for the development of local industries and international marketing. Corporations, universities, colleges, research institutions, and private organizations (with the exception of state-owned enterprises and political groups) are qualified to apply for this subsidy, with the condition that the reward is only offered to the same applicant twice per year.

Furthermore, the *Kaohsiung City Government Local Industry Innovation and R&D Promotion Program* aims to encourage local SMEs to actively engage in R&D by providing additional subsidies above and beyond those already available from the central government. Similarly, the *Kaohsiung City Government Assistance and Care Program for SMEs* also attempts to encourage SMEs in Kaohsiung City to upgrade their R&D capabilities by utilizing academic resources to help solve problems and provide technical consulting and services.

CONCLUSION

The role of the central government in Taiwan and Kaohsiung's industrial evolution looms large. Its long-term and phased plans to foster the emergence and growth of the electronics industry have yielded substantial returns. Over time, FDI-led growth has given way to more locally-driven innovation, as the country has spawned a generation of market-leading firms. This was enabled by a unique combination of high-end infrastructure, specialist research institutes, and universities in and around Taipei.

Despite Kaohsiung's established industrial tradition, the municipality has not witnessed the same evolution. Its firm base is still substantially oriented towards heavy industry, and much of the investment in its EPZs is from foreign-owned MNCs. Although there are technology spillovers from the EPZs to local firms and substantial amounts of R&D is carried out in Kaohsiung, the municipality has yet to see the emergence of local industry leaders. This is despite central government measures to complement the EPZs with universities, high-technology parks, and specialist research institutes.

Progressive measures to decentralize have enabled the municipal government to begin to formulate local-level economic development strategies as well as offer incentives and rebates. However, these have been too recent to enable significant policy innovations to be implemented.

Notes

1 This figure refers to Kaohsiung City only, as the data are from prior to the 2010 administrative reorganization.

2 Telephone interview with Manager of Kaohsiung branch of TEEMA, 24 August 2012.

References

Chan, Jerry. "Taiwan Legislation Guide". *Asialaw Profiles*. 15 August 2012, available at <http://www.asialawprofiles.com/Guide/489/Taiwan-legislation-guide.html>.

Chang, Kuei-Lin. "An Initial Framework of Regional Governance for the Era of Coming 5-Special-Municipality in Taiwan". Conference of County-Municipality Merging & Spatial Development Planning, Taiwan. 12 October 2010.

Chen, C-F. and G. Sewell. "Strategies for Technological Development in South Korea and Taiwan: The Case of Semiconductors". *Research Policy* 25 (1996): 759–83.

Chen, Shih-Yi. "The Evaluation of Kaohsiung City-Harbor in on." *National Policy Commentary* (2007), available at <http://www.npf.org.tw/post/1/94> (accessed 24 July 2010).

Choung, J-Y and H-R Hwang. "National Systems of Innovation: Institutional Linkages and Performances in the Case of Korea and Taiwan". *Scientometrics* 48, 3 (2000): 413–26.

Crook, Steven. "Taiwan's Export-Processing Zones: Shifting Roles Through the Decade". *Taiwan Business Topics* (2010): 29–31.

Department of Statistics, Ministry of Economic Affairs. "Industrial Census Report". 25 July 2011, available at <http://2k3dmz2.moea.gov.tw/gnweb/Publication/wFrmPublicationDetail.aspx?id=15>.

DGBAS. *Commerce, Industry and Services Census 2006 Report*. Taipei: Directorate-General of Budget, Accounting and Statistics, Executive Yuan, 2008.

EPZA. *Export Processing Zone Administration*. Ministry of Economic Affairs. 24 August 2011, availabe at <http://www.epza.gov.tw/index_redir.jsp?>.

Economic Development Bureau, Kaohsiung City Government. *Multifunction Commerce and Trade Park*. Kaohsiung City Investment Invitation Services. 22 August 2012, available at <http://invest.kcg.gov.tw/eng/Model.aspx?CID=1089>.

———. *Privileges for Business Invitation*. Kaohsiung City Investment Invitation Services. 20 July 2011, available at <http://invest.kcg.gov.tw/eng/Model.aspx?CID=1089>.

Finance Bureau Kaohsiung City Government. Available at <http://finance2.kcg.gov.tw/eng/laws_a15.html> (accessed 9 August 2012).

Fujita, Masahisa, Paul Krugman, and Anthony Venables. "The Spatial Economy: Cities." *Regions and International Trade*. Cambridge: MIT Press, 1998.

Government Information Office. "The Green Silicon Island". (2002). Available at <http://www.gio.gov.tw/taiwan-website/ad/win/012/html/silicon.htm> (accessed 15 August 2012).

Hill, E.W. and J.F., A Brennan. "Methodology for Identifying the Driving of Industrial Clusters: The Foundation of Regional Competitive Advantage." *Economic Development Quarterly* 14, no. 1 (2000): 65–96.

Hsu, Teng-Fang. "The Measurement of Manufacturers' Technology Spillover in Taiwan's Export Processing Zone: Two Departments Model Analysis". Teachers' Research Project Report, Tanjen University, 2009.

Hobday, M. "East Asian Latecomer Firms: Learning the Technology of Electronics". *World Development* 23, 7 (1995): 1171–93.

Hu, M. and J.A. Mathews. "National Innovative Capacity in East Asia". *Research Policy* 34 (2005): 1322–49.

Industry-Academy Cooperation Information. Available at <http://www.iaci.nkfust.edu.tw/module/home/TechResearchTeam/technology_ctu.aspx> (accessed 17 August 2011).

Industrial Development Bureau, Ministry of Economic Affairs. "Industrial History of Taiwan". *National Science and Technology Museum* (2009). Available at <http://industry.nstm.gov.tw/english/03_taiwan/10.asp> (accessed 21 June 2011).

International Finance Corporation. *Doing Business 2012: Doing Business in a More Transparent World*. Washington: World Bank, 2012.

Krugman, Paul. *Geography and Trade*. Cambridge: MIT Press, 1991*b*.

———. "Increasing Returns and Economic Geography". *Journal of Political Economy* 99 (1991*a*): 483–99.

Krugman, Paul and Anthony J. Venables. "Globalization and the Inequality of Nations". *Quarterly Journal of Economics* 110 (1995*a*): 857–80.

———. "The Seamless World: A Spatial Model of International Specialization". *NBER Discussion Paper*, no. 5220 (1995*b*).

Kuo, Cheng-Tian. *Global Competitiveness and Industrial Growth in Taiwan and the Philippines*. Pittsburgh: University of Pittsburgh Press, 1995.

Kuo, Shirley W.Y., Gustav Ranis, and John C.H. Fei. "The Taiwan Success Story: Rapid Growth with Improved Distribution in the Republic of China, 1952–1979". Colorado: Westview Press, 1981.

Li, Liang-Jian. "The Industrial Revitalization Strategic Planning of Kaohsiung Area". Commission project of Council for Economic Planning and Development, Kaohsiung. 2009.

Li, Yun-Chieh. "New Budget Allocation Empowers the Local Governance". *National Policy Commentary* (2010). Available at <http://www.npf.org.tw/post/1/7261> (accessed 7 January 2010).

Liu, Yi-Lang. "Review and Challenge of Kaohsiung's Industrial Sector." *Southern Taiwan Alliance of Researchers & Scholars* (2011). Available at <http://tipo.stars.org.tw/Form02.aspx?id=8870&N_Kind=%B2%A3%B7~%B5%FB%AAR> (accessed 21 June 2011).

Marshall, Alfred. *Principles of Economics*. London: Macmillan, 1920.

Ministry of Foreign Affairs. "Science Parks". *Republic of China (Taiwan) Government Entry Point* (2010). Available at <http://www.taiwan.gov.tw/ct.asp?xItem=27510&ctNode=1906&mp=1001> (accessed 30 August 2012).

Small and Medium Enterprise Administration, Ministry of Economic Affairs. "Taiwan's Economic Development". Available at <http://www.moeasmea.gov.tw/ct.asp?xItem=72&CtNode=263&mp=2> (accessed 15 August 2012).

Taiwan Electrical and Electronic Manufacturers' Association. "Introduce to TEEMA". *Taiwan Electrical and Electronic Manufacturers' Association* (2010). Available at <http://www.teema.org.tw/english/about-teema.aspx?unitid=112> (accessed 9 August 2012).

Tseng, Chu-Wei. "Again! Rip-off Among the Fiscal Families of the Government." *National Policy Commentary* (2010). Available at <http://www.npf.org.tw/post/1/8011> (accessed 15 September 2010).

UHDD. *Urban and Regional Development Statistics*. Taipei: Urban and Housing Development Department, 2010.

Urban Development Bureau. *The Comprehensive Development Plan of Kaohsiung City — to Enhance the Industrial Competitiveness and Strategic Planning of Economic Development of the City*. A project executed by National Cheng Kung University. 2006.

Venables, Anthony J. "Equilibrium Locations of Vertically Linked Industries". *International Economic Review* 37 (1996): 341–59.

Wade, Robert. *Governing the Market*. Princeton: Princeton University Press, 1990.

Wang, Audrey. "Taiwan's 2010 Trade Hits Record High". *Taiwan Today* (2011). Available at <http://taiwantoday.tw/ct.asp?xItem=142731&ctNode=453&mp=9> (accessed 15 August 2012).

Wang, Chi-hua. "A New Aspect on Taipei-Industrial Economy Development Issues and Visions". *Taipei Economic Quarterly* 5 (2011): 16–23.

Weiss, Linda. *The Myth of the Powerless State*. Ithaca: Cornell University Press, 1998.

Williams, Jack F. "The Role of Secondary Cities in Rapidly Industrializing Countries: The Example of Kaohsiung, Taiwan". *Challenges to Asian Urbanization in the 21st Century* 75 (2004): 225–41.

Wolfe, D.A. and M.S. Gertler. "Cluster from the Inside and Out: Local Dynamics and Global Linkage". *Urban Studies* 41, nos. 5/6 (2004): 1071–93.

Wu, Ji-hua. *White Paper on the Economic Development of Kaohsiung City*. A commissioned project of the Economic Development Bureau, Kaohsiung City. 2010.

Wu, Lien-shang. "Process of Industrial Development and Transformation of Kaohsiung City". *Kaohsiung City Literature* 22, no. 2 (2009*a*): 1–28.

Yang, Chih-Hai, Kazuyoshi Motohashi, and Jong-Ron Chen. "Are New Technology-based Firms Located on Science Parks Really More Innovative? Evidence from Taiwan". *Research Policy* 38 (2009): 77–85.

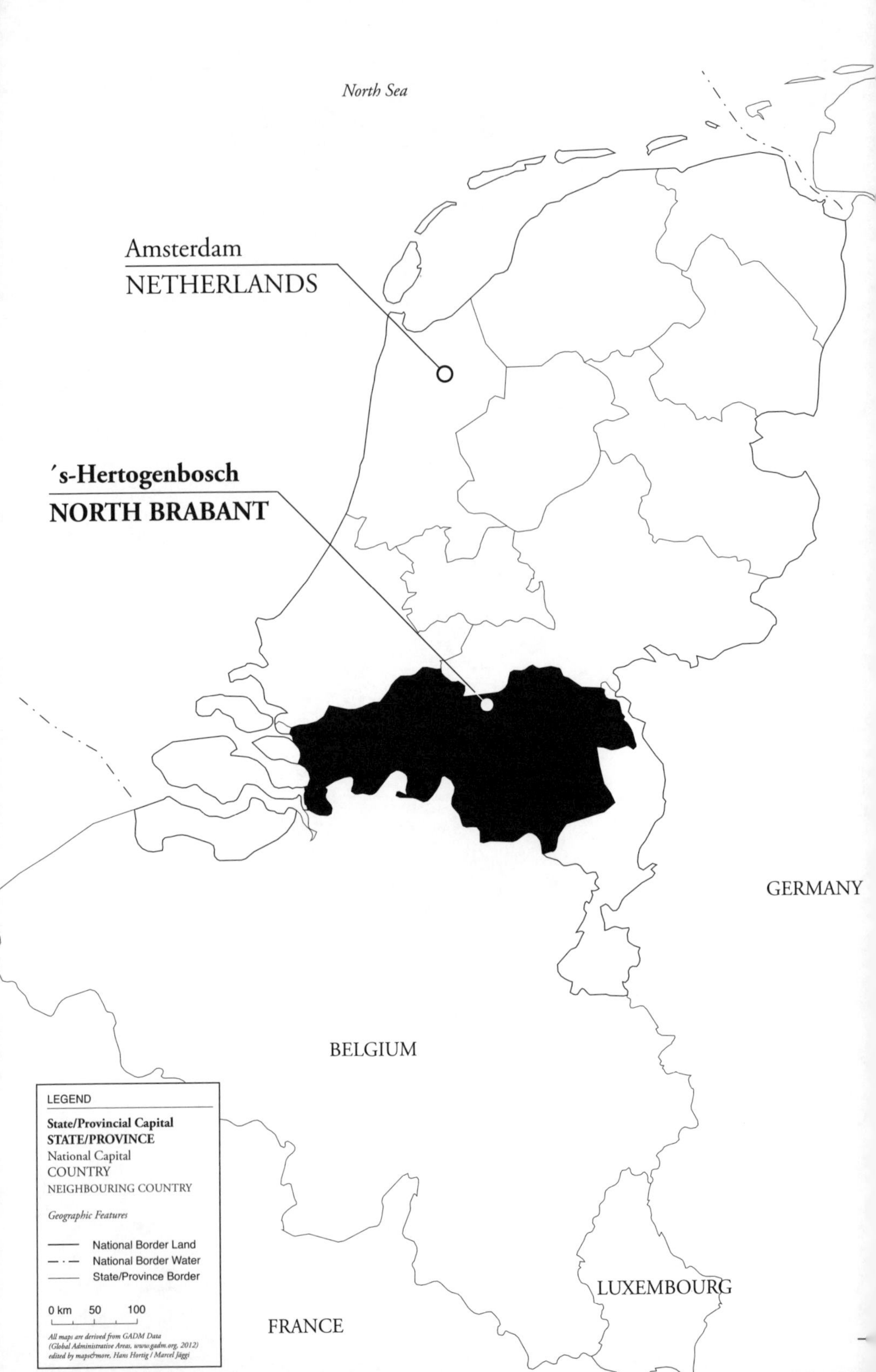
North Sea
Amsterdam
NETHERLANDS
´s-Hertogenbosch
NORTH BRABANT
GERMANY
BELGIUM
LUXEMBOURG
FRANCE
LEGEND
State/Provincial Capital
STATE/PROVINCE
National Capital
COUNTRY
NEIGHBOURING COUNTRY
Geographic Features
National Border Land
National Border Water
State/Province Border
0 km 50 100
All maps are derived from GADM Data (Global Administrative Areas, www.gadm.org, 2012) edited by maps&more, Hans Hortig / Marcel Jäggi

12

SUB-NATIONAL POLICY AND INDUSTRIAL TRANSFORMATION IN NORTH BRABANT, NETHERLANDS

Olaf Merk

INTRODUCTION

North Brabant is a province located in the south of the Netherlands, bordering Belgium. It is the third most populated province of the Netherlands, with 2.4 million inhabitants and home to several medium-sized cities such as Eindhoven, Breda and Tilburg. It is one of the more wealthy provinces in the country with an estimated GRP per capita of €35.665 in 2011 (ING 2011). This is lower than in the western provinces of the Netherlands, where cities such as Amsterdam and Rotterdam are located, but is higher than many other provinces in the Netherlands. It has a relatively low unemployment rate of 5.3 per cent (2010), and relatively high labour productivity and productivity growth.

North Brabant is the leading technological region of the Netherlands, and is also innovative in the European Union (EU) context, as illustrated by its prominent position in the EU Regional Innovation Scoreboard (Hollanders, Tarantola, and Loschky 2009). Within the region of North Brabant, the city-region of Eindhoven plays a particularly important role, as highlighted by recent comparative European studies on knowledge-based regions, typifying it as “star niche player” and ranking it eighth among Europe’s smartest medium-sized cities (Fernandez-Maldonado and Romein 2010).

The leading innovative performance of North Brabant (and Eindhoven City-Region) can be illustrated by a variety of indicators, including high-tech employment, patent applications, research and development (R&D) expenditure, as well as the number of leading firms and universities. Thus, North Brabant is the Dutch region with the highest share of high-tech employment. It made up 5.3 per cent of the labour force in South Netherlands in 2006, more than twice the share of the more affluent West Netherlands, where it comprised 2.2 per cent of the labour force.

The number of patent applications coming from North Brabant is impressive. It has one of the highest scores with respect to patent applications per capita amongst OECD regions and by far the highest in the Netherlands (Figure 12.1). In several sectors, North Brabant is the leading region with respect to patents. For example, in energy efficiency in buildings and lighting

Figure 12.1
Patent Applications Per Capita in OECD Regions, 2006

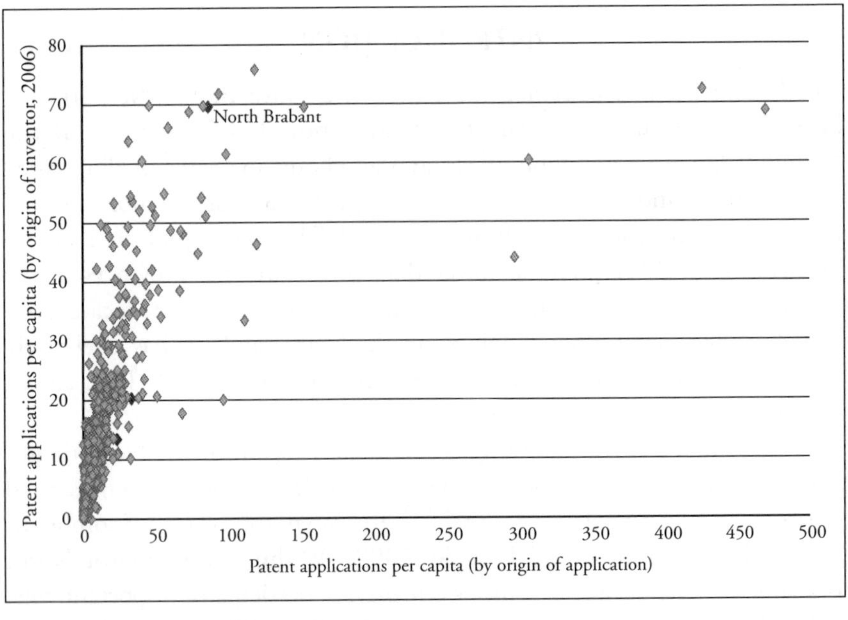

Source: OECD Regional Patent Database.

North Brabant is the best-performing region in the OECD. In North Brabant, like in the Netherlands in general, the business sector plays a key role in generating patents; on average more than 90 per cent of the patents are owned by the business sector in the Netherlands, the third-highest score in the OECD on this indicator. Business-to-business is the most important form of collaboration with regards to patents in the Netherlands, and no other country in the OECD has such a high rate of business-to-business collaboration on patents. The Dutch MNC Philips accounts for a large share of North Brabant's patents, but even if this corporation's patents are not taken into account, North Brabant would still be the leading Dutch province. In the case of optics and semiconductors, a clear but smaller gap remains between the Eindhoven region in North-Brabant and the other regions in the Netherlands (Ponds and Van Oort 2008). MNCs with R&D facilities in the province include: Bosch; Tyco Electronics; Ericsson; and Texas Instruments (Brabant Development Agency 2012).

North Brabant is clearly the leading region in the Netherlands in terms of research and development. The R&D expenditure of South Netherlands (an important part of which is formed by the province of North Brabant) amounted to 2.8 per cent of GDP, almost three times more than the region in the Netherlands with lowest R&D-spending (1.0 per cent). The R&D expenditure share in South Netherlands is also high in comparison with other regions in the OECD (Figure 12.2). A very large share of this R&D expenditure is made by the business sector (2.4 per cent of GDP), which dwarfs the R&D-spending by the business sector in the other Dutch regions. This large business share of R&D spending is also exceptional in international terms. The position of South Netherlands is slightly less dominant with regards to the share of R&D-personnel in the labour force: this share (1.7 per cent in 2006) is more than twice the share of North Netherlands (0.8 per cent), but West Netherlands (1.4 per cent) follows closely. In contrast to other Dutch regions and many regions throughout the OECD, the majority of R&D personnel in South Netherlands (83 per cent) is employed by the business sector (in contrast, in West Netherlands, this figure is 45 per cent).

North Brabant has traditionally been strong in the electronics industry, with an industrial heritage in which Philips Company plays a key role. The conglomerate, number 210 in the Forbes 2000 list of largest companies in the world, was founded in North Brabant in 1891. From its beginnings producing carbon-filament light bulbs, Philips has gone on to produce televisions, compact discs, DVDs, as well as household appliances and

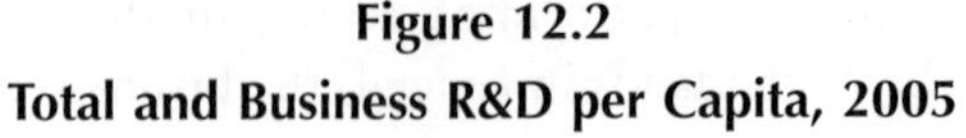

Figure 12.2
Total and Business R&D per Capita, 2005

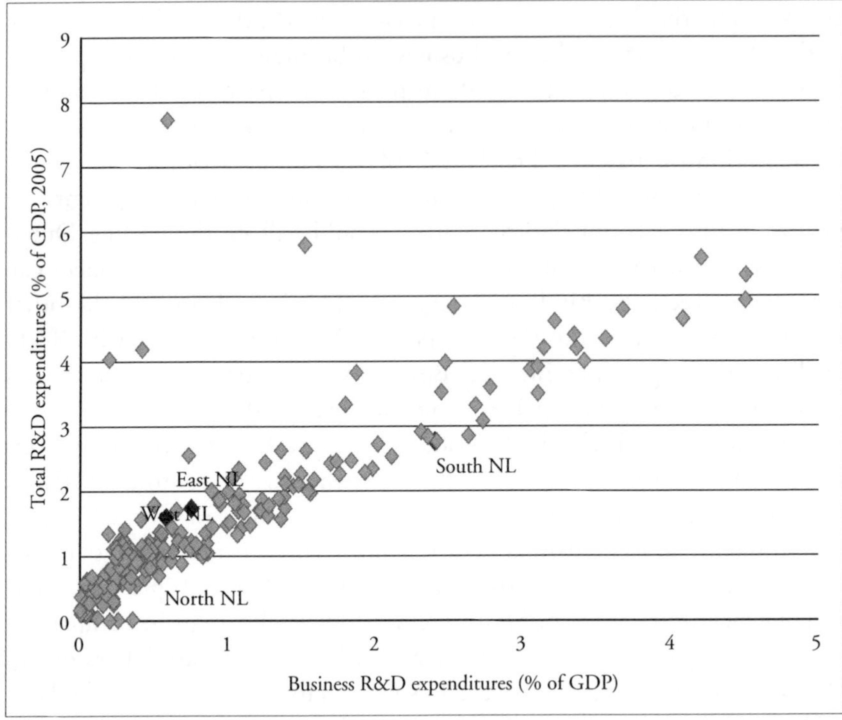

Source: OECD Regional Database.

medical equipment. It is now one of the largest electronics conglomerates in the world, with research and development at the core of its business model.

Although Philips has diversified into a variety of related activities, it still has a large imprint on the province. Its core electronics activities have led to demand for downstream activities such as semi-conductors and several of its spin-offs or joint-ventures have now developed into large companies themselves. Examples are ASML and NXP Semiconductors, both headquartered in the city-region of Eindhoven. These and other firms in the region have a strong export profile. As a result, the region — although only home to 18 per cent of the Dutch population — has exports that represent 35 per cent of the national total.[1]

This industrial heritage is still present, as can be seen by the economic focus of the region. It has still a certain specialization in the electronics sector, at least with regard to the national average. The location quotient of North Brabant is 2.15 in computers, electronic and optical products, and 1.16 in electrical equipment. This means that 115 per cent more people work in the computers, electronic and optical products sub-sector and 16 per cent more people work in the electrical equipment sub-sector in North Brabant than could be expected on the basis of the national average (Table 12.1). In comparison with other European countries, however, this specialization appears much weaker. Location quotients in which Europe as a whole is considered give a much lower location quotient for North Brabant in these sub-sectors. Thus, other regions, particularly from Central Europe, turn out to be much more specialized in these sub-sectors, particularly computers, electronic and optical products, and electrical equipment (Tables 12.2 and 12.3).[2]

Among North Brabant's current economic sectors, several are directly related to electronics, such as: a strong wholesale sector in information and communication equipment as well as other machinery; and developed manufacturing in segments such as pharmaceutical products, motor vehicles and equipment. The strong innovative profile of North Brabant has evidently been stimulated by the large presence of scientific research and development activities. Some of the activities of head offices and management consultancy activities can also be attributed to the presence of high tech firms and their command centres. The strong sectors that have been identified within the high-tech industries are: semi-conductors for consumer electronics, mobile, automotive and lighting; semiconductors

Table 12.1

National and European Location Quotients in Electronics for North Brabant, 2008

	National LQ	European LQ	Employment
Computers, electronic and optical products	2.15	1.12	10,030
Electrical equipment	1.16	0.35	3,802

Source: Elaborations of OECD secretariat on the basis of structural business statistics of Eurostat.

Table 12.2
European Location Quotients in Computers, Electronic and Optical Products, 2008

Region	Country	LQ	Employment
Pohjois-Suomi	Finland	5.58	8,352
Közép-Dunántúl	Hungary	4.94	15,024
Dresden	Germany	3.95	13,241
Southern and Eastern Ireland	Ireland	3.94	19,922
Západné Slovensko	Slovakia	3.62	12,953
Dél-Dunántúl	Hungary	3.34	6,501
Nyugat-Dunántúl	Hungary	3.28	9,232
Freiburg	Germany	3.21	18,530
Thüringen	Germany	3.02	13,347
North Brabant	Netherlands	1.12	10,030

Source: Elaborations of author on the basis of structural business statistics of Eurostat.

Table 12.3
European Location Quotients in Electrical Equipment, 2008

Region	Country	LQ	Employment
Chemnitz	Germany	6.88	17,017
Oberpfalz	Germany	6.27	24,580
Mittelfranken	Germany	5.69	21,542
Kassel	Germany	5.49	20,538
Rheinhessen-Pfalz	Germany	5.37	21,508
Západné Slovensko	Slovakia	4.33	18,701
Unterfranken	Germany	4.12	10,408
Severovýchod	Czech Republic	3.76	22,287
Strední Morava	Czech Republic	3.76	17,722
Detmold	Germany	3.57	23,259
Vzhodna Slovenija	Slovenia	3.40	13,072
Jihozápad	Czech Republic	3.31	15,715
Opolskie	Poland	3.17	6,295
Vest	Romania	3.07	13,244
North Brabant	Netherlands	0.35	3,802

Source: Elaborations of author on the basis of structural business statistics of Eurostat.

Table 12.4
Main Economic Specializations in North Brabant, 2008

	European LQ	National LQ	Employment
Employment activities	3.61	1.03	104,435
Manufacturing of basic pharmaceutical products	2.34	2.73	7,150
Wholesale of agricultural raw materials and live animals	2.28	0.73	4,299
Wholesale of information and communication equipment	2.26	0.91	8,523
Wholesale of other machinery, equipment and supplies	2.20	1.04	16,205
Activities of head offices; management consultancy activities	2.01	0.86	29,363
Scientific research and development	1.87	1.09	19,495
Sale of motor vehicles	1.74	1.02	16,880
Other manufacturing	1.26	1.85	7,259
Manufacturing of beverages	1.11	2.04	2,733
Manufacturing and equipment n.e.c.	0.88	1.64	21,311
Manufacturing of motor vehicles, trailer and semi-trailers	0.53	2.38	9,124
Manufacturing of leather and related products	0.35	3.79	1,236

Source: Elaborations of author on the basis of structural business statistics of Eurostat.

for electronic production equipment, food processing and logistic systems; medical and optical systems; automotive components; and the enabling technology, supply chain and services (Programmacommissie Zuidoost-Nederland 2006).

The leading university of the region is the Technical University of Eindhoven, which has a good reputation and high scores in international university rankings. It ranked as the 114th university (and first Dutch university) in the Times Higher Education Supplement ranking for 2011 and in the 2010 Shanghai University ranking, its score placed it among the 52–75th highest-ranked universities for computer science. Philips is a main sponsor of the Technical University of Eindhoven, and has been at the centre of a close industry-university cooperation in North Brabant, with a large network of suppliers and innovative start-ups that span off from the company. In addition to this university, there is a solid base of polytechnics, technical colleges and vocational schools present in the province.

REGIONAL POLICIES: FROM REGIONAL TO SECTORAL

Regional development policies in the Netherlands evolved during the 2000s, with more emphasis placed on strengthening strong regions, instead of supporting lagging ones. This vision was expressed in the 2004 policy paper "Peaks in the Delta", which formed the start of the phasing out of subsidies to the lagging northern part of the Netherlands and focusing on strengthening strong regions in the Netherlands and their thriving economic activities. The rationale was that strong clusters in regions (peaks) would have to be developed further in order to stimulate growth in the whole of the Netherlands. Southeast Netherlands was defined as one of the peaks, and includes North Brabant and part of another province in the south of the Netherlands, called Limburg. Under the new policy, a programme committee for each peak was established, consisting of local stakeholders from sub-national governments, firms, universities, and research institutes to formulate programmes. These programmes were aimed at stimulating new initiatives that would otherwise not have been developed and were funded for a set period (2006–10). The programme committee for Southeast Netherlands formulated ambitions with respect to commercialization of research, the growth potential of large companies and SMEs, and high value-added employment.

BOX 1

Peaks in the Delta programme for Southeast Netherlands

The strategy for South East Netherlands, as outlined in the Peaks in the Delta programme in 2006, was based on four pillars: excellent clusters (public knowledge infrastructure, cooperation between academia and industry); entrepreneurship; services (skilled workers, accessibility, regional attractiveness); and cross-regional cooperation. The three key economic sectors were defined as: high tech systems and materials; food and nutrition; and life sciences and medical technology. Concrete actions were formulated for these three sectors along five different axes: knowledge; capabilities; cash; open innovation; and knowledge workers. Examples of projects that were developed due to this programme and their aims include:

— *Bèta Technology & Business Accelerator:* accelerate the growth of small innovative firms (5 to 25 employees) in Eindhoven in high-tech sectors by providing office space and research facilities.
— *Biosensing*: develop technology to monitor activity in order to support diabetes patients and for the rehabilitation of orthopaedic patients and athletes.
— *Design Incubator*: finance and coach design start-ups.
— *Embedded Systems Engineering Competence Centre*: development of a post-graduate course for system architects.
— *High Precision Inkjet Printing System:* improve reliability, accuracy and speed of inkjet systems in order to make this technology suitable for the production of printed electronics such as displays, antennas, electronic circuits and solar cells.
— *Intelligent Remote Industrial Services*: develop remote services particularly in operation, diagnosis and maintenance of complex high-tech and production services.

In most cases, the projects were financed by the national government, the provincial government of North Brabant, and the city-region of Eindhoven.

In parallel with this regional economic policy, a national innovation policy called the Key Area Strategy was implemented, which was aimed at stimulating clusters. This policy focused on a set of new businesses such as: creative industries; flowers and food; high tech systems and materials; water engineering; and chemistry. In the high technology sector, embedded systems and nanotechnology were assessed to have great potential; and an innovation programme called Point One was set up for these areas. The aim was to solve the dependence on imports from US and Asia in these areas and to make sure that highly qualified labour in these areas would not be lost. It has been observed that the Key Area programme overlapped in several cases with "Peaks in the Delta" as many of the clusters supported were similar (OECD 2007*a*). Although this might have led to an impression of fragmentation, the coordinating ministry, the Ministry of the Economy assured the coherent implementation of both programmes for which it was responsible (OECD 2010).

Since 2010, regional economic development is considered the prerogative of provinces, and a more sectoral focus has been announced. This approach has been elaborated in a policy document called "Top Sectors", which defines nine top sectors that the national government would like to strengthen. These include: energy; logistics; food processing; hosting MNC headquarters; and high tech. This last sector covers electronics, but also a wider set of related sub-sectors such as nanotechnology, embedded systems, ICT and their applications in all kinds of areas, including health, lighting, logistics, satellites, safety and renewable energies. The approach is one of generic policies to stimulate the economy — reduction of specific subsidies with a simultaneous decrease of the company tax — with specific measures that are sector-specific.

Despite this policy shift, the recognition of the important role of the North Brabant region by the central government has remained strong. The key role of North Brabant has consistently been recognized in government documents, including spatial planning visions (Nota Ruimte), long-term infrastructure investment programmes and policy programmes such as the previously mentioned "Peaks in the Delta" programme. Despite the more sectoral than regional orientation in the "Top Sector" approach, the region of North Brabant (in particular the city-region of Eindhoven) is clearly identified in this policy document as the main spatial concentration of high tech activities, including electronics.

The government that came to power in 2010 (led by Prime Minister Rutte) even strengthened the focus on North Brabant, in line with the special focus on key areas of the previous governments with respect to the Randstad, a conurbation in the western Netherlands that surrounds Amsterdam. Following alarming reports on the lagging competiveness of the region (OECD 2007*a*, Kok Commission 2007), the previous government decided to define a programme in 2007, which involved defining a vision for the region until 2040 and bringing the whole area under the responsibility of one government minister. Such a programmatic approach has now been extended to North Brabant, in line with OECD recommendations made in 2010 regarding the key importance of the region for the country's economy (OECD 2010). As a result, North Brabant's importance for the national economy is now recognized in the coalition agreement of the government in power since 2010 (VVD-CDA 2010). It also has taken the form of a document entitled the "Brainport 2020 Vision", which is an analysis of main challenges for the region that will serve as a basis for a stimulus by the central government. As such, the national government has defined three areas of national importance: the city-region of Amsterdam; North Brabant; and Rotterdam.

Needless to say, many central policies that are officially space-blind have, in effect, spatial consequences. An example is the response to the fiscal and economic crisis of 2008. Although the Netherlands did not explicitly "regionalize" its response to the crisis — as was the case in Australia where national and regional governments agreed on common strategies and priorities — its different elements implicitly supported the strongest sectors in different regions. Elements that were supporting sectors in North Brabant were the extended coverage of: a tax incentive programme which subsidizes R&D staff costs; and the car scrappage scheme, which gave an incentive to buy new cars — many parts of which are produced in North Brabant (Yuill 2009). More generally, substantial parts of education, innovation and infrastructure policies have regional impacts.

SUB-NATIONAL GOVERNMENTS IN THE NETHERLANDS: IMPORTANT PARTNERS FOR THE NATIONAL GOVERNMENT

Sub-national governments have important functions in the Netherlands. Thirty-five percent of total public spending is done by sub-national governments in the Netherlands, which is a relatively high share among OECD countries. They also have important functions in providing public goods and services. As with many other OECD countries, this includes education, social protection and general public services (Figure 12.3). Furthermore, a relatively important budget item of Dutch sub-national governments is economic development (Figure 12.4).

Figure 12.3
Main Spending Items of Sub-national Governments in OECD Countries, 2008
(As Share Of Sub-national Budget)

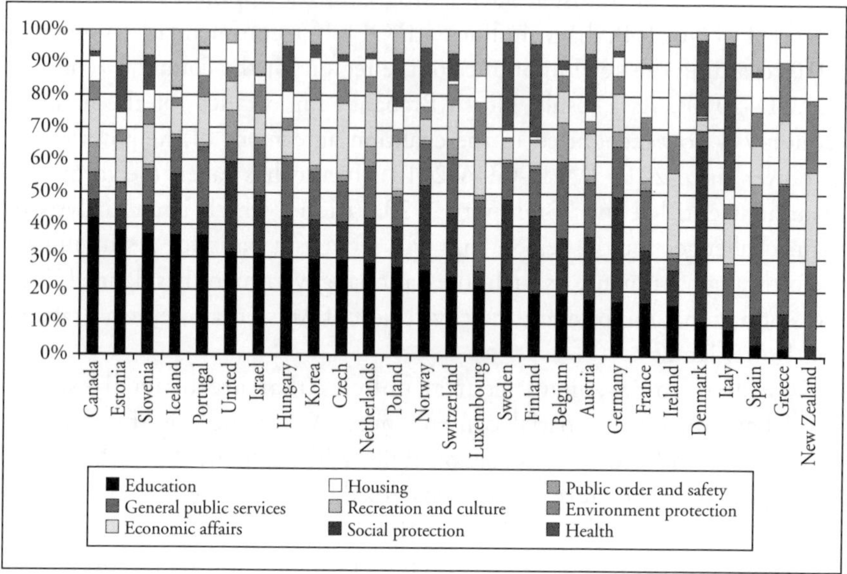

Source: OECD National Accounts Database.

Figure 12.4
Spending on Economic Development as Share of Sub-national Spending in OECD Countries, 2008

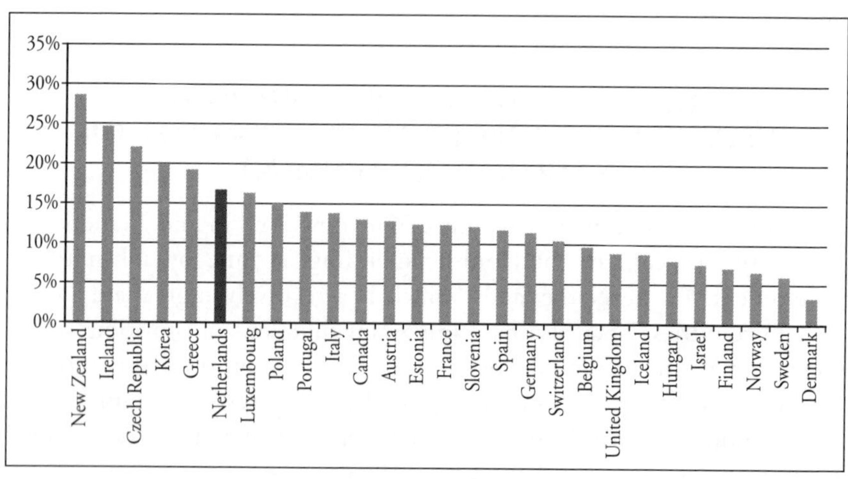

Source: OECD National Accounts Database.

Provinces, with Particular Focus on the Province of North-Brabant

The Netherlands has two tiers of sub-national governments: provinces, of which there are twelve; and municipalities, of which there are some 400. The province is responsible for coordinating policy areas such as spatial planning, environment and regional transportation. Provincial elections take place every four years in which provincial parliaments are elected. Executive power lies with a provincial governor called the Queen's Commissioner, who is selected and appointed by the national government alongside a limited number of representatives from the provincial parliament. The weakness of Dutch provinces can be illustrated by their relatively minor expenditure and staff, as compared to municipalities. In addition their tax revenue is limited, as their only self-generated resources consist of a surcharge on a motor vehicle tax. The rest of their expenditure is financed by grants from the central government. An indication of the limited own-revenue sources of Dutch sub-national governments is provided in Figure 12.5, which illustrates that sub-national governments in the Netherlands have a relatively high share of sub-national expenditures, but a low share of sub-national own-revenues.

Despite these limitations, the province of North Brabant has been able to stimulate economic development within its territory. Since 2011, the province of North Brabant has been run by a coalition consisting of right-wing, centre- (Christian Democrat) and left-wing parties, and headed by a Queen's Commissioner from the Christian Democrats. Despite this new provincial executive, the broad provincial strategy has remained the same with the new administration committed to implementing the same strategic economic agenda called "Agenda for Brabant". In its 2011–15 programme, it pledges to reduce provincial staff from 1400 to 1000 employees. Two of the four last Queen's Commissioners in North Brabant were previously senior national politicians, one being a Prime Minister and the other an Infrastructure Minister. Arguably, this has facilitated provincial-national coordination.

The province of North Brabant has instruments that it can use to foster regional economic development, especially with respect to spatial economic development, attraction and clustering of firms and a role as "broker" between governments, firms and other organizations. Other responsibilities on which the province of North-Brabant is focused include: spatial

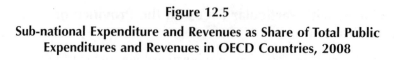

Figure 12.5
Sub-national Expenditure and Revenues as Share of Total Public Expenditures and Revenues in OECD Countries, 2008

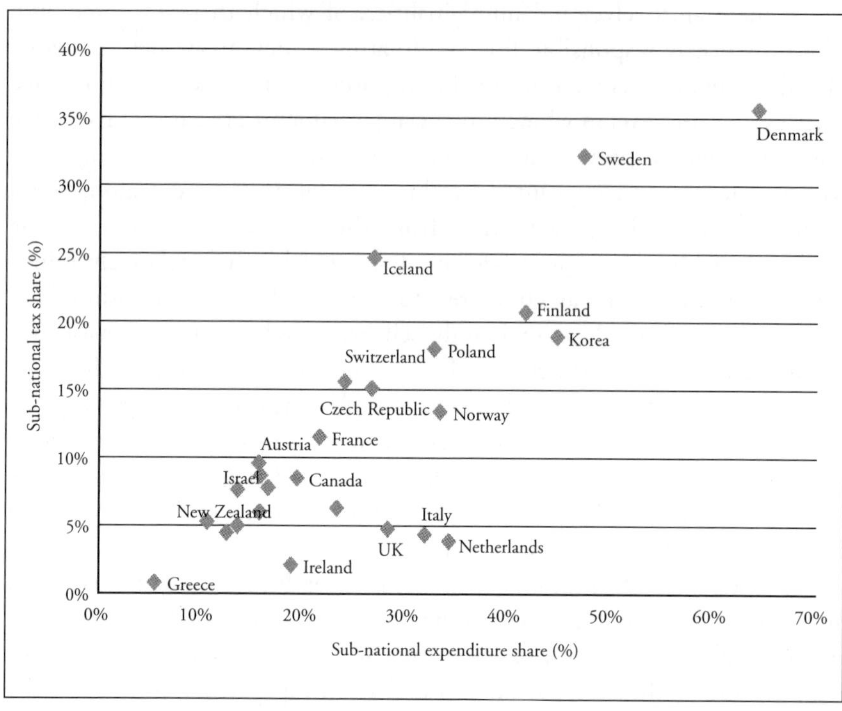

Source: OECD Revenue Statistics and OECD National Accounts Data.

development and planning; the local labour market; rural policy; environment; energy and climate policies; regional accessibility and cultural infrastructure (Province of North Brabant 2010).

One recent thrust of provincial-level policies has been to increase the number of pupils that chose scientific and engineering degrees. In order to achieve this goal, five regional "technical platforms" have been created to contribute to better matching of demand and supply of staff with science and engineering degrees. These platforms organize activities and projects for pupils in primary and secondary education to raise interest in science and engineering. Another goal of regional economic policy in North Brabant has been to improve the skills set of the existing labour force. The instrument put in place to achieve this is the Human Resources Management Service Centre. In such a centre, personalized trajectories for skills upgrading and life-long learning are offered, benefiting from cooperation from: employers' and employees' organizations; social security and employment offices; and vocational education organizations. In addition, provincial governments have been participating in coordination mechanisms called Regional Spatial Coordination Bodies, in which municipalities, province, and water boards (which are functional decentralised government bodies with specific tasks for water management) coordinate: housing development programmes; the development and re-development of industrial estates; and the conservation and development of natural landscapes.

In addition, an important part of the regional economic policymaking is delegated to the North Brabant Development Agency (NBDA). The organization was founded in 1983 by the Netherlands Ministry of the Economy and the province of North Brabant, both of whom are shareholders. NBDA offers four types of core services. The first is assistance for potential international corporations to invest in North Brabant. This involves: organizing fact-finding trips; visiting potential investors abroad; and supporting the selection of production sites and negotiations with national and local authorities. The second is venture capital. NBDA finances innovative and financially healthy companies by providing equity capital and subsidized loans of up to 1.8 million Euros. From 2013 onwards, NBDA will also provide risk capital for innovative business plans via a Development Fund. The third set of services is related to new business development. The NBDA supports R&D programs and initiates and stimulates innovative industrial projects. Moreover, the NBDA offers access to its contacts in the local, national, and European business communities as well as with universities, research centers and financial institutions. The fourth set concerns the development of business and science parks. NBDA is responsible for redeveloping some 1,300 hectares of industrial land, which represents some 25 per cent of the region's land in industrial estates. Examples of successful science parks run by NBDA include the Oss Life Sciences Park and the High Tech Automotive Campus in Helmond.

Municipalities and the City-region of Eindhoven

Municipalities have more extensive revenue sources and responsibilities, particularly in larger cities, such as Amsterdam, Rotterdam and

The Hague. The municipality is responsible for education, environment, social care, housing, local planning, public transport, social affairs and economic development. The municipalities share many areas of responsibility with the national government, but are allowed to act rather independently. They dispose of more money, have effective lobbies with the central government, but also have limited revenue capacity (only the property tax and fees and charges). In some cases the national government has a strong impact on the local agenda, since it imposes the national frameworks, such as legislation, instructions, rules and norms that the municipality must follow.

A major constraint is that municipalities have boundaries that are more restricted than functional areas such as cities and towns. Municipalities in the Netherlands, which on average have 40,000 inhabitants, are not particularly small by OECD standards. However, several municipalities form part of larger functional areas, particularly around large cities. Although it is common in many countries that administrative boundaries do not correspond to the boundaries of functional areas, this might be particularly true in the Netherlands where the combination of density, flat land surface, static housing markets and relatively well-developed transport systems have led to large commuting flows that cross many of these boundaries (Geus et al. 2011).

This means that cross-municipal coordination is necessary, which takes the form of horizontal cooperation between municipalities. Officially, the province has the mandate to invite municipalities to form a city-region. Since 2003, this formation is based on the Joint Agreement Act (in Dutch: *Wet Gemeenschappelijke Regeling-plus*: *WGR-plus-region*). The main responsibilities of a city-region and its regional council are in policy areas such as the environment, housing and regional economic development, but there is a strong emphasis on traffic and transport. The budget is funded by the municipalities. Daily operational work in regional development is usually the work of specific regional development agencies. Although pragmatic cooperative vehicles such as the WGR-plus regions seem to have generated some results, they have received a fair amount of criticism due to their lack of democratic accountability and administrative complexity. The current government has announced plans to abolish these vehicles. The question on city-regional coordination, however, remains unsolved.

In the case of Eindhoven, the innovative centre of the region, the WGR-plus-region took the form of the Eindhoven City-Region, which assembles 21 municipalities including the city of Eindhoven. It has the usual functions of city-regions in the Netherlands (transport, environment, etc), but the Eindhoven City-Region also has acted as programme initiator and promoter, while leaving the programme management to the regional development agency or the urban development agency, *NV Rede*. These agencies provide assistance to potential foreign investors and venture capital to innovative companies, as well as promote R&D programmes and innovative industrial projects in order to help companies speed up the commercialization process. The role of the City-Region becomes clear from several of the programmes it has launched. For instance, the City-Region initiated the Horizon Programme in 2002, entailing a strategic regional action plan, but the implementation was left to an independent programme agency and a steering committee (OECD 2007*b*). Similar procedures were applied for programmes that followed, such as the Brainport programme for economic development, launched in 2004.

Eindhoven's steering committee developed at the City-Region-level, has no formal status, but organizes itself around programmes and projects, such as: Stimulus (1990), a European programme for job creation and strengthening of the industrial fabric; and Horizon (2001), which is focused on industrial innovation, reduction of shortages of skilled labour, diversification of the knowledge industry and international branding. The Brainport programme was established as a successor to the Horizon project and was accompanied by the creation of the Brainport Foundation established in 2005. The Board of the Foundation represents the regional stakeholders with three mayors, the presidents of three knowledge institutes (Eindhoven University, Fontys and DAE), the regional Chamber of Commerce and other business representatives. The Brainport office is in charge of implementing the Brainport Programme and has been considered instrumental in the "reinvention" of Eindhoven (Fernandez-Maldonado and Romein 2010).

On a wider regional scale, the cooperation between sub-national governments in North Brabant took the form of *Brabantstad*, which is a cooperation initiative between the five main cities in the area (Eindhoven, Breda, Helmond, 's-Hertogenbosch, Tilburg), as well as the province of North Brabant. Created in 2001, it acts as a lobby organization towards regional and national government. It also cooperates with other poly-centric networks, such as the *Regionalverband Ruhr* (the cities in the German Ruhr area) and has a multi-annual investment programme for 2008–12 that adds up to €1.4 billion.

Policies are usually made in a partnership between national and sub-national government levels, in order to ensure the coherence of national and sub-national policies. An example is formed by the Area Agendas, which are vehicles for integral area development in the physical-spatial domain that are negotiated between the national and regional governments to identify common priorities for the area. In order for projects to be eligible for multi-year funding from certain budgets (the so-called MIRT budgets), they have to be integrated in these Area Agendas. Eight of these Area Agendas have been established, including one for North Brabant. Similar national-sub-national arrangements can be found in other areas. The private sector is involved in several of these initiatives, including the formulation of regional priorities in the Peaks in the Delta-programme, as well as the Top Sector programmes. Both policies were constructed in such a way that the programme for the region or sector would have to be established by a committee of representatives of business, government and academia.

DETERMINANTS OF SUCCESS[3]

Around two decades ago, the North-Brabant region was confronted with adverse economic developments. Some of the large manufacturing companies in the region, including Philips and the truck manufacturer DAF, had to fire thousands of workers, leading to bankruptcies and lay-offs along the supply chain for hundreds of SMEs. Moreover, in 1997, Philips decided to move its headquarters from Eindhoven to Amsterdam. However, it decided to concentrate the expansion of most of its core R&D activities around Eindhoven. Since the late 1990s, investments by Philips and other companies started to change the region's outlook, developing into the knowledge hub described in the first section of this paper. Investments were channelled into what was initially called the Philips High-Tech Campus, soon changed to High-Tech Campus Eindhoven to reflect a more inclusive strategy (Box 2). This campus has been labelled an outstanding regional initiative for open innovation (Lagendijk and Boekema 2008).

BOX 2
High Tech Campus Eindhoven

The High Tech Campus was initiated by Philips Electronics in order to spatially cluster high-tech research and development. Approximately 700 million euros was invested, and current facilities include: 8,000 sq metres of clean room; 50,000 sq m of laboratories; 185,000 sq m of office space; and an additional 150,000 sq m to be developed. There are 90 companies located at the campus, as well as some 8,000 researchers of around 50 nationalities. Innovation at the Campus is focused on micro/nano systems, lifetech, infotainment, embedded software and embedded systems.

Key determinants of the transformation of the North Brabant economy have been: a turnaround driven by mobilising key actors and stakeholders in the region, continuity in policy programs and goals over several cycles resulting in a shift from exogenous based programmes to internal ones, a common voice, and open innovation value chains with a strong involvement of the private sector.

A Turnaround Driven by Mobilising Key Actors and Stakeholders in the Region

The crisis in the 1990s, which was accompanied by the relocation of Philip's headquarters to Amsterdam induced a sense of common urgency among the different stakeholders in the region. A critical event was the coming together of three leaders in the region, the Mayor of Eindhoven (Rein Welschen), the President of the Technical University of Eindhoven (Henk de Wild), and the President of the Chamber of Commerce of Eindhoven (Theo Hurks) combining efforts of the private, public and educational sector. The crisis offered the opportunity and need for these parties to start cooperating. An important element for success was the strong personality of all three representatives, resulting in a very cooperative and trustworthy relationship, which helped mobilize a large number of stakeholders from the three communities.

By the end of 2005, these initial measures were formally institutionalised in the Brainport Programme. Since 2010, the Brainport Programme is carried out by Brainport Development and its board composition is based on the regional triple-helix model comprising: four mayors (of the municipalities of Eindhoven, Veldhoven, Helmond, Waalre); the presidents of the four knowledge institutions (Technical University Eindhoven, FONTYS Polytechnic University, University of Tilburg and the institute for vocational education ROC Eindhoven); and four business representatives in the region (Philips, the VDL industrial corporation specialised in buses and related intermediate products, the employers' association from the provinces North Brabant and Zeeland BZW and the regional Chamber of Commerce). The Brainport programme has been very influential in the region's overall development and still maintains a strong spirit of cooperation among the three communities resulting in a wide arrange of programs and initiatives. Many provinces from the Netherlands and regions elsewhere regard the regional triple-helix institutional arrangement as a best practice.

Continuity in Policy Programs and Goals over Several Cycles Resulting in a Shift from Exogenous Based Programmes to Internal Ones

After the crisis, representatives from the three communities (private, public and academia) established a plan over 1993–95, which was formalised into a regional development plan in 1996. This plan was quite defensive with the purpose of getting through the crisis and was heavily driven by external resources. The programme also known as "stimulus" combined resources from the EU (both from the European Social Fund and Structural Funds), the national government, the provinces and the 21 municipalities. Funds were allocated to enhance the triple-helix structure supporting high-technological business development with close links to higher educational facilities. At the outset, a focus of stimulus has bringing low-skilled unemployed workers back into the workforce.

The external sources were gradually phased out over time, especially with the Horizon Programme established in 2000. Horizon was a strategic action plan to strengthen the economic structure of the Eindhoven region over the medium term, and foster a change from an industrial region to a top-technology region. It was based on a triple-helix governance arrangement giving emphasis to people, technology, and focusing on international competitiveness. The Programme was also market-driven, focusing on internal private resources rather than external ones but providing a continuation with the previous established goals and objectives.

The Brainport Navigator 2013 programme established in 2005 was also a joint initiative from the triple-helix partners defining projects in four domains: technology (more public and private R&D, better linkages); business (start-ups, attracting FDI, new business on the axes of societal challenges and technological excellence); people (more students, life-long learning, attracting knowledge workers); and basics (living climate, cultural facilities, accessibility).

Finally, the Brainport 2020 strategy and action programme, ordered by the national government, was established by the regional triple helix in 2011. This programme focuses on the same domains as the Brainport Navigator, but involves more parties, both from the regional and national level. All the different programs and initiatives over this period have a continuity and consistency in goals, all based on strengthening links and aligning goals among the private, public and educational communities in the region.

A Common Voice

Through the close links between the private and public sectors as well as academia, the region has managed to reach a common voice, a common agenda and a common goal in the different policy areas becoming quite efficient in its bargaining position with respect to other regions and the central government. The same message has been given by the Chairman of Philips Nederland and by the Mayor of Eindhoven in different negotiations.

The Region's Strong Open Innovation Value Chains with Strong Private Sector Involvement

Innovation in the region was previously based on a closed model, mainly driven by Philips. Its loss of international competitiveness brought with it a crisis to the region . Subsequently, a period of restructuring ensued, with the adoption of the concept of open innovation, establishing the first knowledge campus, and creating many spin-offs to existing business operations (such as ASML, NXP, FEI, Atos Origin, Liquavista, Panalytical, VDL ETG, Assembleon, Atos Origin, Keytec, M&T and many others). This change of approach, and subsequent initiatives have been quite successful making the region one of the main knowledge hubs in the OECD area.

Main Policy Challenges

There is a broad consensus on the main challenges ahead for the development of the high-tech region of North Brabant. As part of the different policy trajectories (such as *Peaks in the Delta, Key Area Strategy, Top Sectors, Brainport 2020*), a variety of documents has been produced that deal with main challenges for the region. There is a remarkable similarity in the challenges that are identified, which can be summarized under four headings: skilled labour; public R&D spending; accessibility; and regional attractiveness.

- There is a scarcity of labour with the requisite technical skills and education. Studies on future labour market frictions indicate possible bottlenecks in this respect. This scarcity is linked to: the relatively limited number of university and polytechnic students in technology and exact sciences in the Netherlands; brain drain in some specific exact science fields (e.g. mathematics); and relatively limited attraction of the Netherlands for foreign high tech-workers.
- North Brabant has a high share of R&D spending, but this is mostly private spending. In contrast, the public R&D rate is actually below other regions in the Netherlands. This might be a bottleneck for those innovations for which a market does not appear to exist immediately. Well-developed publicly-funded R&D might also be of particular importance for knowledge transfer to the SME sector.
- Accessibility of the region is increasingly becoming problematic, with growing congestion and saturated railway lines, as large infrastructure investment has been focused on the Randstad region (western Netherlands), where the largest part of the population in the Netherlands lives. This has been considered problematic for the international connectivity of high-tech workers.
- The region does not have much international appeal. Its assets are hardly known, it does not figure in the renowned international city-rankings and even nationally, the cities in the region are not considered to be the most attractive places to live: only 's-Hertogenbosch makes it into the top 10 ranking of the fifty largest municipalities in the Netherlands (Marlet and Woerkens 2011).

These challenges can, to some extent, be tackled by the sub-national governments in the area. The area where sub-national governments could have some room to manoeuvre is regional attractiveness; local and regional governments can influence this with spatial planning decisions, housing development, cultural and other activities and city and region marketing.

Most of the challenges, however, refer to framework conditions on which the national government has more influence. The lack of exact science students might be related to the way students' bursaries and tuition fees are regulated. For example, students have to pay back bursaries if they take too long time to study, and this gives an incentive to choose less-technical subjects. In addition, the Netherlands does not give lower tuition fees for technical studies, as is the case in Germany. The entry procedures for foreign knowledge workers have been improved, but are still quite bureaucratic. With respect to public R&D, it is also up to the national government to set priorities. This includes spending more public money on R&D in high tech systems and materials, as proposed by the High Tech Top Sector-committee and would certainly increase the share of public R&D in North Brabant (Topteam High Tech Systemen en Materialen, 2011). Infrastructure decisions that would improve the accessibility by road and rail of the cities in North Brabant would also imply a role for the national government. In short, it is clear that many of the challenges for the high-tech and electronics sector in North Brabant are in fact challenges where the national government has an important role to play.

A particular challenge, on which less has been published, is cross-border cooperation. This is relevant because of the proximity of North Brabant to Belgium and Germany. The potential for this cooperation is evident in terms of knowledge and innovation. There are already considerable links between the regions in this field. There is, for example, an established cooperation between the universities of Eindhoven, Leuven (Belgium) and Aachen (Germany), via the so-called Eindhoven Leuven Aachen Triangle (ELAT) organization. The aim of this organization is to create links between these institutes and firms and clusters in the area; the project is supported by the European Union. In order for this cooperation to develop further, there is a need to remove barriers to deepen cooperation between the three countries. At present, they face bottlenecks in different areas and have differences in fiscal rules, tax regimes and social security rules. This makes it difficult to mobilise resources, labour and knowledge amongst the different locations. In particular, the movement of foreign workers is currently inflexible and subject to different national regulations.

The links between scientific and corporate actors also becomes clear from co-patent patterns; that is the number of patent applications of an actor (firm, inventor) in one region with actors in other regions. These patterns can give an indication of innovative links between regions. The strongest co-patent links with foreign regions of North Brabant are with

Flanders/Belgium (in which for example the earlier-mentioned University of Leuven is located) and Nordrhein-Westfalen/Germany (the state of Aachen university).

There are also cross-border links that imply wider economic synergies, e.g. with respect to logistics. North Brabant happens to be at the crossroads between two of Europe's largest seaports (Rotterdam and Antwerp), the third largest cargo airport (Amsterdam Schiphol airport) and one of the largest European manufacturing centres (Rhine Ruhr area). As a result, North Brabant is increasingly used as a linking pin between the logistics hubs and the manufacturing hinterlands: both the ports of Antwerp and Rotterdam use the area as an extended gate for their operations and a variety of European distribution centres has seen birth in North Brabant. This unique geographical position might have advantages for the high-tech sector — and the electronics sector in particular — in that it facilitates exports and decreases transport costs. The presence of a massive flow of goods could also bring opportunities for value-added services providers in North Brabant.

CONCLUSION

North Brabant is a good example of a region where the traditional focus on electronics has spilled over to many other highly innovative areas. Although the region is still to some extent specialized in electronics, other regions in Europe, in particular in central Europe, show much higher specialization rates in electronics. North Brabant remains strongly specialized in a variety of high-tech sectors and industries related to these, such as whole trade and research and development.

The policy concern in the Netherlands is now less with electronics as such, but with high-tech, in which electronics is included. Despite apparent shifts in policy focus, government support is relatively consistent over time. The high-tech sector, mainly located in North Brabant, continues to be facilitated by national initiatives. The challenges identified over time in the different programmes tend to be similar and focus on the availability of skilled workers, public R&D, accessibility and regional attractiveness. Despite a fair amount of decentralization in the Netherlands, most of the challenges related to the high tech sectors in the Netherlands are not exclusive sub-national challenges, but strongly dependent on national frameworks in many cases. This requires adequate mechanisms for national-sub-national policy coordination.

Notes

1 These numbers refer to Southeast Netherlands, which not only includes North Brabant but also part of the province of Limburg.

2 Countries included in the calculation of the European location quotient are: Belgium, Bulgaria, Czech Republic, Denmark, Germany, Estonia, Ireland, Spain, Italy, Cyprus, Latvia, Lithuania, Hungary, Netherlands, Austria, Poland, Portugal, Romania, Slovenia, Slovakia, Finland, Sweden, and the United Kingdom.

3 This section draws on the case study of Zuid-Nederland in OECD (forthcoming).

References

Brabant Development Agency. Available at <http://www.foreigninvestments.eu/smart-solutions/r-d.html> (accessed 18 July 2012).

Brainport 2020. *Top Economy, Smart Society*. Eindhoven: Secretariat Brainport, 2011.

Geus, A.J. de, R. Gradus, and O. Merk. "Krachtig Uit De Crisis". *Economische Statistische Berichten*, [Economic Statistical Messages], 96, no. 4601 (2011): 10–12.

Fernandez-Maldonado, A. and A. Romein. "The Role of Organisational Capacity and Knowledge-based Development: the Reinvention of Eindhoven". *International Journal of Knowledge-Based Development* 1, nos. 1/2 (2010): 79–96.

Hollanders, H., S. Tarantola and A. Loschky. "Regional Innovation Scoreboard 2009". INNO Metrics Thematic Paper, Brussels: European Commission, DG Enterprise, 2009.

ING. "Economische Kloof Tussen Regio's Groeit". Kwartaalbericht Regio's [Regional Quarterly], Amsterdam: ING Bank Economic Research Desk, 2011.

Kok Commission. "Advies Commissie Versterking Randstad". The Hague: Staatsuitgeverij, 2007.

Lagendijk, A. and F. Boekema. "Global Circulation and Territorial Development: Southeast Brabant from a Relational Perspective". *European Planning Studies* 16, no. 7 (2008): 925–39.

Marlet, G. and C. van. Woerkens. *Atlas voor Gemeenten 2011*, [Atlas for Municipalities 2011]. Utrecht: VOC Uitgevers, 2011.

OECD. *Territorial Review of Randstad-Holland.* Paris: OECD Publishing, 2007*a*.

———. *Globalisation and Regional Economies; Can OECD Regions Compete in Global Industries?* Paris: OECD Publishing, 2007*b*.

———. *Place-based Policies in the Netherlands*. Paris: OECD Publishing, 2010.

———. *Growing Lagging Regions*. Paris: OECD Publishing, forthcoming.

Ponds, R. and F. Van Oort. "Spatial Patterns of Innovation in Science-based Technologies in the Netherlands". *Tijdschrift voor Economische en Sociale Geografie* 99, no. 2 (2008): 238–47.

Programmacommissie Zuidoost-Nederland. *Pieken in Zuidoost-Nederland; Uitzicht op de Top* [Peaks in South-East-Netherlands: View at the Top]. The Hague: Ministry of the Economy, 2006.

Province of North Brabant, *Agenda van Brabant; Traditie en Technologie* [Agenda of Brabant: Tradition and Technology] 's-Hertogenbosch: Province of North Brabant, 2010.

Topteam High Tech Systemen en Materialen. *Holland High Tech; Advies Topteam High Tech Systemen en Materialen.* [Holland High Tech; Advice of Topteam High Tech, Systems and Materials]. The Hague: Ministry of the Economy, 2011.

VVD-CDA. "Vrijheid en verantwoordelijkheid: Regeerakkoord VVD-CDA" [Freedom and responsibility: coalition government agreement between VVD and CDA]. The Hague: State Publishing House, 2010.

Yuill, D. *The Netherlands: Regional Dimensions of the Economic Crisis.* EoRPA Paper 09/1 Glasgow: European Policies Research Centre, University of Strathclyde, 2009.

SECTION V

Conclusion

13

SUB-NATIONAL GOVERNMENTS AND INDUSTRIALIZATION: SOME CONCLUSIONS

Francis E. Hutchinson

INTRODUCTION

The previous chapters have contained numerous examples of dynamism, entrepreneurship, and policy innovation from state and provincial governments in Asia and beyond. Referring to the central question of whether sub-national governments can be "architects of growth" — to the extent that they design and implement policies to address needs regarding their industrial sectors — the answer is "yes".

However, given their subordinate position *vis-á-vis* their national counterparts, sub-national governments never assume this role under circumstances of their own choosing. Furthermore, even if there is political will to prioritize industrialization, their restricted range of responsibilities and limited financial wherewithal — as well as the nature of their entry into the electronics sector — may preclude effective outcomes.

Consequently, the questions that then arise are: when do sub-national governments take on the role of "architects"; to what extent can they become the prime drivers of industrialization; and which of the strategies and policy measures that are available to them are productive and under what conditions?

In order to shed light on these questions, this chapter will draw out common themes and conclusions from the ten cases. This will be done in five parts.

The first section will analyse the "birth" of the electronics sector in the various cases. New industries emerge in a given location for a range of reasons, of which government initiative is but one. Subsequently, once an industry is mature, traded and untraded externalities favour established centres, making it harder to establish agency on the part of the sub-national government concerned. Thus, the moment when an industry first emerges in a given location offers a unique opportunity to evaluate whether the respective state or provincial government was aware of its existence and/or involved in its emergence.

The subsequent two sections will explore how the respective sub-national governments supported the development of their electronics sectors following their emergence and consolidation. This will be done along two axes — vertical and horizontal.

As regards vertical dynamics, this includes: how the various sub-national governments were affected by the governance structure within which they had to operate; the extent to which centrally-pursued regional policy worked in their favour; and the manner in which state and provincial leaders managed relations with the centre to secure resources and pursue their own aims.

In terms of horizontal dynamics, this encompasses analyzing the nature and extent of ties between the assorted sub-national governments and their private sector. This will compare and contrast the mechanisms through which state and provincial governments communicated with local and international firms in order to establish needs and implement policies, and worked with them to improve capabilities and establish or strengthen inter-firm networks.

The fourth section will evaluate the policy frameworks developed by the various sub-national governments to support their industrialization drives. The key questions will be: whether they sought to go beyond competing on price to attract investment and attempted to construct a locationally-rooted competitive advantage; and whether they explicitly sought to foster sectoral transformation through broker, demand-side, and market-complementing measures.

The fifth section will assess the success of the various cases in fostering structural and sectoral transformation.

BEGINNINGS

As put forward in the introduction, when a given economic activity first emerges in a new location, there are few or no agglomeration economies. As a result, a nascent industry can be, and often is, geographically dispersed. This is, according to Storper and Scott, an "open window of locational opportunity" (2003). Over time, as firm clusters grow, supporting services develop, technological learning takes place, and agglomeration economies begin to accrue. Path dependence then sets in as successful regions enjoy more spillover effects, attract or grow more firms, and pull ahead of competing locations. The "window of opportunity" then closes, making it harder for late-comer regions to break into the sector.

Thus, this section will ask what enabled the electronics sector to emerge in the various cases when the "window" was still open. This will involve looking at: how the international electronics sector was developing when each local industry emerged; whether it was present in other parts of the country at that point in time; and the extent to which the relevant sub-national government was a catalyst in its development.

The earliest case to develop was North Brabant in the Netherlands. The sector's incipience in the province can be traced back to the establishment of Philips, a local privately-owned firm that began producing light bulbs in 1891. The corporation began to produce radios in the 1930s and consolidated its position following the World War II with a number of leading-edge consumer electronics products. While the provincial government subsequently played a role, in this case the local industrial tradition was key to the emergence of the electronics sector.

The rest of the cases developed later, once the electronics sector had begun to internationalize. Although some Asian countries had developed small, local, and protected electronics industries in the 1950s, their development really only took off once they adopted export-oriented industrialization. This became possible from the 1960s onwards, when technological and organizational developments enabled complex production tasks to be broken down into basic steps and coordinated from afar (Frobel et al. 1980).

Outside of Japan, the Newly Industrialized Countries (NICs) of Singapore, Taiwan, and Korea were — along with Hong Kong — the first entrants into the sector in Asia. In the first three cases, these countries' incursion into the industry was the result of conscious central government initiative that identified the electronics sector as being of strategic importance. Thus, Singapore, Taiwan, and Korea all sought foreign direct investment in their industrial sectors and began producing radio components and semiconductors for the American and Japanese markets (Hobday 1995; Pempel 1999).

The emergence of the electronics sector in Gumi, Taiwan, and Kaohsiung, Korea, was the result of conscious central government policy. Indeed, administrative control and economic planning in both countries at that time was very centralized, precluding much agency at the sub-national level. In Korea, the national government sought to promote a more decentralized pattern of industrialization in the 1960s and 1970s. To this end, it designated Gumi as a preferred location for the electronics sector and the city's industrial sector began to develop in the late 1970s — some ten years after the first MNC, IBM, arrived in Korea (Evans 1995). However, in Gumi's case, relatively few firms were MNCs and most were branch plants of Korean conglomerates, or *jaebols*, who were obeying central directives to locate operations there.

The pattern in Taiwan was similar, in that the central government designated Kaohsiung as a key exporting location because of its developed infrastructure. To this end, it established the country's first export processing zone in the city in 1966 and the first investments from MNCs there were roughly coterminous with those occurring elsewhere in the country.

In the early 1970s, the electronics sector entered its second stage of internationalization, which lasted until the late 1980s. The NICs began to move into more technologically complex and skill-intensive areas of manufacturing. MNCs began to transfer more sophisticated tasks and set up regional procurement centres in Singapore and Hong Kong. For their part, Korea and Taiwan successfully achieved sectoral transformation in the late 1980s, through developing indigenous manufacturing companies and entering higher value-added segments of the consumer electronics and semiconductor market (Hobday 1995).

Other East Asian countries such as Malaysia, Thailand, and the Philippines then moved into the more labour-intensive sectors vacated by the NICs. In addition to semiconductor firms, disk drive and personal computer manufacturers also began to locate operations in East Asia (Mckendrick et al. 2000; Sturgeon and Lester 2004). As offshore production became more established and East Asia provided the requisite levels of manpower and political stability, MNCs started collaborating with local firms and setting up facilities to take advantage of lower costs. Furthermore, companies began to create an intricate set of production networks across countries in the region, seeking to leverage different capabilities and prices (Yusuf 2004).

It was in this context that the electronics sector emerged in Cebu in the Philippines, Johor in Malaysia, and the Eastern Seaboard Provinces in Thailand.

Of these three cases, the Philippine province provides the clearest example of sub-national agency. While Cebu had a tradition of outward orientation, an established manufacturing tradition, and even an export processing zone built in 1979 by Marcos, it was local initiative from 1986 onwards to: ally with the private sector; procure international capital to invest in infrastructure; and market the province overseas. Indeed, Cebu's emergence as a site of electronics production occurred more than three decades after the first consumer electronics manufacturers established facilities southeast of Manila to target the domestic market (Reyes-Macasaquit 2010).

Johor's case is one of serendipity coupled with a modicum of sub-national government initiative. The state's electronics sector began to develop in the late 1970s, as rising labour costs in Singapore encouraged many smaller firms to relocate operations to the neighbouring territory. Relative to other parts of Malaysia, Johor was a late entrant to the electronics sector, as Kuala Lumpur and Penang had attracted their first international investments in 1965 and 1971, respectively. Johor's development really came in the mid-1980s, when the 1985 Plaza Accords led to an appreciation of the Singaporean dollar against the U.S. dollar. Following this, a substantial number of Singaporean firms and MNCs based in Singapore began to relocate their operations to Johor, which offered substantially cheaper land and labour costs. While the Johor state government had developed a number of industrial parks in the 1970s to absorb the spillover from Singapore, it had not specifically targeted the electronics sector. It was only in the wake of this influx of investment that the state government began to develop its planning capabilities to target the electronics sector more effectively.

In contrast to the two cases above, the emergence of the electronics sector in Thailand is not due to any agency by provincial governments. The sector developed in the country in a number of stages, beginning first with semiconductor production in late 1970s, computer peripherals in the 1980s, and particularly disk drives in the late 1980s (Jomo 1997). The bulk of the electronics sector is concentrated in the northern part of the Bangkok metropolitan region; however there has been some development of the electronics sector in the eastern seaboard provinces, particularly Rayong, Chonburi, Chachoengsao, and Prachinburi. That said, this has been due to centrally-mandated locational incentives as well as the proximity of these provinces to two major ports and the country's main airport (Webster 2006).

In the 1990s, the electronic sector entered another stage. Aided by trade liberalization, market deregulation, and advances in information technology, global production networks[1] became more established (Yusuf 2004). Shortening production and production cycles meant that locating engineering, design, and even some R&D closer to manufacturing sites was increasingly advantageous.

It was in this context that Tamil Nadu's electronics sector emerged, providing another clear case of sub-national initiative. From the 1960s onwards, both central and state governments in India had been directly involved in the electronics sector through state-owned enterprises (Sridharan 1996). However, domestic production was uncoordinated and local private sector participation was limited. International investment began to arrive in the country in the late 1980s and was further facilitated by the 1991 liberalization measures. Central government investments in research and development facilities had favoured Karnataka, which emerged as the country's leading site for the software sector. With regard to electronics manufacture for the domestic market, Uttar Pradesh was the leading state (Heitzman 2004; ESC 2011). Spurred on by competition from Karnataka as well as neighbouring Andhra Pradesh, the Tamil Nadu state government adopted an aggressive approach to attracting foreign investment, first in the auto sector and, subsequently, consumer electronics and telecommunications equipment. This drive, accompanied by consistent investment in infrastructure and its formidable human resource base allowed Tamil Nadu to overtake the other states to emerge as the pre-eminent electronics manufacturer for export.

The post-2000 era marks another era, characterized by heightened competition. Leading MNCs have focused more on research and development, concept development, and marketing. Consequently, tasks such as manufacturing and, increasingly, detailed design work are being outsourced. Contract electronics manufacturers (CEMs) have emerged in this space, as service providers who manufacture, assemble, and provide designs to lead firms (Sturgeon and Lester 2004).

This has entailed greater competition — as well as heightened requirements regarding technological competence and geographical reach — for local firms catering to MNCs. In addition, many industry segments are now very difficult to enter, due to technological standards set by lead firms. Furthermore, the removal of trade barriers has entailed less protection for small firms, who are immediately exposed to international competition. And, stricter intellectual property rights have limited the possibilities of reverse engineering, which has been a standard recourse for "late-comer" firms to break into new sectors (Joseph, this volume).

Da Nang and Chengdu have both emerged in this more competitive context. Da Nang, for its part, displays considerable sub-national government initiative — albeit under central government control. In the mid-1980s, the Vietnamese government implemented a series of pro-market reforms, opening up the country to FDI and revitalizing its industrialization drive. The bulk of the Vietnam's manufacturing and particularly electronics, has been concentrated in Hanoi and Ho Chi Minh City. In 1996, the approval of foreign direct investment projects was devolved to municipal and provincial governments. And, in 1997, Da Nang's status was changed to a municipality under the supervision of the Prime Minister, allowing its government greater financial resources and autonomy. In this context, the municipal government has emerged as prime mover in seeking to develop its electronics sector, spurring rapid growth since 2002.

Despite the institutional similarities between China and Vietnam, as well as important decentralization measures implemented in the former country, sub-national initiative is not very prevalent in Chengdu's case. Its incursion into the electronics sector is better explained by central government initiative. The clustering of state-owned defence enterprises as well as research institutes and universities built up a core of skilled labour. Central changes in regional priorities towards the centre and west of the country, accompanied by investments in infrastructure and the prioritization of electronics production for the local market catalyzed Chengdu's industrial development (van Grunsven and Wang, this volume).

The cases thus show a variety of beginnings, ranging from conscious policies to serendipity. In some instances, such as Cebu, Tamil Nadu, and Vietnam, the sub-national government was the initiator of the state or province's industrialization drive. In other cases, such as North Brabant and Johor, fortune — in the way of industrial tradition or proximity — played a big role. And, in Gumi, Kaohsiung, Chengdu and, of course, Singapore, central government decisions were key.

The next sections will compare and contrast how the various state and provincial governments sought to foster the industry once it had emerged.

VERTICAL LINKAGES

State and provincial governments can assume a variety of roles *vis-á-vis* their industrial sectors. But their choice set is bounded by: the overall governance structure within which they operate; implicit or explicit regional preferences displayed by the centre; and their ability to secure resources from the national government. This section will therefore explore: how much autonomy each country's governance structure afforded sub-national governments; how regional policies helped or hindered the cases under study; and what tactics sub-national governments used to secure support from the centre.

The Southeast Asian cases display a wide range of vertical dynamics. Over the past two decades, the Philippines, Thailand, and Vietnam have enacted important decentralization measures. However, the results have not always directly benefitted state and provincial governments, with resources and responsibilities often bypassing them in favour of the lowest tier of government. Malaysia presents an interesting contrast as, over the same time period, it has centralized important functions and revenue sources. In all cases, the sub-national governments sought to cultivate ties with the centre to obtain financing and resources. However, where they differ is in the methods used to attain their goals and the end results of these efforts.

In 1991, the Philippines implemented a far-reaching decentralization reform that empowered the country's local governments. The decentralization measures included transferring responsibilities for: public services such as public works, environment and natural resources, and education; regulation of land management and building processes; and revenue-raising, including new sources of taxation revenue as well as greater flexibility regarding joint ventures, bond issuance, and loans. Interestingly, this new financial structure has benefited city governments more than municipalities or provinces, as they have received the bulk of revenue. In Cebu's case, these reforms came just after the province's leaders began to address local economic issues, entailing more room for manoeuvre and resources. While the reforms benefited city governments more than provincial governments, this was not a hindrance for Cebu, as the leaders of the provincial government and the city government of Cebu came from the same family and were influenced by the same regional identity. As a result, the decentralization measures strengthened the provincial government indirectly.

During the Marcos era, provincial-central relations and regional policy were not to Cebu's benefit. The province was associated with the opposition, and regional policy explicitly favoured northern Luzon. However, following the end of the Marcos era, Cebu benefited from well-articulated ties to the national leadership based on the province's economic importance and the influence of powerful local dynasties. While he served only one term as Governor of Cebu, Lito Osmeña, who is credited with ushering in the most important reforms for the province's economic take-off, subsequently served as Chief Economic Adviser to President Ramos.

This facilitated considerable amounts of resources from the centre to be invested in developing the province's infrastructure, as well as preferential access to international project financing.

Thailand presents a counterexample, in that it enacted important decentralization measures that did not result in significant autonomy at the sub-national level. Historically a highly centralized country, Thailand enacted decentralization measures in 1997, some two decades after the electronics sector first began to develop in the country. The measures entailed the creation of provincial administrative organizations, which are led by elected officials. However, their effectiveness is essentially neutralized by a parallel government structure comprised of provincial governors, who are delegates of the central government appointed by the Ministry of the Interior. Furthermore, while these new administrative organizations do manage funds, their autonomy is compromised by central regulations, which mandate set expenditure patterns. Last, a considerable portion of the funds are destined, not to provincial governments, but to local governments. However, regional policy has favoured the Eastern Seaboard provinces to the extent that the national government has sought to promote the dispersal of economic activity outside of Bangkok. This region benefited from central investment in planning, infrastructure, and expedited investment (Unger 1998).

Vietnam also enacted a number of measures that have altered the structure of its governance structure but, in contrast to Thailand, these did increase agency and autonomy at the provincial level. Over the past four decades, the central government has increased the number of its provinces and municipal governments substantially. In Da Nang's case, its elevation from a city to a municipality in 1997 implied a drastic increase in its autonomy and access to resources. This increased level of agency, coupled with Da Nang's status as an economic hub for the central part of the country, allowed it to compete with Vietnam's two prime centres, Hanoi and Ho Chi Minh City. In addition, devolving the approval of FDI projects to provinces and municipalities entailed unprecedented responsibility for provincial cadres, as well as heightened competition for investment. Furthermore, Da Nang's status as a municipality under direct supervision by the Prime Minister, as well as the combined administrative and political power of the municipality's first Mayor, enabled reforms to be implemented quickly. These enabled the municipality to dramatically improve the local business context prior to the emergence of the electronics industry.

Malaysia presents a marked contrast to the other cases. Created as a federation, state governments had significant amounts of autonomy well before the emergence of the electronics sector. Of key importance, they had control over land management and public utilities. However, successive rounds of expropriation, altered incentive structures, privatization, and "organizational duplication" by the central government have seen sub-national autonomy decrease substantially over time (Hutchinson, forthcoming).

Despite Johor being a bastion of support for the coalition in power at the national level, federally-pursued regional policies have not been to its favour, as central government priorities have favoured a concentration of resources in the capital, Kuala Lumpur. In addition, Johor's leaders have been forced by internal party discipline to implement policies pursued by the centre — even those that are not in the state's interest. Thus, Johor has seen a decrease in its competitive advantage in a number of areas, due to opaque privatization policies favoured by the national leadership.

Despite possessing multi-levelled governance frameworks, India and China were both centralized to a high degree up until the late 1970s and early 1980s, respectively. In China, fiscal reforms undertaken in 1978 provided sub-national governments with incentives to foster economic growth in their constituencies, and in subsequent years provincial and municipal governments have pursued a range of economic strategies. However, the central government still retains a very significant degree of influence, through: designating areas for export processing zones; investing in infrastructure; influencing labour flows; and controlling access to its domestic market. In recent years, regional policies have gravitated away from the eastern coastal states towards the interior. While Remick (2002) and Segal and Thun (2001) have shown a range of innovative strategies adopted by provincial leaders to foster their industries, in Chengdu's case, its leaders seem to have remained wedded to following central dictates (van Grunsven and Wang, this volume).

In India, the central government began to relax its control of the economy in 1980. The 1991 reforms took this process further, with the repeal of central regulations revealing an underlying layer of state-level prerogatives (Kohli 1989; Jenkins 1999). This allowed India's state governments to pursue a range of industrialization strategies. While the Tamil Nadu state government did run on a platform of cultural sub-nationalism and anti-central discourse in the pre-reform period, this changed following the 1991 reforms. A series of local-level social changes as well as the need to court investment in the face of increasing competition from neighbouring provinces, led the state government to abandon this approach and focus on providing a business-friendly environment and maintaining good relations with the central government. These dynamics predated and underlaid the emergence of the electronics sector in the state (Ilavarasan and Hutchinson, this volume).

The vertical dynamics at play in Korea and Taiwan also offer interesting parallels. Both countries maintained a centralized system of government until relatively recently, with key decentralization measures taking place well after the emergence of their respective electronics sectors. In both cases, leadership positions in sub-national governments were made elective, as opposed to appointive, in the mid-1990s. Prior to that time, Gumi and Koahsiung benefited from regional policies adopted by the centre regarding the location of industries outside the capital region. Subsequently, the national governments have sought to further promote the dispersal of industry, which has been to Gumi and Kaohsiung's benefit.

The governance situation in Holland has been much more fluid than any of the other cases. Regarding provincial prerogatives, much as in the Philippines and Thailand, provincial governments do not receive the bulk of revenue destined to sub-national governments — this is received by local governments. However, because municipal boundaries are very porous, it often is more effective for planning to occur collectively among several municipalities or at a regional level. Over time, the central government's approach to regional policy has evolved considerably, changing from seeking to equalize growth between regions to supporting leading ones — which is to North Brabant's benefit. Regarding sub-national and national ties, as seen in the Philippines and Korea, the provincial government has benefited from close ties to the national leadership.

This survey of the cases paints a picture that is more nuanced than originally anticipated. Governance structures changed in a number of ways with a number of unexpected outcomes. In most cases, decentralization measures were implemented, with important revenue-raising and implementation responsibilities devolved or delegated to sub-national governments. In some cases such as Tamil Nadu and Cebu, they occurred prior to, or co-terminously with, provincial or state-level industrial drives. In other cases, such as Kaohsiung and Gumi, they occurred well after the fact. Often, new states, provinces, or municipalities were created, entailing reconfigurations of power relations and, frequently, increased autonomy. This was most marked in Vietnam, where Da Nang benefited from new financial and administrative powers. However, provincial governments were not always the beneficiaries of decentralization processes. In the Philippines and Thailand, such measures bypassed provincial governments and benefited local governments. In some cases, various levels of governments worked together to achieve specific goals. In the Netherlands and Korea, collective efforts between various tiers of government were visible. And, in Cebu's case, close ties between the provincial and city governments allowed the limited purview of the former to be circumvented. Malaysia stands out as the outlier, as it has centralized rather than decentralized power over time.

Regional policies also differed considerably across countries and over time. Gumi, Kaohsiung, and Thailand's Eastern Seaboard Provinces were consistent beneficiaries of central government planning and investment decisions, often to promote the dispersal of economic activity. In contrast, China's regional policy evolved over time, away from focussing on the coastal region to the west and interior of the country, which was to Chengdu's benefit. Tamil Nadu and North Brabant also benefited from their central governments' change to seeking to support high-performing states and regions, rather than seeking to lessen intra-territorial income disparities. For its part, Cebu did not consistently benefit from regional policies, but its industrialization was coterminous with a more equitable regional policy.

Similarly, national/sub-national ties took a variety of forms. With the exception of Tamil Nadu and Cebu during the 1970s and 1980s, no sub-national government adopted an opposition or anti-centre position. Rather, the most frequent dynamic was close personal ties between provincial and national leaders. And, as far as Tamil Nadu and Cebu are concerned, their economic and political fortunes improved markedly once national and sub-national priorities were more closely aligned. However, close ties did not always result in beneficial outcomes. In Johor's case, the ruling coalition's domination at the national and state level meant led to provincial leaders accepting national-level decisions that curtailed their autonomy. And, in Chengdu's case, consistent central government support did not perceptibly translate into greater autonomy or innovation at the sub-national level.

HORIZONTAL LINKAGES

The literature on industrial transformation in Japan, South Korea, and Taiwan and elsewhere places a premium on well-established state-business relations. The consistent flow of up-to-date information to be fed into policy-making, as well as the ability of governments to encourage firms to undertake new and more technologically-demanding activities, have been pinpointed as key elements in the successful pursuit of industrial transformation. Furthermore, the domestic and international private sectors can constitute important sources of capital, bolstering transfers from the centre and thereby increasing autonomy for aspiring state and provincial governments.

With regard to the Southeast Asian cases, Cebu stands out as a case of very good government-private sector ties. This stands to reason, as, while the Philippines is characterized by a high degree of primacy, it also has very powerful rural elite with extensive economic interests (Sidel 1999). As one of the country's economic centres, Cebu has a number of political families that overlap with the business community, and the provincial governor that initiated Cebu's forms was tied to the local entrepreneurial class. These close links as well as a business-friendly outlook enabled coordination and communication between the provincial government and the local private sector, as well as a number of joint ventures to market the province overseas. This pattern of interaction was also applied to the international business community through the formation of a business association for MNCs.

While also oriented towards attracting international capital, Johor presents a markedly different approach with regard to domestic entrepreneurship. In line with national-level priorities, the state government sought to foster the emergence of a new business class from the Malay community. To this end, Johor's ample, flat, and fertile land enabled the state government to raise funds and promote a bevy of operations in various service sectors, notably real estate, tourism, restaurants, and health care. As a result, the local, largely Chinese manufacturing community has been bypassed, and communication between the state government and this section of the private sector is inexistent. To a large extent, this "oversight" was made possible by the ample rewards available to the state government in the agriculture sector.

As with the Philippines, Thailand also suffers from a high degree of primacy and has a tradition of provincial entrepreneurship (Hewison and Thoungyu 2000). However, as with Johor, the potential rewards from the agricultural sector have acted as a disincentive for both the local private sector as well as provincial governments to seek to promote more technologically-complex activities such as those in manufacturing (Unger, this volume). As a result, there have been no policy innovations at the sub-national level to overcome the rather distant relations between the public and private sector in manufacturing.

While provincial governments in Vietnam have pursued FDI-led industrialization aggressively, as seen elsewhere in Southeast Asia, the local private sector has not received the same level of attention. This has been compounded by a historically rather difficult context for the local private sector, which has only changed recently. For example, the Enterprise Law enacted in 2000 has seen a rapid increase in the number of registered local firms. However, many of these are micro- and small-firms, meaning that the base of firms that have the scale and resources to begin manufacturing is relatively under-developed. Most of the larger firms are state-owned enterprises that have proven formidable rivals to local firms for market space and access to capital (Hakkala and Kokko 2007).

Turning to China and India, beyond their FDI-focus, Chengdu and Tamil Nadu have quite different state-local business relations. With regard to the first country, provincial and municipal governments have adopted a variety of strategies with regards to promoting industrialization. However, contrary to the municipal governments of Shanghai and Beijing, who pursued their own strategies towards their sector of choice (Segal and Thun 2001), the Chengdu government seems to have remained focussed on attending to central government needs as well as those of international firms. As a result, ties with the local private sector are very under-developed.

As with Chinese provinces, Indian states have pursued a variety of development strategies and, consequently, have distinct government-business relations. In the pre-reform period, Tamil Nadu lagged behind other states in reaching out to the local private sector. Following 1991, the Tamil Nadu state government has adopted a more business-friendly stance. This has included partnering with local private sector associations to build technological capabilities and to market the state overseas. And, very importantly in the Indian context, Tamil Nadu has managed to secure consistent support for urban industries without generating the level of anti-urban bias that has crippled reform efforts in Karnataka and Andhra Pradesh.

Turning now to the industrialized countries, Singapore differs from the others in that it has not traditionally focussed on fostering the acquisition of domestic technological capabilities. Although this is changing of late (Toh, this volume), in the past the country has focussed on attaining a better position within global production chains by meeting MNC requirements insofar as skilled labour, infrastructure, and supplier services (Felker 2003). However, Singapore does have quite an extensive range of state-business relations, particularly in strategic niches of the electronics sector. This includes: government-linked firms establishing joint ventures with leading MNCs; government-owned equity in firms in key upstream activities such as wafer fabrication; as well as joint ventures with large, local firms. Furthermore, there are a number of active business associations, including the Singapore Business Federation of which all companies whose paid up capital is above a certain threshold are members.

With regard to Korea, the horizontal dynamics in Gumi seem to have been an extension of those at the central level. Much of the industrial activity in the city is carried out by affiliates of *jaebols*, who have located production facilities there at the behest of the central government. This seems more in accordance with Amsden's (1989) concept of "reciprocity" — where the private sector complies with governmental policies and targets in return for support and credit — rather than any unique local dynamic. That said, the recent emergence of a local business association bodes well for the future.

Interestingly, the Kaohsiung case seems to have more in common with those from Southeast Asia than its national counterpart. To a greater extent than the industry found in and around Taipei and the Hsinchu Science-based Industrial Park, the electronics sector in the city seems to centre around foreign-owned MNCs, rather than local firms. While there is a branch of the main business association in the electronics sector, it would appear that most of the communication between public agencies and local firms takes place in the north of the country, with little in the way of unique forms of social capital in Kaohsiung itself.

In comparison to the other cases, North Brabant is striking in the variety of forms of state-business communication as well as the frequency with which they change. In addition, many forms of consultation go beyond the government and the private sector to include academia and social groups. Consultation also takes place at the municipal, provincial, and city-region level. While some may be organic, it is also likely resources made available at different levels encourage the formation of many of these initiatives.

What can be said about the patterns observed at the sub-national level? In contrast to national-level efforts in Korea and Taiwan, the cases from Southeast Asia, China, and India focused to a much greater extent on FDI-led industrialization than on fostering indigenous entrepreneurship. Indeed, even the Gumi and Kaohsiung cases show interesting variants on their national counterparts. As a result, more attention was placed on attracting capital and attending to the needs of the international private sector. Consequently, the linkages and communication with the local private sector are less extensive than one would expect.

Several structural reasons may account for this. First, the centralization of political and economic power in capital cities would mean a consequently smaller pool of locally-based entrepreneurs. In addition, relative to national capitals, provincial capitals offer less access to banking and finance, constituting a critical bottleneck for aspiring entrepreneurs. Indeed, in some cases, the critical mass of potential industrialists may simply not be there. Furthermore, the agricultural sector in provincial areas is likely to be proportionately more important than in national capitals, thus constituting an alternative source of rents for local entrepreneurs.

The later entrants to the electronics industry also face additional hurdles. First, there has been less time for local firms to emerge in response to locally-present demand from MNCs, and heightening competition has also raised the bar for successful entry. In the event that their electronics sectors develop later than in other locations in the same country, local supplier firms may be crowded out by large domestic firms from elsewhere. This has occurred to a certain extent in Tamil Nadu and Chengdu, where large supplier firms from established industry centres in the north and east of the country, respectively, have captured a significant amount of the locally-present demand in these emerging clusters.

POLICY OBJECTIVES AND FRAMEWORKS

The previous sections set out the context within which the electronics sector emerged, and analysed the vertical and horizontal strategies that the various sub-national governments used to pursue their goals. This section will look at the various types of policies that they implemented in order to foster the development of their electronics sectors.

As regards the Southeast Asian cases, Cebu had a very well-developed and explicit policy platform of catalyzing its growth through linking to the global economy (and bypassing Manila) by building up its infrastructure base and attracting foreign direct investment. In order to do this, the provincial government mobilized funds from international financial institutions, issuing bonds, and selectively selling assets. These funds were then invested in improving infrastructure as well as marketing the province overseas. This was complemented by mobilizing central resources for upgrades in public services and utilities. In addition, the provincial government, relevant local governments, and largest corporations active in Cebu jointly funded an investment promotion centre to market the island to overseas investors. However, despite this business-friendly approach, the provincial government did not engage with local firms to: link up with MNCs; overcome collective action dilemmas; or map or develop their technological capabilities. The local private sector also does not seem to have collaborated consciously to boost capabilities on its own.

Despite their dramatically different governance contexts, the Da Nang municipal government shares some similarities with its counterpart from Cebu. It also explicitly targeted the "primate" cities — in this case Hanoi and Ho Chi Minh — as comparators and competitors. The province also sought to compete on cost with these locations, particularly insofar as land was concerned. As with Cebu, it placed significant emphasis on increasing taxation revenue to invest in infrastructure, industrial parks, and marketing. In addition, Da Nang sought to capitalize on its high-quality regulatory environment for business, and has sought to develop its human resource base through partnering with international education providers. In contrast to Cebu, however, the provincial government does not have extensive ties with the local private sector, due, in large part, to the country's tradition of central planning. This has precluded any significant attempt to bolster local capabilities or foster inter-firm networks.

The Johor state government mirrored its national counterpart's twin goals of promoting industrialization and creating a new business class. However, these policy goals were pursued separately. On one hand, the state government sought to develop its industrial sector largely through competing on price relative to Singapore. To this end, it created a network of industrial parks to cater to the outflow of manufacturing activities from its neighbour. However, despite official pronouncements to the contrary, very little was done to broker contacts between MNCs and local firms, or bolster local technological capabilities. Conversely, considerable efforts were placed into creating a new cohort of business-people, including brokering, demand-side, and market-complementing measures. However, these initiatives did not include those ethnic groups not targeted by national policies, nor did they focus to a significant extent on the manufacturing sector.

Due to the structural considerations outlined above, Thailand's provincial governments and provincial administrative organizations have attempted little in the way of seeking to support industrialization. However, the legislation concerning the roles of the elected provincial administrative organizations does allow some leeway to address industrialization-related issues. That said, local cultural expectations are that these organizations address more short-term issues such as law and order, cultural affairs, sports, and maintaining public facilities. As a result, sub-national governments have implemented very little of import to industrialization in general and the electronics sector in particular.

Chengdu and Tamil Nadu both explicitly targeted industrialization as a policy priority. Both also capitalized on their infrastructure and abundance of cost-effective, skilled labour. Chengdu complemented this with tax incentives for priority activities, such as integrated circuit design and software development services, as well as deductions for training activities. However, the city government did not seek to bolster local inter-firm networks or technological capabilities in any significant way. One partial exception was the decision to co-locate firms in similar industries in the same areas, which could promote the development of inter-firm networks.

Tamil Nadu had an explicit goal of outcompeting the surrounding states of Andhra Pradesh and Karnataka. However, mirroring Cebu and Da Nang, the state government was only able to embark on its industrialization drive thanks to a commitment to fiscal prudence. Sustained investment in infrastructure, coupled with a well-developed human resource base, and an investor-friendly approach enabled Tamil Nadu to attain leadership in the export-oriented electronics sector. Unlike most other cases, the state had actually developed a range of organizations and policies to develop the electronics sector before it embarked on its FDI-led drive. Thus, the state did implement a range of market-complementing measures. This included: a university; joint-ventures with private businesses in strategic areas; as well as the founding, in conjunction with a leading business association, of a technology development promotion centre for local firms. However, the predominance of international firms in Tamil Nadu's economy indicates that overall business-friendly initiatives were more important than attempts to boost local capabilities. However, the state government's large-scale procurement of electronics goods and personal computers could, should they contain local content requirements, prove powerful tools for supporting local industries in the future.

The cases from industrialized countries provide some interesting contrasts. Singapore, in particular, provides many examples of successful and consistently-evolving policy-making. While there are certain aspects of policy that are not applicable to sub-national governments, nevertheless a chronological study of what the city-state did and when provides many insights that are relevant for aspiring sub-national regions. While much of its industrialization drive also consisted of industrial parks, tax incentives, and marketing, it is also striking how early an emphasis was placed on attracting and retaining value-added activities. Thus, benchmarks for qualifying for incentives were continually raised, agglomeration economies were encouraged, and perceived market failures were targeted. Clusters and inter-firm networks were encouraged in the 1980s, as were linkages between local firms and MNCs. During the 1990s, this was complemented by a range of specialist research organizations as well as programmes to support high-tech start-ups.

Gumi provides an interesting case of infrastructural investment and physical planning being carried out by the national government, with the provincial and city governments focussing on more technologically-intensive policies. In recent years, following decentralization, the provincial and city governments have collaborated to bolster local capabilities for research and development in key industries. This has included an electronics and IT research institute, as well as rolling out support programmes for local firms, business incubators, and research efforts. This has also included substantial work to foster agglomerations, establish business networks, and encourage inter-firm linkages, including fostering cooperative research and development among SMEs.

As in the case of Gumi, the Kaohsiung municipal government has followed central directives regarding strategic industries, and the central government has remained the main player insofar as infrastructure and industrial parks are concerned. In recent years, the Kaohsiung municipal government has implemented several broker and market-complementing policies. This includes loans for local firms, credit guarantees, university-industry endeavours, as well as incubation centres.

As regards North Brabant, the provincial government has focused to a greater extent on broker functions due, in part, to its relative paucity of funds. This has involved promoting science and technical education, upgrading the skills of the workforce, and establishing skills development centres. The provincial government has also worked with municipal governments to coordinate efforts. However, the most hands-on role was played by the North Brabant Development Agency (NBDA), jointly founded by the national government and the North Brabant provincial government. The NBDA markets the province overseas and runs a network of industrial parks. The agency also provides venture capital to local firms, supports R&D, and helps local firms expand operations overseas.

What can be said about discernable patterns across the various state and provincial governments? In many cases, they began their incursion into the manufacturing sector in textiles, seeking to capitalize on relatively low costs for land and labour. Subsequently, they sought to mobilize investment, either through improving their taxation collection system or tapping new sources of revenue. These funds were then usually invested in infrastructure, often to improve connectivity with the wider economy, thus bypassing the national centre. It is in this respect — rather than attempts to foster their local private sectors — that the various sub-national governments have shown the most innovation, whether it be managing oil palm estates in Johor or attracting financing from the Asian Development Bank in Cebu.

This was usually followed by a sustained attempt to market the territory to international investors, which was often matched by a drive to improve the local business climate. Da Nang, Cebu, and Tamil Nadu were among the most aggressive in seeking to offer investors a business-friendly climate. Da Nang, in particular, was remarkable for being able to offer investors a clear legal framework and well-developed regulatory agencies a few years after its establishment as a municipal government.

It is past this point that government policies begin to diverge. In the cases from industrializing countries, broker policies were relatively under-utilized. Policy pronouncements aside, there was relatively little conscious targeting of specific industry activities. With the exception of very low skill items such as textiles and garments, state and provincial governments were largely agnostic as to the specific nature or the technological intensity of investments coming in. Technology mapping and value chain analyses were not carried out with firms to identify weaknesses or gaps in value chains. Nor were collective planning exercises carried out. In addition, initiatives to build local or regional innovation systems were not prevalent. Thus, while strategic investments were made in specialized institutes, relatively little was done in order to bring the various actors in the local economy together to build or strengthen the local innovation system.

Demand side policies were relatively rare. The exceptions to this could be Tamil Nadu and Chengdu, due to the sheer size of their domestic markets. However, neither has used procurement policies to support their local electronics sector directly.

Market-complementing measures were more common. States and provincial governments established facilities such as industrial training institutes and specialized universities, and also set up joint-ventures in strategic areas. In addition, tax incentives for research and development were used relatively frequently.

When seen as an ensemble, with the exception of Johor, the Southeast Asian cases did not seek to work with local firms. Perhaps because of their later entrance into the electronics sector or their FDI-focused investment drive, they perceived their roles to be limited to providing a conducive environment for international businesses. This can also be seen to a significant extent in Chengdu and Tamil Nadu as they, too, have largely focused on FDI-led export-oriented industrialization.

The cases from industrialized countries tended to be more ambitious insofar as policies were concerned, particularly with regard to local firms. While demand-side policies were not visible, most had a range of broker and market-complementing measures in place. In particular, provincial and municipal governments themselves engaged in technical, research-intensive, and contact-intensive work with the private sector.

STRUCTURAL VERSUS SECTORAL TRANSFORMATION

Having put these policy frameworks in place, to what extent were the various sub-national governments able to promote industrial transformation? With regard to structural transformation, all the cases witnessed an important and far-reaching change in their economic structures. Either as part of nation-wide attempts, or a sole-orchestrated push to pursue export-oriented industrialization, the state and provincial governments received considerable amounts of electronics-related FDI, with ensuing changes in their firm bases and labour forces.

However, this is only part of the process as simple export-oriented production cannot guarantee sustained economic growth. In many cases, electronics production remained in enclaves, with few linkages to the domestic economy. In addition, an exclusive reliance on labour-intensive processes leaves the given region open to competition on price from other locations.

The challenge for aspiring regions is thus to pursue sectoral transformation. This involves encouraging flagship firms to carry out more technologically-intensive and value-added tasks, fostering inter-firm networks, and working with local firms to boost technological capabilities.

It is in this aspect that there has been more limited success. To borrow Markusen's typology, this book has found significantly more examples of satellite platforms and, to a lesser extent, hub-and-spoke clusters driven largely by MNC branch plants, as opposed to Italianate districts sustained by dynamic local firms.

The electronics sectors in the rather more mature manufacturing locations of Cebu, Johor, and Thailand are still overwhelmingly dependent on MNCs and confined to relatively labour-intensive sectors. Largely led by Japanese flagship firms and their suppliers, Cebu is very focused on the assembly of semiconductors and electronics components. Johor's is comprised of Japanese, Singaporean, and American firms engaged in CEM, as well as the manufacture of passive components and consumer electronics. Inter-firm linkages are very under-developed, as most firms are linked to production chains centring on regional headquarters in Singapore. Thailand, for its part, is focussed on hard disk drives, as well as semiconductors, and integrated circuits. Ownership is largely foreign, and tasks are still focussed on assembly operations.

This is not to say that there are no signs of progress. Some MNCs are beginning to establish R&D facilities in Cebu, which is also seeing the emergence of some local firms in downstream sectors such as plastics, metal stamping, and semiconductor packaging. Although it is lagging behind Singapore and Penang, Johor is also seeing investments from MNCs in new, more technologically-intensive sectors such as solar power and medical devices. In addition, some local firms are climbing the value chain, by moving from the plastics sub-sector into CEM and precision engineering. There is some evidence of local supplier firms in Thailand, but these are largely confined to electrical appliances.

For its part, the electronics sector in Da Nang is much younger and exhibits even more of the traits of a satellite platform, comprised almost entirely of Japanese, South Korean, and Singaporean firms. In contrast to the other cases, production is more geared towards industrial electronics.

Thus, insofar as industrializing Southeast Asia is concerned — despite some investments in more technologically intensive areas, and some linkages between MNCs and local supplier firms — production is still largely of an enclave variety and firm dynamics are more reminiscent of satellite platforms. However, over time, linkages between MNCs and local firms could thicken, with the firm groupings acquiring more characteristics of a hub-and-spoke structure.

Chengdu and Tamil Nadu, despite their later entry into the sector, have had more success at pursuing sectoral transformation. Chengdu has received entire production chains of investment in relatively sophisticated subsectors, such as flat panel displays and communication equipment. While the cluster is largely MNC-led, there are a number of important local players. However, many of these are from other provinces in China who have moved to cater to the emerging demand. Tamil Nadu's firm dynamics are similar, in that there are large flagship firms who are accompanied by chains of supplier firms, and there are a significant number of larger, local firms who are from other parts of the country. This relatively different structure may be a product of these territories' relatively later incursion into the electronics sector. Thus, technological standards may have been set by anchor firms, precluding a more gradual emergence of local supplier firms. This may have been exacerbated by the arrival of large local firms in downstream sectors from more mature industry locations. That said, the degree of inter-firm linkages is encouraging, and these territories are closer to the hub-and-spoke ideal type.

Despite their greater progress towards the acquisition of local technological capabilities, the cases from industrialized countries yield interesting insights. For its part, while Singapore has: excelled in encouraging MNCs to locate high value-added tasks in its territory; built a formidable array of supporting institutions such as high-end research institutes — often in joint ventures with industry leaders; and has a number of large electronics firms, the country is still quite reliant on foreign-owned MNCs for technology (Wong and Singh 2008).

Despite their similarities, Gumi and Kaohsiung offer interesting contrasts. While both have had to compete with their national capitals for resources and attention from industry leaders, Gumi has had markedly more success at fostering a dynamic local firm base. This process took more than two decades, but the large number of branch plants present in the city has led to a network of spinoffs and supplier firms. As a result, there are a substantial number of linkages between firms present in the city, which is now an important location for the production of patents. In addition, a substantial number of local SMEs have linkages with firms outside the city, indicating a potential base of export activity. In contrast, despite the national government's considerable investments in Kaohsiung's knowledge infrastructure, the municipality has more in common with the satellite platforms from Southeast Asia than it does with Gumi. Much of the more sophisticated manufacturing in municipality is confined to MNCs based in EPZs, and, unlike Taipei, Kaohsiung has not witnessed the growth of a generation of local firms.

North Brabant has come into its own insofar as high value-added activities and innovation are concerned, with a large share of employment in high-tech sectors, lots of patents generated by firms, and extensive business-to-business links. However, this has not always been the case, and the province's electronics sector underwent considerable upheaval when its largest and most important anchor firm, Philips, restructured and moved from a closed innovation paradigm to a more open and flexible arrangement. Since then, the corporation has actively encouraged more spinoffs, which has thickened the base of local firms considerably. This resembles a transition from a hub-and-spoke structure to one more reminiscent of the 'Italianate' firm groupings.

CONCLUSION

The initial expectation was to find numerous examples of agency, innovation, and entrepreneurship at the sub-national level. Mini-developmental states were thought to be at work behind the scenes of many of these cases, single-mindedly pursuing industrial transformation and engaged in constant communication with the local private sector to this end.

The picture that emerges from this collection of cases is substantially different. There are certainly cases of dynamic sub-national governments and even of (qualified) success insofar as sectoral transformation is concerned. However, the picture itself is rather more nuanced, with even the most promising regions benefiting from doses of luck, support from national governments at specific points in time, and even collaborative efforts between various layers of governments.

Catalysts for new industries in these regions were often serendipitous, such as a long-running industrial tradition or a national policy change that consciously or inadvertently benefited the state or province in mention. That said, there were some clear cases of sub-national agency, with Cebu perhaps being the clearest. However, state or provincial governments sometimes became more dynamic in the wake of missed opportunities, as in the case of Tamil Nadu.

The influence of national governments looms large. Most sub-national governments had a clear strategy of nurturing ties or leveraging influence with national leadership, and there are no examples of regional autarky resulting in positive outcomes. Tamil Nadu and Cebu did certainly distance themselves from the national leadership at key points, but their subsequent industrialization occurred following *rapprochement*. On the other hand, close ties did not automatically result in unwavering support, as the case of Johor demonstrates.

Counter-intuitively, decentralization did not always benefit state or provincial governments. In some cases, such as North Brabant and Cebu, they were bypassed and, in others, such as the Thai Eastern Seaboard Provinces, their room for manoeuvre was curtailed. Yet, in cases such as Cebu and North Brabant, multi-levelled government investment drives enabled provincial governments to retain control. Thus, the influence of sub-national governments changed over time, and was not strictly determined by the formal division of responsibilities. Rather, sub-national leaders evaluated partnerships and linkages strategically, making them in order to magnify their influence.

The state and provincial governments studied did certainly approach the private sector. However, the majority chose to focus on attracting and retaining foreign direct investment rather than seeking to work with their local business groups. While this could be a product of limited formal responsibilities and scarce resources on the part of sub-national governments, it could also be due to a relatively limited base of entrepreneurs due to structural issues. In certain cases, such as Tamil Nadu and Chengdu, it is also possible that local entrepreneurs were crowded out by larger firms from other parts of the country.

With regard to policy frameworks, there were a variety of approaches. In all cases, the state and provincial government concerned sought to go beyond merely providing low-cost land and labour. However, the development of a vibrant local firm base and the fostering of local technological capabilities did not figure frequently on the radars of most governments. Rather, innovation and persistence were seen in the areas of broadening the revenue base, investing in infrastructure, and providing a more conducive environment for business.

Note

1 GPNs are defined as 'inter- and intra-firm relationships through which the firm organizes the entire range of its business activities for R&D, production definition and design, supply of inputs, manufacturing, distribution, or support services (Borrus et al. 2000, p. 1).

References

Amsden, A. *Asia's Next Giant: South Korea and Late Industrialization.* New York: Oxford University Press, 1989.

Borrus, M. Ernst, D. and Haggard S. "Introduction: cross-border production networks and the industrial integration of the Asia-Pacific region". In *International Production Networks in Asia: Rivalry or Riches*, edited by M. Borrus, D. Ernst, and S. Haggard. London: Routledge, 2000.

ESC. "Statistical Yearbook 2010–11". New Delhi, Electronics and Computer Software Export Promotion Council, 2011.

Evans, P. *Embedded Autonomy: States and Industrial Transformation*. Princeton: Princeton University Press, 1995.

Felker, G. "Southeast Asian Industrialisation and the Changing Global Production System". *Third World Quarterly* 24, no. 2 (2003): 255–82.

Frobel, F., Heinrichs, J., and Kreye, O. *The New International Division of Labour: Structural Unemployment in Industrialised Countries and Industrialisation in Developing Countries.* Cambridge: Cambridge University Press, 1980.

Hakkala K. and A. Kokko. "The State and Private Sector in Vietnam". EIJS Working Paper 236. Stockholm: Stockholm School of Economics, 2007.

Heitzman, J. *Network City: Planning the Information Society in Bangalore.* Delhi: Oxford University Press, 2004.

Hewison, K. and M. Thongyou. "Developing Provincial Capitalism: A Profile of the Economic and Political Roles of a New Generation in Khon Kaen, Thailand". In *Money and Power in Provincial Thailand*, edited by R. McVey. Singapore: Institute of Southeast Asian Studies/Silkworm Books, 2000.

Hobday, M. "East Asian Latecomer Firms: Learning the Technology of Electronics". *World Development* 23, no. 7 (1995): 1171–93.

Hutchinson, F.E. "Malaysia's Federal System: Overt and Covert Centralization" in *Journal of Contemporary Asia*, forthcoming.

Jenkins, R. *Democratic Politics and Economic Reform in India.* Cambridge: Cambridge University Press, 1999.

Jomo, K.S. *Southeast Asia's Misunderstood Miracle: Industrial Policy and Economic Development in Thailand, Malaysian, and Indonesia.* Boulder: Westview Press, 1997.

Kohli, A. "Politics of Economic Liberalization in India". *World Development* 17, no. 3 (1989): 305–28.

McKendrick, D.G., Doner, R.F., and S. Haggard. *From Silicon Valley to Singapore: Location and Competitive Advantage in the Hard Disk Drive Industry*. Stanford: Stanford University Press, 2000.

Pempel, T.J. "The Developmental Regime in a Changing World Economy". In *The Developmental State*, edited by M. Woo-Cumings. Ithaca: Cornell University Press, 1999.

Remick, E. "The Significance of Variation in Local States: The Case of Twentieth Century China". *Comparative Politics* 34 (2002): 399–419.

Reyes-Macasaquit, M-L. "Case Study of the Electronics Sector in the Philippines: Linkages and Innovation". In *Fostering Production and Science and Technology Linkages to Stimulate Innovation in ASEAN*, edited by P. Intarakumnerd. ERIA Research Project Report, No. 7-4 2009. Jarkarta: ERIA, 2010.

Segal. A. and Thun E. "Thinking Globally, Acting Locally: Local Governments, Industrial Sectors, and Development in China". *Politics and Society* 29, no. 4 (2001): 557–88.

Sidel, J.T. *Capital, Coercion and Crime: Bossism in the Philippines.* Stanford: Stanford University Press, 1999.

Sridharan, E. *The Political Economy of Industrial Promotion: Indian, Brazilian, and Korean Electronics in Comparative Perspective 1969–1994.* Westport: Praeger, 1996.

Storper, M. and Scott, A.J. "The Wealth of Regions: Market Forces and Policy Imperatives in Local and Global Context". *Futures* 27, no. 5 (2003): 505–26.

Sturgeon, T.J. and Lester, R.K. "The New Global Supply-Base: New Challenges for Local Suppliers in East Asia". In *Global Production Networking and Technological Change in East Asia*, edited by S. Yusuf, M.A. Altaf, and K. Nabeshima. Washington D.C.: World Bank/Oxford University Press, 2004.

Unger, Danny. *Building Social Capital in Thailand.* New York: Cambridge University Press, 1998.

Webster, Douglas. "Supporting Sustainable Development in Thailand: A Geographic Clusters Approach". Bangkok: National Economic and Social Development Board/World Bank, 2006.

Wong, P-K. and A. Singh. "From Technology Adopter to Innovator: Singapore". In *Small Country Innovation Systems: Globalization, Change and Policy in Asia and Europe*, edited by C. Edquist and L. Hommen. Edward Elgar: Cheltenham, 2008.

Yusuf, S. "Competitiveness through Technological Advances under Global Production Networking". In *Global Production Networking and Technological Change in East Asia*, edited by S. Yusuf, M.A. Altaf, and K. Nabeshima. Washington D.C.: World Bank/Oxford University Press, 2004.

INDEX